Lukas Schmidt-Mende, Jonas Weickert
Organic and Hybrid Solar Cells
De Gruyter Graduate

Also of interest

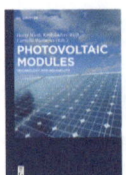
Photovoltaic Modules: Technology and Reliability
Wirth, Weiß, Wiesmeier, 2016
ISBN 978-3-11-034827-9, e-ISBN 978-3-11-034828-6

Physics of Energy Conversion
Krischer, Schönleber, 2015
ISBN 978-1-5015-0763-2, e-ISBN 978-1-5015-1063-2

Nanomaterials in Joining
Charitidis (Ed.), 2015
ISBN 978-3-11-033960-4, e-ISBN 978-3-11-033978-7

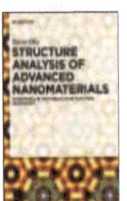
Structure Analysis of Advanced Nanomaterials: Nanoworld by High-Resolution Electron Microscopy
Oku, 2014
ISBN 978-3-11-030472-5, e-ISBN 978-3-11-030501-2

Green: A Systemic Approach to Energy
Schlögl, Robert (Editor-in-Chief)
ISSN 1869-876X, e-ISSN 1869-8778

Lukas Schmidt-Mende,
Jonas Weickert

Organic and Hybrid Solar Cells

An Introduction

DE GRUYTER

Authors
Prof. Dr. Lukas Schmidt-Mende
University of Konstanz
Faculty of Physics
Universitätsstr. 10
78464 Konstanz
Germany
Lukas.Schmidt-Mende@uni-konstanz.de

Dr. Jonas Weickert
University of Konstanz
Faculty of Physics
Universitätsstr. 10
78464 Konstanz
Germany
jonas.weickert@googlemail.com

The **cover image** shows an AFM topographic height profile of an anodized aluminum oxid (AAO) membrane electrochemically processed in phosphoric acid used in the Schmidt-Mende group for the preparation of nanostructured organic and hybrid solar cells. The shown AFM image of a self-organized AAO membrane was recorded by Dr. Thomas Pfadler with a Bruker Multimode8 AFM in tapping mode, equipped with a Nanotools MC60 high precision probe. The authors thank Nanotools Munich GmbH for access to their laboratories.

ISBN 978-3-11-028318-1
e-ISBN (PDF) 978-3-11-028320-4
e-ISBN (EPUB) 978-3-11-038851-0

Library of Congress Cataloging-in-Publication Data
A CIP catalog record for this book has been applied for at the Library of Congress.

Bibliographic information published by the Deutsche Nationalbibliothek
The Deutsche Nationalbibliothek lists this publication in the Deutsche Nationalbibliografie; detailed bibliographic data are available on the Internet at http://dnb.dnb.de.

© 2016 Walter de Gruyter GmbH, Berlin/Boston
Typesetting: Lumina Datamatics
Printing and binding: CPI books GmbH, Leck
Cover image: Dr. Thomas Pfadler
♾ Printed on acid-free paper
Printed in Germany

www.degruyter.com

Contents

Preface —— IX

1 Introduction —— 1
1.1 General Introduction —— 1
1.2 Brief History —— 3
1.2.1 Inorganic Solar Cells —— 3
1.2.2 Organic and Hybrid Solar Cells —— 4

2 Semiconductors and Junctions —— 6
2.1 Introduction to Inorganic Semiconductors —— 6
2.1.1 Electronic States in Inorganic Semiconductors —— 6
2.1.1.1 Energy Bands in Solids —— 6
2.1.1.2 Electrons, Holes, and Polarons —— 8
2.1.1.3 Band Structure —— 10
2.1.1.4 Direct and Indirect Bandgaps —— 17
2.1.1.5 Density of States —— 19
2.1.2 Doping of Inorganic Semiconductors —— 22
2.1.2.1 Intrinsic Semiconductors —— 22
2.1.2.2 n-Type Semiconductors —— 23
2.1.2.3 p-Type Semiconductors —— 25
2.2 Introduction to Organic Semiconductors —— 26
2.2.1 Carbon Hybridization and Organic Molecules —— 26
2.2.1.1 Carbon Orbitals and Orbital Hybridization —— 26
2.2.1.2 Single, Double, and Triple Bonds —— 28
2.2.1.3 Structural Formulas —— 30
2.2.1.4 Mesomerism —— 33
2.2.2 Electronic States in Organic Semiconductors —— 34
2.2.2.1 Conjugated Molecules —— 34
2.2.2.2 Molecular Orbitals —— 37
2.2.2.3 Optical Transitions —— 45
2.2.2.4 Excitons and Exciton Diffusion —— 60
2.2.3 Charge Transport in Organic Semiconductors —— 67
2.2.3.1 Charge Hopping in Disordered Systems —— 67
2.2.3.2 Macroscopic Parameters of Charge Transport —— 78
2.2.3.3 Doping of Organic Semiconductors —— 79
2.3 Junctions —— 81
2.3.1 Inorganic p–n Homojunctions —— 81
2.3.2 Inorganic Heterojunctions —— 89
2.3.3 Schottky Junctions —— 92

2.3.4	Organic–Organic Heterojunctions —— 93
2.3.5	Organic–Inorganic Heterojunctions —— 95

3 Working Mechanisms of Organic and Hybrid Solar Cells —— 97

3.1	Basic Principles of Solar Cells —— 97
3.1.1	Energy Conversion and the Solar Spectrum —— 97
3.1.2	Shockley–Queisser Limit —— 101
3.1.3	Photocurrent and Spectral Response —— 102
3.1.4	Characteristics of a Solar Cell —— 106
3.1.5	Parasitic Resistances and the Generalized Shockley Model —— 108
3.2	Inorganic Junction Solar Cells —— 113
3.2.1	p–n Junction Solar Cells —— 114
3.2.2	Inorganic Heterojunction Solar Cells —— 116
3.3	Organic Junction Solar Cells —— 117
3.3.1	Charge Separation in Organic Solar Cells —— 117
3.3.1.1	Donor–Acceptor Junctions for Exciton Splitting —— 117
3.3.1.2	Charge Transfer States —— 118
3.3.1.3	Detailed Energetic Picture of Charge Transfer States —— 124
3.3.2	Flat Heterojunctions —— 130
3.3.3	Bulk Heterojunctions —— 132
3.3.4	Mixing of Polymers and Fullerenes —— 135
3.3.4.1	Pure Materials —— 135
3.3.4.2	Mixed Films —— 143
3.4	Hybrid Junction Solar Cells —— 150
3.4.1	Dye-Sensitized Solar Cells —— 151
3.4.1.1	Liquid Electrolyte Dye-Sensitized Solar Cells —— 151
3.4.1.2	Solid-State Dye-Sensitized Solar Cells —— 156
3.4.2	Inorganic–Organic Hybrid Solar Cells —— 158
3.4.2.1	Metal Oxide–Polymer Bulk Heterojunctions —— 159
3.4.2.2	Dye-Sensitized Metal Oxide–Polymer Hybrid Solar Cells —— 160
3.4.2.3	Hybrid Solar Cells with Inorganic Absorbers —— 163
3.5	Perovskite Solar Cells —— 165
3.5.1	History of Perovskite Solar Cells —— 166
3.5.1.1	Perovskite Solar Cell Architectures —— 167
3.5.2	Organo-Metal Halide Perovskite —— 168
3.5.2.1	Structure and Phase Transitions —— 169
3.5.2.2	Band Positions and Engineering —— 170
3.5.3	Layer Preparation Methods —— 171
3.5.3.1	One-Step Preparation Method —— 172
3.5.3.2	Two-Step Preparation Method —— 175
3.5.3.3	Flexible Perovskite Solar Cells —— 176

3.5.4	Device Physics —— 177	
3.5.4.1	Perovskite Solar Cell Efficiency Measurements —— 179	
3.5.4.2	Photoexcited States and Relaxation Dynamics —— 181	
3.5.4.3	Excitons and Exciton Binding Energy —— 182	
3.5.4.4	Charge Transport and Carrier Diffusion Length —— 183	
3.5.4.5	Interfaces in PSCs —— 184	
3.5.4.6	Ferroelectricity —— 189	
3.5.5	Stability —— 190	
3.5.6	Outlook —— 191	

4 Characterization Techniques —— 193

- 4.1 Characterization of Solar Cell Components —— 193
- 4.1.1 Absorption Spectra of Photoactive Materials —— 193
- 4.1.2 Emission Spectra of Photoactive Materials —— 195
- 4.1.3 Ellipsometry —— 198
- 4.1.4 Quality of Transparent Contacts —— 202
- 4.1.4.1 Sheet Resistance —— 202
- 4.1.4.2 Transmission —— 204
- 4.1.4.3 Haacke's Figure of Merit for Transparent Contacts —— 206
- 4.2 Steady-State Device Characterization —— 208
- 4.2.1 Total Absorption Measurements —— 208
- 4.2.2 Current Density–Voltage Measurements —— 212
- 4.2.3 External Quantum Efficiency Measurements —— 218
- 4.2.4 Light Intensity–Dependent Current–Voltage Measurements —— 221
- 4.3 Time-Resolved Characterization Techniques —— 226
- 4.3.1 Transient Photocurrent and Photovoltage Decay Measurements —— 226
- 4.3.2 Photo-CELIV Measurements —— 230
- 4.3.3 Impedance Spectroscopy —— 233
- 4.3.4 Transient Absorption Spectroscopy —— 238
- 4.3.5 Time-Resolved Photoluminescence Spectroscopy —— 243

5 Fabrication and Device Lifetime —— 248

- 5.1 Processing —— 249
- 5.1.1 Processing Steps —— 250
- 5.1.1.1 Substrate —— 250
- 5.1.1.2 Transparent Conductive Electrodes —— 252
- 5.1.1.3 Active Semiconducting Film —— 254
- 5.1.1.4 Counter Electrode —— 255
- 5.1.1.5 Encapsulation Layer —— 255
- 5.1.2 Coating Technologies —— 257

5.1.3 Module Design —— 262
5.2 Stability and Lifetime —— 265

6 Conclusion and Outlook —— 268

Bibliography —— 269

Index —— 289

Preface

Over the past years the field of organic and hybrid solar cells has attracted many researchers and the progress has been tremendous. Whereas the organic electronics with organic light-emitting diodes and organic field-effect transistors has been seen as a promising technology with market readiness, the organic and hybrid solar cells have remained for a long time a niche field of research due to their low efficiencies and short lifetimes. However, over the past few years the efficiencies have increased drastically and also research on the lifetime of organic and hybrid solar cells has shown promising results. Until 2012 organic bulk heterojunction and dye-sensitized solar cells have been the focus of research. Since first reports of 10% efficient perovskite solar cells came out, many researchers have started to work on these exciting solar cell materials. As perovskite solar cells have been achieved with efficiencies of >20%, in terms of efficiency these cells are already competitive to inorganic semiconductors, even though reproducibility and lifetime issues need to be solved. Indeed, with these perovskite solar cells, the field of organic and inorganic solar cells seems to merge somehow with the field of inorganic solar cells as these cells are solution processed like the organic solar cells but in many aspects resemble the inorganic solar cells in their behavior. It remains exciting whether other new materials with similar properties will be found in future.

This book is based on a lecture given at the University of Konstanz for master's students with background in solid-state physics. We have not yet found such an introductional textbook on organic and hybrid photovoltaics and hope that this book will close that gap and be the starting point for everyone who is interested in this topic. We kept it short to focus only on the general ideas and methods. The reader is advised to consult the literature for deeper insights into the topics. The book is designed to give the interested reader an overview and to discuss the most relevant topics. It will be very helpful as a starting point when working on organic and hybrid solar cells. We also hope that the more advanced researcher finds it useful when consulting it for different aspects in this field of research and hopefully as a source of inspiration. The reader should have some basic background in solid-state physics, even though the most important aspects concerning semiconductor physics are briefly introduced. The book is divided into six chapters. It starts with a short general introduction giving a brief historical overview. In Chapter 2 we describe semiconductors and junctions, including an introduction to inorganic semiconductors, before discussing organic semiconductors. Chapter 3 is devoted to the working mechanisms in organic and hybrid solar cells. In this chapter the basic principles of solar cells are described and then the different organic and hybrid solar cell architectures and their working mechanisms are discussed in more detail, including bulk heterojunction, dye-sensitized, hybrid, and perovskite solar cells. In Chapter 4 we discuss the most relevant characterization techniques to investigate organic and hybrid solar cells, which includes steady-state as well as time-resolved characterization techniques. Finally in Chapter 5 we discuss some issues concerning fabrication and device lifetime, the important factors for commercialization of organic and hybrid solar cells. We finish the book with a brief conclusion and outlook.

Certainly the book focuses on many important and interesting aspects. It is a very opinionated selection of topics. There are many exciting topics that could have additionally been included. However, that would have extended the book considerably and might be less appropriate for an introduction into the field.

We realized that writing a book is far more time consuming than writing a research or review article, thesis, or book chapter as it is much longer and contains a variety of aspects. Not in all of them we are experts or do research on, which means that we had to collect a lot of information and summarize the most important aspects in a concise way. The book is certainly not perfect and we are very happy to get inputs from the readers on what could or even should be added or improved in future.

Finally we thank everyone who contributed to this book in one or another way by supporting us, especially all members of the "Hybrid Nanostructures" group at the University of Konstanz for their fruitful discussion on various topics. We also thank the students who listened to our lecture, gave us important feedback, and helped us correct the script. Finally we thank our families for their continuous support and encouragement.

Lukas Schmidt-Mende & Jonas Weickert
Konstanz, 2016

1 Introduction

1.1 General Introduction

Currently the main source of our energy supply is delivered through not only fossil fuels, such as oil, gas, and coal, but also nuclear power using ^{235}U. As it becomes more and more apparent that these sources are limited and environmentally problematic due to the large CO_2 emission or the danger of radioactive pollution, there is a strong need for alternative sources to deliver the needed energy.

A combination of increasing energy demands and thereby the increasing CO_2 emission, which is a consequence of combustion of fuels, leads to global warming. In Figure 1.1 the temperature anomalies and CO_2 emissions over the years are plotted and a strong correlation between the emissions and the global warming is found.

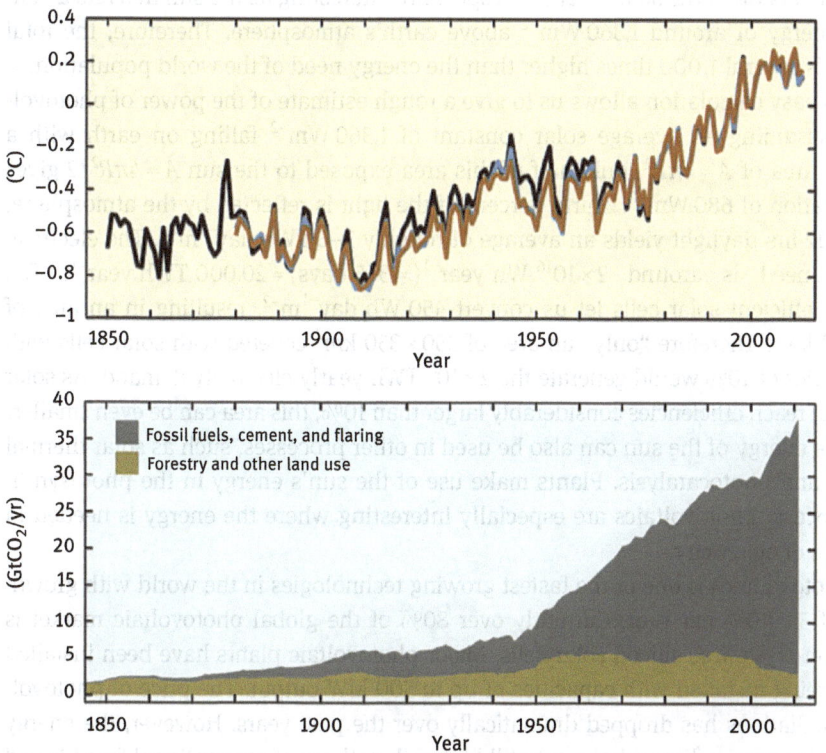

Fig. 1.1: Temperature anomalies and CO_2 emission. Top: Annually and globally averaged combined land and ocean surface temperature anomalies relative to the average over the period 1986 to 2005 with different data sets indicated by colors. Bottom: Global anthropogenic CO_2 emission. Quantitative information of CH_4 and N_2O emission time series from 1850 to 1970 is limited Figure adapted from [1].

In addition to these environmental issues we see a limitation of fossil fuels, which will make them more and more expensive. If we want to avoid these effects, we need to save energy and simultaneously find alternative energy sources that are renewable. Currently the largest renewable energy source are hydropower stations. Other renewable energy sources are wind power, biomass energy, geothermal power, biofuel production, tidal power, and solar power. A finding by the International Energy Agency states that renewable power could increase from just over a fifth of global electricity generation today to nearly a third by 2030. This would be a bigger share when compared to coal, gas, or nuclear power. Photovoltaics are expected to have the largest potential to grow significantly.

Renewable energy sources will be successful only if they are delivered at a competitive price compared to conventional energy sources based on fossil fuels. As the price of fossil fuels will rise due to larger demands, it will make renewable energy sources more attractive.

Conversion of the solar energy is especially interesting as the sun delivers a constant energy of around 1,360 Wm^{-2} above earth's atmosphere. Therefore, the total energy is several 1,000 times higher than the energy need of the world population.

An easy calculation allows us to give a rough estimate of the power of photovoltaics. Assuming an average solar constant of 1,360 Wm^{-2} falling on earth with a surface area of $A = 4\pi R^2$, only half of this area exposed to the sun $A = 4\pi R^2/2$ gives a irradiation of 680 Wm^{-2}. Thirty percent of the light is reflected by the atmosphere, and 6–12 hrs daylight yields an average of roughly 3–6 kWh day^{-1}m^{-2}. The electrical energy need is around 2×10^{16} Wh year^{-1} ($=365$ days) $= 20.000$ TWh year^{-1}. Ten percent efficient solar cells let us convert 450 Wh day^{-1}m^{-2}, resulting in an area of 1.2×10^5 km^2. Therefore "only" an area of 350×350 km^2 covered with solar cells with efficiencies of 10% would generate the 2×10^4 TWh yearly electricity demand. As solar cells can reach efficiencies considerably larger than 10%, this area can be even smaller.

The energy of the sun can also be used in other processes, such as solar thermal energy and photocatalysis. Plants make use of the sun's energy in the photosynthesis process. Photovoltaics are especially interesting where the energy is needed in the form of electricity.

Photovoltaics is one of the fastest growing technologies in the world with growth rates of 35–40% per year. Currently over 80% of the global photovoltaic market is based on crystalline silicon solar cells. Major photovoltaic plants have been installed or are being installed with capacities of up to 500 MW output. The price of photovoltaic installations has dropped dramatically over the past years. However, the energy costs of inorganic photovoltaics are still higher than those of conventional fossil-based energy sources. Organic solar cells allow a simple and cheap fabrication technique as they usually can be processed from a solution and at low temperatures. Even a roll-to-roll fabrication on plastic substrates seems to be feasible. It is expected that this can lower the cost considerably, if efficiency and lifetime become comparable to inorganic solar cells.

1.2 Brief History

In this chapter we briefly describe the history of inorganic, organic, and hybrid solar cells, listing only the most important milestones that have been achieved so far with respect to photovoltaics. It is interesting that some findings have been made by different groups in the same year, indicating that in these cases the research development promoted such findings or in other words the time was just ripe.

1.2.1 Inorganic Solar Cells

The word *photovoltaics* is a combination of the Greek word for light $\varphi\omega\varsigma$ and the name of the Italian physicist and chemist Alessandro Volta, the inventor of the battery. The photovoltaic effect was first recognized in 1839 by the French physicist Alexandre Edmond Becquerel (the father of Antoine Henry Becquerel, known for winning the Nobel Prize for the radioactivity). While working in his father's lab at age 19, he experimented with silver chloride in an acidic solution and illuminated it while keeping it connected to platinum electrodes, which generated a voltage and current. Over 30 years later, in 1873, Willoughby Smith and Joseph May discovered that selenium changes its electrical resistance under illumination, which triggered research on this effect and in 1877 led to the observation of the photovoltaic effect in solid-state selenium by Williams G. Adams and Richard E. Day [2]. In 1883, the first photovoltaic device was developed by Charles Fritts. He coated selenium with a thin layer of gold to form a junction, which gave a solar cell performance of ~1%. Only few years later, in 1887, Edward Weston received the first patents (US389124 and US389125) on solar cells.

Many years later, in 1946, Russell Ohl from Bell Labs received the first patent on modern junction semiconductor solar cells (US2402662, "Light-sensitive device"). This can be seen as the starting point of the modern solar cell. At Bell Labs, Chapin, Fuller, and Pearson further developed this into a diffused p–n junction solar cells by diffusing dopants, which had an efficiency of 6% with a remarkable stability [3]. The cell was remeasured in 2005 – almost 50 years later – and still it showed an efficiency of 5.1%. Hoffman Electronics Corporation pioneered the fabrication of solar cells and improved its efficiency. Still the costs of these cells were about $250 per Watt, a multiple of the cost of the energy delivered by a coal plant. These high costs have always been the major hurdle to purchase photovoltaics onto the mass market. The breakthrough of solar technology came with the advent of satellites, where solar cells could deliver the energy required to power them. The processing had been further improved, which allowed to cut down the production costs considerably. With the growing silicon industry, silicon from feedstock became largely available and silicon even with minor imperfections could still be used for solar cells, although useless for the silicon electronics. Today crystalline silicon is the main material and gives highest efficiencies of over 25%. GaAs solar cells are close to 30% efficiency, and the most efficient multijunction solar cells now reach values of over 45%. However, these high efficiencies also come with a high production cost, making the price exorbitant and uninteresting for day-to-day applications.

1.2.2 Organic and Hybrid Solar Cells

In this subsection we now focus on the development of organic and hybrid solar cells. It gives an overview over the most important device architectures that have been developed up to now, which will be discussed in detail in chapter 3.

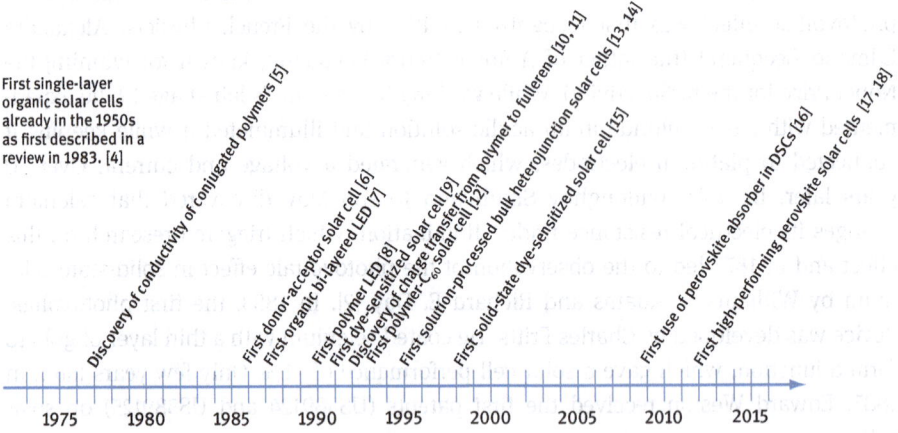

Fig. 1.2: **Milestones in organic and hybrid solar cell research.** In this timeline we have added the most important discoveries that have led to the state-of-the art organic and hybrid solar cells over the last few decades.

In Figure 1.2, we have plotted the most important milestones that have been achieved Interestingly, research on organic solar cells started very early sometime in the 1950s as described by a review on organic solar cells from 1983 written by Chamberlain [4]. Here single-layer organic solar cells have been investigated. An important milestone was the discovery of conductivity of conjugated polymers by Heeger, MacDiarmid, and Shirakawa in 1977 [5]. In 2000, they have been awarded the Nobel Prize in Chemistry for their discovery. The first modern donor–acceptor solar cell was then introduced by Tang in 1986 using a copper–phthalocyanine in combination with a perylene derivative – materials that are still under investigation. The solar cell exhibited an impressive efficiency of about 1% under AM2 illumination [6]. A tear later, the same author discovered the organic LED [7]. This was followed in 1990, by the discovery of polymer LEDs [8]. In the 1990s, the field developed rapidly. In 1991 O'Regan and Grätzel reported about the first dye-sensitized solar cell based on mesoporous TiO_2 [9]. In 1992, the groups of Yoshino and Wudl reported about electron transfer from polymer to fullerene, which is the basic working mechanism for organic donor–acceptor solar cells [10, 11]. This was followed by the first polymer–C_{60} solar cells in 1993 [12] and the first solution-processed bulk heterojunction solar cells in 1995 [13, 14]. In 1998, the first solid-state dye-sensitized solar cells were reported by Bach et al. [15]. This has triggered a lot of research in hybrid solar cells. All these cell types have been improved considerably since their first discovery. The newest milestone is the use of methylammonium lead iodide perovskite, which was first introduced as a sensitizer in a liquid electrolyte dye-sensitized solar cell architecture [16]. The real breakthrough of perovskite solar cells came in 2012 when the groups of Snaith and Grätzel/Park reported perovskite solar cells with efficiencies around 10% [17, 18]. Now this type of solar cell has further developed in terms of efficiency reaching more than 20%. Also we find a variety of different solar cell architectures leading to high-efficiency perovskite solar cells. The device physics is quite different from organic and dye-sensitized solar cells. Therefore, we see perovskite solar cells as a new additional branch of hybrid solar cells, which show remarkable properties resembling in many aspects the properties of inorganic solar cells.

Even though in the described timeline (Figure 1.2) only some important milestones are reported, it can be clearly seen that they cover the full range with the newest important results only a few years back. We expect that this timeline will need updating in future as more – not yet been discovered – milestones will add to it. This is a very vibrant field of research, and we expect many exciting further developments and achievements over the next decade.

2 Semiconductors and Junctions

To understand the function of a solar cell, a device with a complicated multilayer structure, it is imperative to take a closer look at the properties of the constitutive parts. For organic and conventional inorganic solar cells, there are semiconductors that absorb and convert light and transport charge carriers to external metal electrodes.

This chapter gives a brief overview of inorganic semiconductors and introduces the concept of doped semiconductors. We will then move on to a more general introduction to organic semiconductors. The final section of the chapter discusses semiconductor–semiconductor and semiconductor–metal junctions, which play a central role in the working mechanisms of photovoltaic devices.

For further reading on inorganic semiconductors, see, e.g. Hunklinger [19], Seeger [20], Kittel [21], Grundmann [22], and many others. Fundamentals of organic semiconductors are discussed in more detail, e.g. in Pope and Swenberg [23], Brütting [24], and Köhler and Bässler [25]. A more comprehensive overview of different electronic junctions can be found in Milnes [26] and Mönch [27].

2.1 Introduction to Inorganic Semiconductors

A semiconductor is an ensemble of atoms. It therefore exhibits properties like mass density, melting point, and stiffness, which arise from a collective behavior of the constituent elements. For a photovoltaic device, which converts photon energy into electric energy, mainly the optoelectronic properties are of interest.

2.1.1 Electronic States in Inorganic Semiconductors

2.1.1.1 Energy Bands in Solids
Within an atom, electrons can occupy only discrete electronic states (orbitals) as schematically depicted in Figure 2.1. For two atoms in close proximity the electronic states combine and form molecular orbitals. These molecular energy levels lie slightly below and above the original energy level of a single atom and effectively every energy level is split into two new levels. For every atom that is added to the solid the energy levels split accordingly, so that for N atoms every orbital splits into N states. For large numbers of atoms the energy splitting between two adjacent levels becomes very small, which can be separated by quantum mechanically forbidden energy regions where no states are located.

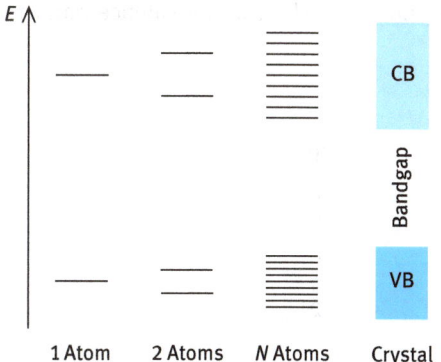

Fig. 2.1: Energy level splitting in a solid. When two or more atoms are brought together to form a molecule or a solid, the original energy levels of electrons split up so that for large numbers of atoms energy bands are formed eventually.

In semiconductors, the energetically highest band occupied by electrons is called the *valence band* (VB), and the energetically lowest unoccupied band is called the *conduction band* (CB). The energy gap between the valence band and the conduction band is called the *bandgap*. The bandgap of semiconductors plays an important role in photovoltaics, since electrons can be excited from the valence band to the conduction band upon photon absorption. If the photon carries more energy than the bandgap, the electron is initially elevated to a higher energetic state, from which it rapidly relaxes down to the lowest energy state in the conduction band via the so-called thermalization (this relaxation is not accompanied by the emission of a photon, but the excess energy of the electron induces vibration of the semiconductor crystal). This process typically occurs within femtoseconds (fs) and is schematically depicted in Figure 2.2. Since there are no states available within the bandgap, the excited electron does not directly relax back to its ground state but remains in the valence band. Its potential energy can be harvested by building a solar cell out of the semiconductor. Compared to the fs timescale of thermalization the lifetime of this excited state is very long and can be on the order of μs or ms.

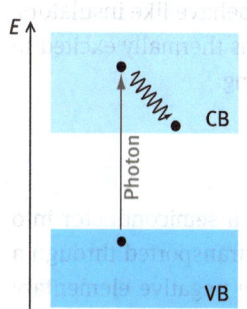

Fig. 2.2: Thermalization upon photon absorption in a semiconductor. In the case that an electron is excited from the valence band to a higher energy state in the conduction band, it rapidly relaxes to the lower edge of the conduction band.

Depending on the bandgap we distinguish between conductors (metals), semiconductors, and insulators (Figure 2.3). This terminology originates from the fact that only electrons in bands that are almost empty (like the conduction band) can move freely since there are many empty states to move in, whereas the valence band is fully occupied by electrons.

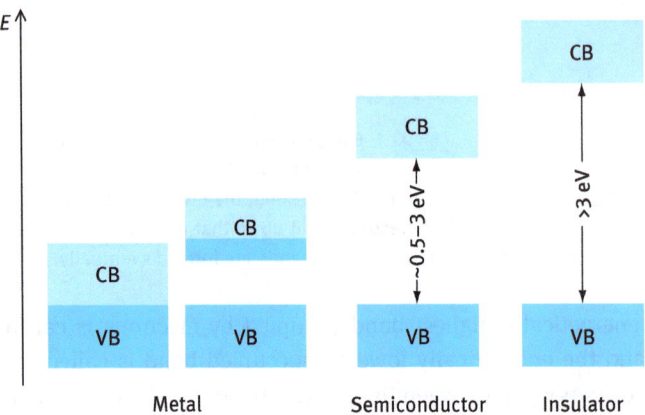

Fig. 2.3: Schematic illustration of the energy bands of metals, semiconductors, and insulators. Due to energetic proximity of bands or half-filled bands, metals show high conductivities. Solids with bandgaps between 0.5 and 3 eV are referred to as semiconductors, whereas wider bandgap materials are insulators.

In metals, the highest occupied band and the lowest unoccupied band are energetically very close or bands are only partly filled so that the thermal energy suffices to excite the electrons into a (partly) empty band. Therefore, the bands in metals are often not referred to as conduction or valence band. For increasing bandgaps, the probability of finding electrons with sufficient thermal energy to reach the conduction band as well as the intrinsic conductivity of the solid decreases. The solids with bandgaps of roughly between 0.5 and 3 eV are usually referred to as semiconductors. These materials exhibit a certain conductivity at room temperature, although much lower than metals. At low temperature semiconductors behave like insulators. If the bandgap increases further, there are virtually no electrons thermally excited to the conduction band and the material is electronically insulating.

2.1.1.2 Electrons, Holes, and Polarons

Since a solar cell is a device that converts light shining onto a semiconductor into electric current, we are especially interested in how charge is transported through a semiconductor. In addition to electrons, which carry a single negative elementary charge, the so-called holes also play an important role as charge carriers in semiconductor physics. Let us consider a single electron in a semiconductor that is removed from its ground state in the valence band to an empty state in the conduction band as shown in Figure 2.4. This electron excitation can either be due to interaction with a photon or occur thermally, when the electron gets sufficient kinetic energy from crystal vibrations. As mentioned in the previous section, this electron can relatively freely move through the semiconductor as long as it remains in the conduction band since there are many unoccupied electron states. If a driving force like an external electric field is applied to the semiconductor, electrons in the conduction band can carry a current from the cathode to the anode of the external source.

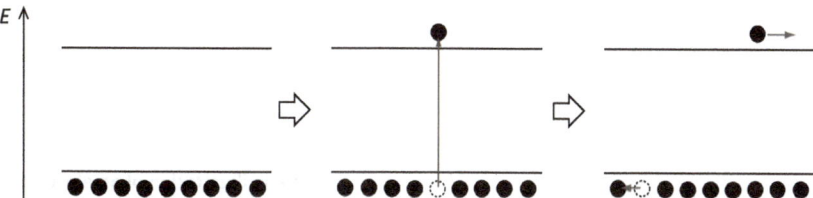

Fig. 2.4: **Electrons and holes in semiconductors.** As discussed in the previous section, in the ground state the valence band of a semiconductor is occupied by electrons, whereas the conduction band is empty. After an electron is excited to the conduction band it can move through the semiconductor since there are free states available. On the other hand, the hole remaining in the valence band can be occupied by adjacent electrons. A sequence of several of these processes gives a net movement of a positive charge, a so-called hole.

However, besides the electron in the conduction band we have to consider the situation in the valence band. The excited electron leaves behind a single unoccupied electron state in the valence band, which can be subsequently occupied by one of the neighboring electrons. In turn once an electron occupies this empty state, it leaves behind a new unoccupied state. This iterative movement of electrons leads to a net movement of the unoccupied state in the opposite direction. This process can be described by a quasiparticle with one positive elementary charge, which is called an *electron hole* or most commonly just a *hole*. Even though in most cases holes can move through the valence band only at a lower mobility compared to electrons in the conduction band, consideration of the hole current is important when dealing with semiconductors. There are semiconductor devices in which the majority of charges transported are holes.

Now that both electron and hole transport have been discussed, we further have to consider the environment of these charges when moving through a crystal. Every semiconductor is built of atoms, either neutral as in the case of pure materials like crystalline silicon or ions as in the case of crystals like NaCl. These atoms can be polarized in the presence of a charge as exemplarily depicted in Figure 2.5 for the case of an electron in an ionic crystal. Note, however, that the mechanism is the same in any polarizable medium. The electron attracts and repels the positively and negatively charged ions, respectively, which leads to a polarization of the electron's environment. When the electron moves through the crystal at velocities well below the phase velocity of light (which is the case for normal charge transport), this polarization moves together with the electron. Depending on the material, this leads to a certain screening of the charge of the electron modifying its physical properties. It is therefore helpful to treat the electron in combination with its induced polarization as a quasiparticle, which is called an *electron polaron*. In a similar fashion there are *hole polarons* for holes in polarizable solids.

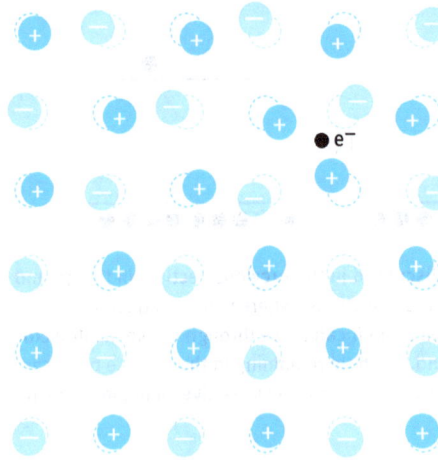

Fig. 2.5: Schematic illustration of an electron polaron in a crystal. A conduction band electron in a crystal attracts and repels positive and negative charges from the atoms constituting the crystal. The induced polarization of the electron environment travels through the crystal together with the electron. The combination of the electron with this polarization is called a polaron.

Mathematically polarons are complicate to handle, and there are only approximate solutions for a Hamiltonian describing a polaron when using Schrödinger's equation. The effective mass of a polaron depends on the coupling of the charge carrier to the polarizable environment, which is quantified through Fröhlich's coupling constant α. Depending on the material, α varies over several orders of magnitude. Depending on the magnitude of α, Also the effective mass and the energy of a polaron are also experimentally found to differ. Generally, the effective mass of a polaron is higher than the effective mass of the charge carrier when neglecting the polarization of its environment.

2.1.1.3 Band Structure

We now take a closer look at the energy states in the valence and conduction bands of a crystal that can be occupied by electrons. The energetic location of these states depends on the momentum of the electrons in the crystal. We therefore have to consider the periodicity of the crystal and in particular the reciprocal crystal lattice since there is a simple relation between the coordinates in reciprocal space and the momentum as outlined in this section.

A crystal is characterized by its translational symmetry, i.e. moving through the crystal reproduces its properties with a certain periodicity. The first step to mathematically handle a crystal is the introduction of a point lattice. These points are spatially distributed with respect to the translational symmetry of the crystal and mark points of identical crystal properties like the relative position of anions and cations composing the crystal with respect to these lattice points. This is exemplarily shown in Figure 2.6 for a two-dimensional crystal with two different ions.

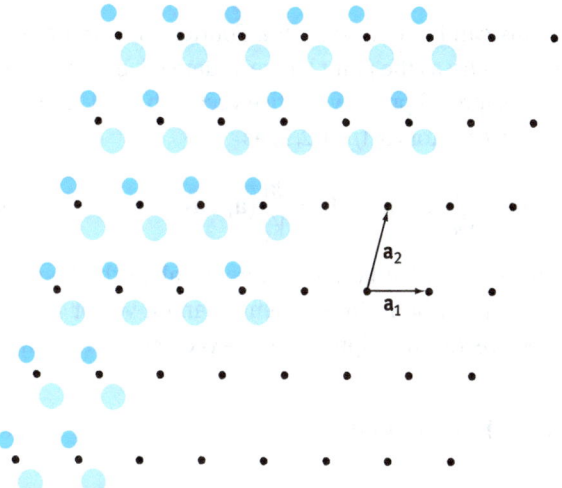

Fig. 2.6: Schematic representation of a two-dimensional point lattice. The point lattice (represented by black dots) is defined by lattice vectors $\mathbf{a}_1$ and $\mathbf{a}_2$ and exhibits the same symmetry as crystal properties like the position of ions (represented by gray circles), etc.

For any quantity $\mathfrak{U}$ describing a certain property of the crystal this can be expressed as

$$\mathfrak{U}(\mathbf{r}) = \mathfrak{U}(\mathbf{r} + \mathbf{R}), \qquad (2.1)$$

where $\mathbf{r}$ denotes an arbitrary position vector and $\mathbf{R}$ is a so-called translation vector. In three dimensions this translation vector is given as

$$\mathbf{R} = n_1 \mathbf{a}_1 + n_2 \mathbf{a}_2 + n_3 \mathbf{a}_3, \qquad (2.2)$$

where the n_i ($i = 1, 2, 3$) are integers and $\mathbf{a}_1$, $\mathbf{a}_2$, and $\mathbf{a}_3$ are the (not necessarily orthogonal) primitive lattice vectors, which define a three-dimensional coordinate system with the symmetry of the point lattice. The n_i being integers implicates that $\mathbf{R}$ is a transition through the crystal to an environment of identical properties since the relative position of $\mathbf{r} + \mathbf{R}$ in the crystal with respect to the point lattice is the same as the position $\mathbf{r}$.

Using the primitive lattice vectors $\mathbf{a}_i$ ($i = 1, 2, 3$) it is possible to construct volume repeat units, which can assemble the whole crystal. Most straightforward is the unit cell, which is spanned by the three primitive lattice vectors and has the volume

$$V_u = (\mathbf{a}_1 \times \mathbf{a}_2) \cdot \mathbf{a}_3. \tag{2.3}$$

We can now introduce a reciprocal lattice using the reciprocal lattice vectors $\mathbf{b}_1$, $\mathbf{b}_2$, and $\mathbf{b}_3$, which need to fulfill the condition

$$\mathbf{b}_i \cdot \mathbf{a}_j = 2\pi \delta_{ij}, \tag{2.4}$$

where δ_{ij} is the Kronecker delta. This can be derived from a Fourier transform of the crystal property $\mathfrak{U}$ with the same symmetry as the point lattice as derived, e.g. in Hunklinger [19]. From this condition the reciprocal primitive lattice vectors $\mathbf{b}_i$ ($i = 1, 2, 3$) can be determined from the primitive lattice vectors $\mathbf{a}_j$ ($j = 1, 2, 3$) as

$$\mathbf{b}_1 = \frac{2\pi}{V_u}(\mathbf{a}_2 \times \mathbf{a}_3), \quad \mathbf{b}_2 = \frac{2\pi}{V_u}(\mathbf{a}_3 \times \mathbf{a}_1), \quad \mathbf{b}_3 = \frac{2\pi}{V_u}(\mathbf{a}_1 \times \mathbf{a}_2). \tag{2.5}$$

The space spanned by the reciprocal primitive lattice vectors is called the **k**-space since the $\mathbf{b}_i$ ($i = 1, 2, 3$) have the dimension of an inverse length and a vector in this space is a wave vector **k**. Using these vectors a reciprocal lattice vector can be defined analogous to Equation (2.2) as

$$\mathbf{G} = h_1 \mathbf{b}_1 + h_2 \mathbf{b}_2 + h_3 \mathbf{b}_3 \tag{2.6}$$

for integers h_i ($i = 1, 2, 3$).

The reciprocal space is of particular interest when dealing with the momentum of a particle since the momentum is given as $\hbar \mathbf{k}$. Many processes like electron transport and lattice dynamics are easier to describe in **k**-space so problems in solid-state physics are often addressed in reciprocal space rather than in real space.

An important concept in this respect is the first Brillouin zone, which has been developed by Leon Brillouin. The first Brillouin zone is a primitive cell of the reciprocal space. Magnitudes that exhibit the same symmetry as the reciprocal point lattice are unambiguously described by their properties within the first Brillouin zone, which is why for many problems it is sufficient to look at the first Brillouin zone only.

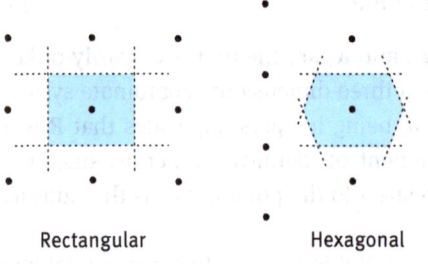

Fig. 2.7: Construction principle of the first Brillouin zone for a rectangular and a hexagonal reciprocal point lattice. The first Brillouin zone is the area defined by the perpendicular bisectors between the center point and the adjacent points.

The first Brillouin zone can be constructed from an inverse point lattice as exemplarily shown for two-dimensional lattices in Figure 2.7. For a certain point that serves as the center, the perpendicular bisector is drawn for each connection between the center point and each of the adjacent reciprocal lattice points. The area (volume in three dimensions) defined by these bisectors is the first Brillouin zone. Commonly, the center of the Brillouin zone is referred to as Γ-point.

In real crystals the Brillouin zone is three dimensional as shown in Figure 2.8 for a face-centered cubic (fcc) crystal structure. Other points on the surface of the first Brillouin zone are also labeled with letters as commonly used in group theory. As will be shown later these characteristic points are helpful when drawing energy band diagrams since energies are often represented with respect to connections between these different points as x-axis.

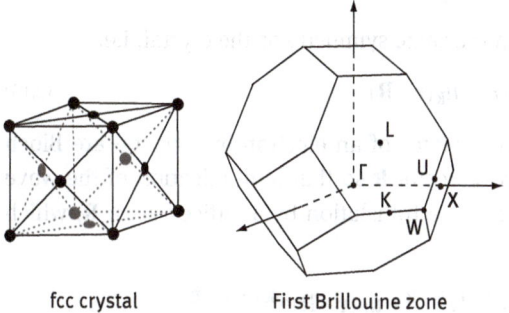

fcc crystal   First Brillouine zone

Fig. 2.8: Sketch of the face-centered cubic (fcc) crystal structure and the corresponding first Brillouin zone. Following the nomenclature of group theory, important points of the Brillouin zone are labeled. Connections of these points are typically used as x-axis in band diagrams.

Since we are still interested in the structure of the energy bands in a crystal with respect to the momentum of the electron we now have to solve Schrödinger's equation for an electron in the crystal considering the symmetry of the crystal lattice or rather of the Brillouin zone since the momentum can be easily treated in **k**-space.

For an electron in a potential $V(\mathbf{r})$ the time-independent Schrödinger equation reads

$$\left(\frac{-\hbar^2}{2m}\Delta + V(\mathbf{r})\right)\Psi_\mathbf{k}(\mathbf{r}) = E_\mathbf{k}\Psi_\mathbf{k}(\mathbf{r}), \qquad (2.7)$$

where m is the mass of the electron, $\Psi_\mathbf{k}(\mathbf{r})$ the wave function of a particular state, and E the respective energy of this state. The notation with the wave vector **k** is used as index since **k** is a good quantum number, which means that every quantum mechanical state is unambiguously described by **k**. We can now take advantage of the symmetry properties of the crystal when considering the following: since the environment at a certain point **r** in the crystal is the same as at a point **r** + **R** as discussed before, there is a priori no argument for why the occupation of an electronic state at point **r** should be favored over occupation of an electronic state at point **r** + **R**.

The fact that the environment of a certain point in the crystal is invariant under translation to an equivalent position in the point lattice implicates that we get the following expression for the potential:

$$V(\mathbf{r}) = V(\mathbf{r} + \mathbf{R}), \tag{2.8}$$

where again **R** denotes a translation vector. Furthermore, when we state that the occupation of electronic states exhibits the same invariance, we have an heuristic argument that the electron wave function is a so-called Bloch wave

$$\Psi_\mathbf{k}(\mathbf{r}) = u_\mathbf{k}(\mathbf{r}) e^{i\mathbf{k} \cdot \mathbf{r}}. \tag{2.9}$$

Here, $u_\mathbf{k}(\mathbf{r})$ is a periodic function with the same symmetry as the crystal, i.e.

$$u_\mathbf{k}(\mathbf{r}) = u_\mathbf{k}(\mathbf{r} + \mathbf{R}). \tag{2.10}$$

The Bloch theorem states that the eigenstates of an electron in a crystal are Bloch waves. This means that there is a wave vector **k** so that multiplication of the wave function with a phase factor $e^{i\mathbf{k} \cdot \mathbf{R}}$ equals a translation by a lattice vector **R**, which can be easily seen from

$$\begin{aligned}\Psi_\mathbf{k}(\mathbf{r}) e^{i\mathbf{k} \cdot \mathbf{R}} &= u_\mathbf{k}(\mathbf{r}) e^{i\mathbf{k} \cdot \mathbf{r}} e^{i\mathbf{k} \cdot \mathbf{R}} = u_\mathbf{k}(\mathbf{r} + \mathbf{R}) e^{i\mathbf{k} \cdot \mathbf{r}} e^{i\mathbf{k} \cdot \mathbf{R}} \\ &= u_\mathbf{k}(\mathbf{r} + \mathbf{R}) e^{i\mathbf{k} \cdot (\mathbf{r} + \mathbf{R})} = \Psi_\mathbf{k}(\mathbf{r} + \mathbf{R}).\end{aligned} \tag{2.11}$$

It is further possible to show that two Bloch waves are identical if the wave vector **k** is shifted by a reciprocal lattice vector **G** from Equation (2.6), i.e.

$$\Psi_\mathbf{k}(\mathbf{r}) = \Psi_{\mathbf{k}+\mathbf{G}}(\mathbf{r}). \tag{2.12}$$

Accordingly, the same holds for the energy eigenstates

$$E_\mathbf{k} = E_{\mathbf{k}+\mathbf{G}}. \tag{2.13}$$

This has the important implication that for an electron in a crystal its wave function can always be solved within the first Brillouin zone since for wave vectors **k'** there can always be found a reciprocal lattice vector **G'** with

$$\mathbf{k'} + \mathbf{G'} = \mathbf{k}, \tag{2.14}$$

where **k** is in the first Brillouin zone. The wave functions of the electron as well as the respective eigenenergies exhibit the same periodicity as the reciprocal lattice. Accordingly, when we want to understand the band structure of the crystal, we only have to examine it in the first Brillouin zone.

To illustrate that we take a closer look at a simplified model of the band structure of a crystal. In this model we assume quasi-free electrons in the crystal, i.e. we assume $V(\mathbf{r}) = 0$ for all **r**. However, the symmetry of the crystal and implicated properties discussed so far should still hold, which is why the situation is referred to as empty lattice.

The Schrödinger equation then simply reads

$$\frac{-\hbar^2}{2m}\Delta\Psi_\mathbf{k}(\mathbf{r}) = E_\mathbf{k}\Psi_\mathbf{k}(\mathbf{r}), \tag{2.15}$$

which has the energy eigenvalues

$$E_\mathbf{k} = \frac{\hbar^2}{2m}|\mathbf{k}|^2 \tag{2.16}$$

for a plane wave

$$\Psi_\mathbf{k}(\mathbf{r}) = u_0 e^{i\mathbf{k}\cdot\mathbf{r}} \tag{2.17}$$

with constant u_0. Because of Equation (2.13) we see that the periodicity of the reciprocal lattice leads to

$$E_\mathbf{k} = E_{\mathbf{k}+\mathbf{G}} = \frac{\hbar^2}{2m}|\mathbf{k}+\mathbf{G}|^2, \tag{2.18}$$

which means that the energy eigenvalues for an electron in an empty lattice are parabolas in the $\mathbf{k}$-space shifted by lattice vectors $\mathbf{G}$. This is illustrated in Figure 2.9(a) for a one-dimensional lattice with primitive lattice parameter a and accordingly reciprocal primitive lattice parameter $g = 2\pi/a$, where reciprocal lattice vectors are integer multiples of g. Due to the symmetry of the lattice the Schrödinger equation can be solved within the first Brillouin zone (Figure 2.9(b)), i.e. for wave vectors

$$\frac{-\pi}{a} \leq k \leq \frac{\pi}{a}. \tag{2.19}$$

To get the eigenenergies the parabolas from Figure 2.9 are shifted by integer multiples of g. The resulting plot gets us toward a more precise band diagram, since we see that (1) the allowed energies for an electron in the crystal depend on the $\mathbf{k}$-vector, i.e. on the momentum of the electron, and (2) there are multiple allowed energies for an electron with wave vector $\mathbf{k}$, i.e. electrons with identical momenta can occupy states at different energies. Furthermore, some states are degenerated since different energy eigenvalue parabolas meet, e.g. at the borders of the Brillouin zone.

For three-dimensional crystals it is analogously possible to reduce the electronic states to the first Brillouin zone; however, the resulting $\mathbf{k} - E_\mathbf{k}$ plot is getting more complex. For more complicated crystal structures the energy eigenstates are also degenerated, i.e. projections from different locations in the $\mathbf{k}$-space into the first Brillouin zone end in identical points in the diagram.

In real crystals this degeneracy is lifted due to a nonvanishing potential $V(\mathbf{r}) \neq 0$. This is exemplarily shown in Figure 2.9(c) for a weak potential $V(\mathbf{r})$, which can be expanded in a Fourier series after reciprocal lattice vectors $\mathbf{G}$ as

$$V(\mathbf{r}) = \sum_{\mathbf{G}} V_{\mathbf{G}}(\mathbf{r})e^{i\mathbf{G}\cdot\mathbf{r}} \quad (2.20)$$

with coefficients $V_{\mathbf{G}}$. For a weak potential we can assume that only coefficients $V_{\mathbf{g}}$ play a role, i.e. only for the smallest reciprocal lattice vector $\mathbf{G} = \mathbf{g}$. This means that only the potential in the neighboring Brillouin zones is taken into account. One can then show that the originally degenerated eigenenergies from the empty lattice split at the border of the Brillouin zone by an energy of $2|V_{\mathbf{g}}|$ as indicated in Figure 2.9(c).

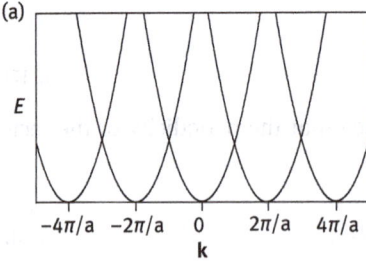

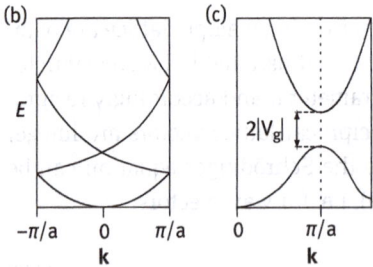

Fig. 2.9: Eigenenergies for an electron in an empty one-dimensional lattice with lattice constant a. (a) In k-space the eigenenergies are parabolas shifted by lattice vectors $n \cdot 2\pi/a$ for integers n. (b) Eigenenergies reduced to the first Brillouin zone. (c) Upon introduction of a nonvanishing potential the degeneracy from projections into the first Brillouin zone is lifted and a forbidden energy region can form, which is referred to as bandgap.

For more complex potentials the splitting of energy levels is more complex accordingly, but in the same manner there is the possibility that forbidden energy regions are forming, i.e. the crystal exhibits a bandgap. The band structure of a real crystal (AlN in this case) is shown in Figure 2.10. Since the band structure is three dimensional for a real three-dimensional crystal, the energy bands are typically represented along paths between significant points in the Brillouin zone as indicated in Figure 2.8 for an fcc lattice. AlN has a hexagonal wurtzite structure so that the shape of the Brillouin zone is different than in a fcc crystal and accordingly other letters from group theory are used to label points in the Brillouin zone. As apparent from the figure, certain energy regions can be accessed by electrons depending on their momentum so that energy bands are formed. AlN exhibits a bandgap of 6.2 eV at the Γ-point and even wider bandgaps for other direct transitions from the valence to the conduction band (the upper edge of the valence band is set to energy 0 in the figure).

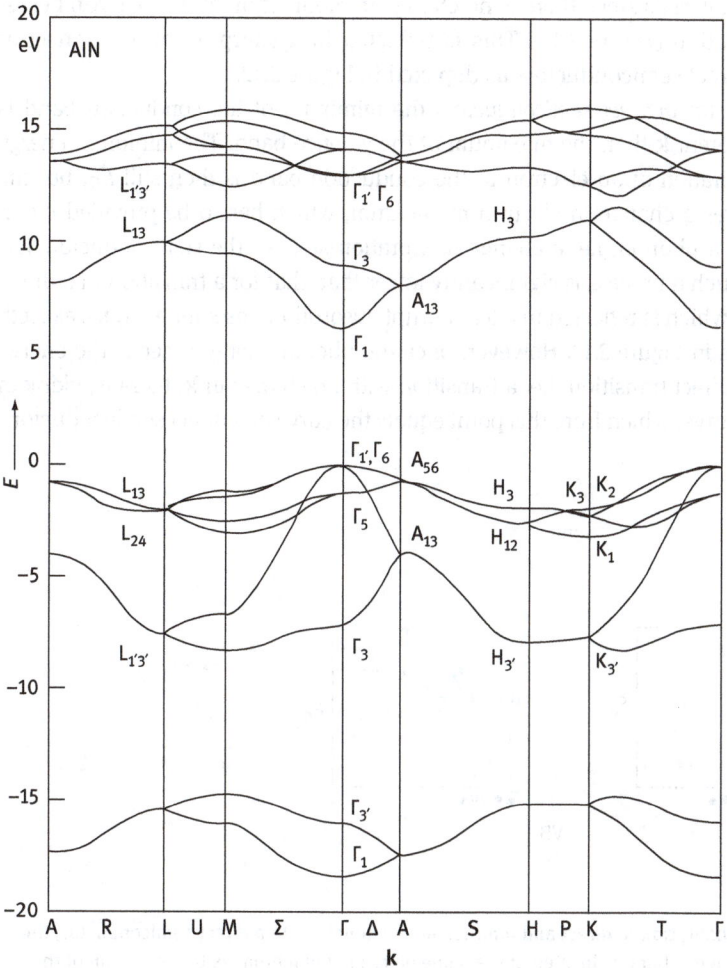

Fig. 2.10: Energy band diagram for an AlN crystal. Energy bands are shown along paths between significant points in the Brillouin zone. Figure adapted from [28].

2.1.1.4 Direct and Indirect Bandgaps

In the previous section we saw that the band edge energy depends on the momentum of the electron, i.e. on the wave vector **k**. Instead of the general term *bandgap* that we used in the previous sections it is now more precise to use the term *fundamental bandgap* for the energy E_G, which is the energy difference between the minimum of the conduction band and the maximum of the valence band for all **k**. For some semiconductors these extrema occur at the same. These semiconductors are called *direct semiconductors*. Optical transitions, i.e. the excitation of electrons from the conduction to the valence band upon interaction with a photon of energy E_G, are likely in these materials since there is no change in momentum of the electron necessary upon excitation (Figure 2.11). This is reflected in a sharp absorption onset at energy E_G for direct semiconductors as depicted in Figure 2.12.

In contrast, for *indirect semiconductors* the minimum of the conduction band is located at a different **k** than the maximum of the valence band. The minimum energy required for excitation of an electron to the conduction band is then still E_G, but the transition requires a change in electron momentum, which has to be provided by an interaction with a phonon, i.e. a change in vibration state of the semiconductor. The probability for such a process is significantly lower than that for a transition in a direct semiconductor, which is reflected in a less abrupt absorption onset for energies exceeding E_G as shown in Figure 2.12. However, once the photon energy exceeds the energy necessary for a direct transition, i.e. a transition with no change in **k**, there is a kink in the absorption curve, which from this point equals the curve of a direct semiconductor.

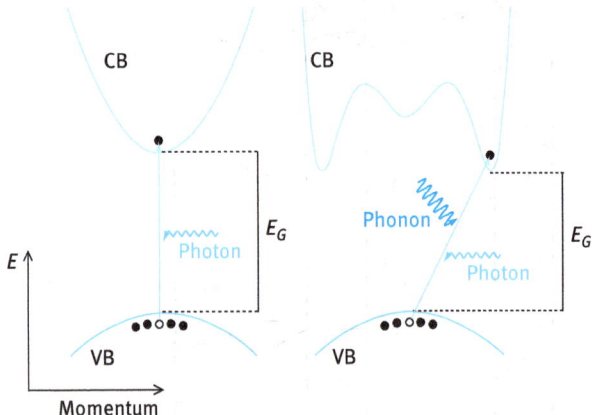

Fig. 2.11: Photon absorption in direct and indirect semiconductors. In a direct semiconductor, the maximum of the valence band is located at the same electron momentum as the minimum of the conduction band. A photon with energy E_G can therefore directly excite an electron to the conduction band. In contrast, for an indirect semiconductor a photon of energy E_G can only excite an electron in combination with a momentum transfer with a phonon since the conduction band minimum is located at a different electron momentum than the valence band maximum.

For application in photovoltaic devices direct semiconductors are more attractive since efficient light absorption for photons exceeding E_G can be achieved by relatively thinner absorber layers compared to indirect semiconductors. Nevertheless, crystalline silicon is the most commonly used semiconductor for solar cell applications due to its favorable bandgap energy with respect to the solar spectrum and charge transport properties, although it is an indirect semiconductor. GaAs is the most common direct semiconductor material for solar cells, currently providing the highest efficiencies (>30%) for single junction solar cells.

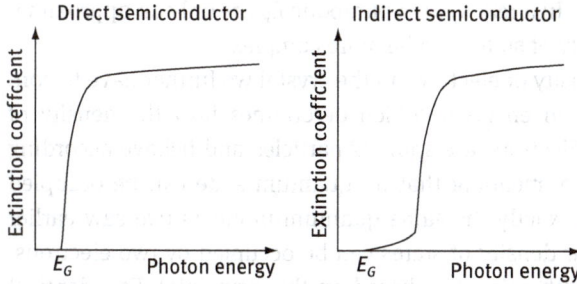

Fig. 2.12: Sketch of absorption spectra of a direct and an indirect semiconductor. A prompt absorption onset is typical for direct semiconductors, whereas a weak absorption over a certain energy region and a sudden onset at a higher energy is a typical footprint of an indirect semiconductor.

2.1.1.5 Density of States

So far we have discussed the structure of allowed energies in **k**-space, which lead us to band structures of crystals. Knowing the band structure of a crystal we can thus say which energies and momenta electrons can carry when occupying states in the crystal. However, for many problems it is mainly interesting to know how electron energies are distributed. We therefore have to discuss the density of electronic states and the distribution function, which determines the occupation of available states.

First, we look at the density of states in **k**-space. As we saw in the Schrödinger equation (Equation (2.7)) every quantum state is unambiguously described by a wave vector **k** and for a three-dimensional problem the components k_x, k_y, and k_z are good quantum numbers. Thus, in a crystal region of volume $L \times L \times L$, there are $L^3/(2\pi)^3$ quantum states, which follow from the condition that in each crystal direction $k_{x,y,z}L/(2\pi)$ has to be an integer for a wave function confined within L in this direction. Due to the Pauli exclusion principle and two possible spin states, each of these quantum states can be occupied by two electrons, so there are $2/(2\pi)^3$ electron states per unit volume. With this, we can write down an expression for the density of states in **k**-space:

$$g(\mathbf{k}) = \frac{2}{(2\pi)^3} d^3\mathbf{k}. \quad (2.21)$$

Since we are interested in the density of states with respect to the energy of the electron and not its momentum, we then have to transform this using the dependency of the energy eigenstates $E_\mathbf{k}$ on the wave vector **k** from Equation (2.16). In the simplest case of parabolic eigenenergies (isotropic around E_0) as discussed for quasi-free electrons in an empty lattice, one can show that this leads to an energy density of states (from now on referred to only as density of states) of

$$g(E) = \frac{(2m^*)^{3/2}}{2\pi^2 \hbar^3} \sqrt{E - E_0} \quad (2.22)$$

with effective electron mass m^*. In a real system parabolic E_k are only an approximation and the shape of the density of states can be more complex.

When interested in the density of electrons in the crystal we further have to consider the distribution of electron energies, which determines how the density of states is filled with electrons. Electrons are spin 1/2 particles and behave according to the Pauli exclusion principle, meaning that no quantum state can be occupied by two or more particles with exactly the same quantum numbers (we saw earlier that every quantum state in our density of states can be occupied by two electrons, one with spin +1/2 and one with spin −1/2, based on this principle). For identical particles with half-integer spins (fermions) the energy distribution is given by the Fermi–Dirac statistics, which accordingly also applies for electrons in a crystal:

$$f_e(E) = \frac{1}{\exp((E-E_F)/k_BT) + 1}. \tag{2.23}$$

Here k_B is Boltzmann's constant, T the absolute temperature, and E_F the Fermi level or electrochemical potential. At $T = 0$ the electrochemical potential is called the Fermi energy. Thermodynamically, E_F is the energy required to add one electron to the otherwise unchanged system. It can also be understood as the energy that has a probability of 1/2 of being occupied at thermodynamic equilibrium. It is important to note that E_F can be a hypothetical energy level, i.e. there is not necessarily an electron quantum state with the energy E_F existent in the crystal. For example, for a direct semiconductor at 0 K the Fermi level is located exactly at $(E_C - E_V)/2$, i.e. in the center of the bandgap, where no quantum state is located.

The shape of the Fermi–Dirac statistics for increasing absolute temperature is shown in Figure 2.13. While at 0 K there is a sharp transition from occupation probability 0 to 1 at $E = E_F$; the transition is broadened at increasing temperatures. Due to their thermal energy, k_BT, particles can occupy higher energetic states (broadening toward higher energies) while lower energetic states remain unoccupied (broadening toward lower energies).

It should be noted that the Fermi level itself exhibits a temperature dependence, which for a non-doped semiconductor (see next section) reads as

$$E_F(T) = \frac{E_C - E_V}{2} + \frac{3}{4} k_B T \ln\left(\frac{m_h^*}{m_e^*}\right), \tag{2.24}$$

where E_V and E_C are the edge energies of the valence and conduction bands and m_e^* and m_h^* the effective electron and hole mass, respectively (see common solid-state physics literature for a derivation, e.g. Hunklinger [19]). With equal effective masses of electron and hole corresponding to the same bending of valence and conduction bands, the position of the Fermi level E_F remains in the center of the bandgap. If the masses differ, then the Fermi level shifts with temperature. The temperature shift is small compared to the size of the bandgap.

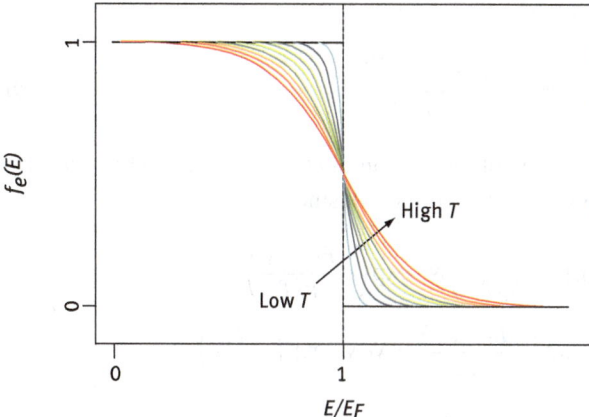

Fig. 2.13: Fermi distribution function $f_e(E)$ for various temperatures. $f_e(E)$ defines the probability for the occupation of a state with energy E by an electron. For low temperatures there is a sudden drop in $f_e(E)$ at the Fermi energy E_F, whereas the edge is broadened at elevated temperatures.

In order to calculate the density of electrons n_0 in a crystal we can now combine the density of states (Equation (2.22)) and the probability distribution function (Equation (2.23)):

$$n_0 = \int g(E) f_e(E) \, dE. \qquad (2.25)$$

It is further possible to find an expression for the electron and hole densities in the conduction and valence bands, respectively (for a derivation, see, e.g. Würfel [29]). In short, the expression for the Fermi–Dirac statistics can be simplified for $E_C \gg E_F + k_B T$ since the term $+1$ in the denominator of the Fermi distribution function can be neglected. With the definition of the so-called effective density of states of the conduction band

$$N_C = 2\left(\frac{m_e^* k_B T}{2\pi \hbar^2}\right)^{3/2} \qquad (2.26)$$

we then find a simplified expression for the electron density in the conduction band:

$$n_0 = N_C \exp\left(-\frac{E_C - E_F}{k_B T}\right). \qquad (2.27)$$

In a similar fashion we can write the hole density in the valence band as

$$p_0 = N_V \exp\left(-\frac{E_F - E_V}{k_B T}\right) \qquad (2.28)$$

with effective density of states of the valence band

$$N_V = 2\left(\frac{m_h^* k_B T}{2\pi \hbar^2}\right)^{3/2}. \quad (2.29)$$

This leads us to the important statement that the product of electron and hole density in a semiconductor is independent of the Fermi level, since

$$\begin{aligned} n_0 p_0 &= N_C \exp\left(-\frac{E_C - E_F}{k_B T}\right) N_V \exp\left(-\frac{E_F - E_V}{k_B T}\right) \\ &= N_C N_V \exp\left(-\frac{E_C - E_V}{k_B T}\right) = N_C N_V \exp\left(-\frac{E_G}{k_B T}\right), \end{aligned} \quad (2.30)$$

where $E_G = E_C - E_V$ is the bandgap energy (the difference between valence and conduction band edges). The product $n_0 p_0$ being independent of the Fermi energy has the important implication that upon doping of semiconductors, $n_0 p_0$ is invariant. While the density of either electrons or holes can be changed via shifting the Fermi level, as will be discussed in the next section, this accordingly results in a reduction of the density of the other carrier type.

2.1.2 Doping of Inorganic Semiconductors

2.1.2.1 Intrinsic Semiconductors

Up to this point we were treating semiconductors as crystals featuring identical repeat units that determine the optoelectronic properties of the material. The discussed properties apply for a pure semiconductor with a negligible amount of defects like missing atoms or crystal impurities. For many applications, however, foreign atoms, the so-called dopants, are introduced in such a crystal on purpose in order to change its properties.

Before we discuss doped semiconductors we quickly recall the properties of pure *intrinsic* semiconductors. If a semiconductor features only a negligible amount of impurities, the Fermi level is generally located relatively far away from both the valence and the conduction bands, so it lies well in the bandgap as depicted in Figure 2.14. In this situation only few electrons are thermally excited to the conduction band and accordingly only few holes are located in the valence band since $n_0 = p_0$ in intrinsic semiconductors. Therefore, the conductivity of intrinsic semiconductors is low up to moderate temperatures and only increases for elevated temperatures, when the Fermi–Dirac distribution is broadened around E_F and the Fermi level is shifted closer to the conduction band. However, for any temperature $T \neq 0$ there are at least a few electrons elevated from the valence to the conduction band where they can move freely (and so can the remaining holes in the valence band). Therefore, even an intrinsic material is always conducting to a certain extent at temperatures above 0. It is further important to note that in intrinsic semiconductors current is always carried by both electrons and holes since their densities are equal ($n_0 = p_0$).

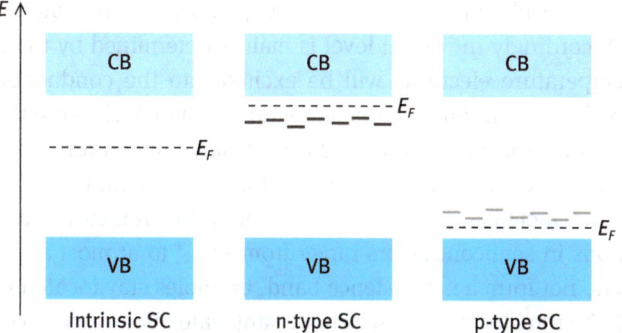

Fig. 2.14: Schematic band diagrams of intrinsic and doped semiconductors. In an intrinsic semiconductor there are no electronic states in the bandgap and the Fermi level is located in the bandgap relatively far away from valence and conduction bands. Thermal excitation of electrons to the conduction band is therefore unlikely. In the case of n-doping there are additional occupied electron states introduced into the semiconductor by impurities or crystal defects. These states are located close to the conduction band and thermal excitation of electrons from these states to the conduction band is possible. These electrons can carry an electric current. For p-doping, empty electron states are introduced close to the valence band, which can be thermally populated by electrons from the valence band. Accordingly, the holes remaining in the valence band can carry a current.

As mentioned before for many electronic applications like transistors or solar cells, it is imperative to adjust the conductivity of semiconductors as well as the majority carrier type, i.e. the type of charges (electrons or holes) that mainly carry a current through the semiconductor. Both properties can be tuned by controlled introduction of dopants into the crystal lattice as will be discussed in the following sections.

2.1.2.2 n-Type Semiconductors

First we look at the introduction of foreign atoms with additional valence electrons compared to the host material, which leads to n-doping of the semiconductor. The principle of n-doping is exemplarily shown for a Si crystal in Figure 2.15. The pure Si crystal is an intrinsic semiconductor. Each Si atom is covalently bound to four other Si atoms so that all valence electrons are involved in atomic bonds. It is now possible to introduce weakly bound electrons by exchanging single Si atoms with atoms of higher valency, e.g. with P atoms. Only four of the five valence electrons of P participate in bonds with the neighboring Si atoms, resulting in a weak binding energy of the fifth electron. Due to its low binding energy thermal excitation of this electron suffices to remove it from the potential of the P atom so that it can move through the semiconductor.

In an energy band picture the introduction of P atoms into the Si semiconductor results in additional electron states slightly below the conduction band as schematically shown in Figure 2.14. At 0 K all of these states are occupied by the fifth valence electrons of the P atoms. Accordingly the Fermi level is mainly determined by these states. With increasing temperature electrons will be excited into the conduction band and the Fermi level will rise. In n-doped semiconductors it is much closer to the conduction band than in an intrinsic semiconductor. At increasing temperature more and more electrons overcome the energy gap between the dopant states and the conduction band, resulting in an increasing conductivity of the semiconductor. Since typical dopant concentrations in semiconductors range from ~10^{-9} to at most a few percent, the dopant states do not form a new valence band, but holes stay located in one state and cannot easily be translocated to another dopant state due to a comparably large distance between these states in the crystal. Therefore, the current in an n-type semiconductor is mainly carried by electrons in the conduction band, while only very few holes reside in the valence band. In an n-type semiconductor electrons are therefore referred to as majority carriers, while holes are the minority carriers.

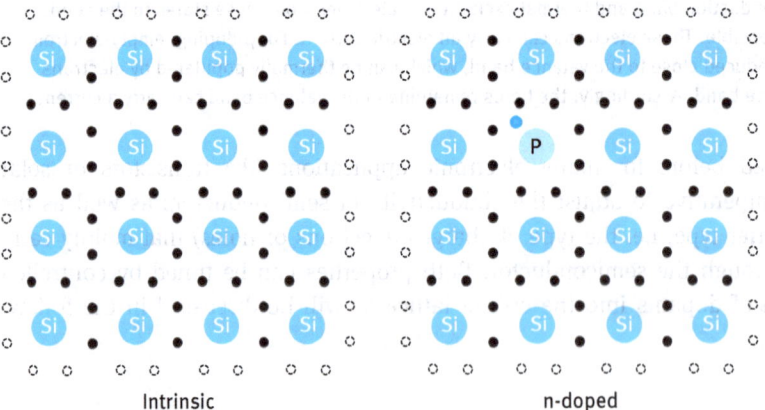

Intrinsic   n-doped

Fig. 2.15: n-type doping of crystalline silicon. Substitution of single Si atoms with group V atoms like P introduces weakly bound electrons to the crystal. Since only four of the five valence electrons of the P atom participate in the crystal bonds, translocation of the fifth electron can happen easily due to its low binding energy. The crystal becomes conductive with the current being carried by electrons.

As we saw in the previous section the product of the electron and hole density $n_0 p_0$ is independent of the Fermi level and in an intrinsic semiconductor $n_0 = p_0$. We can therefore introduce an intrinsic carrier density n_i with $n_i^2 = n_0 p_0$. Upon doping of a semiconductor the product of the charge densities remains constant so we can write

$$np = n_i^2 \qquad (2.31)$$

for electron and hole densities n and p, respectively. This relation will be used for the calculations of the current through p–n junctions in Section 2.3.1.

2.1.2.3 p-Type Semiconductors

Analogous to n-doping it is possible to introduce foreign atoms with less valence electrons into the semiconductor crystal. For a Si crystal atoms of valency III are p-dopants as exemplarily shown in Figure 2.16 for B atoms in a Si crystal. Due to its valency B can participate only in three atomic bonds with the neighboring Si atoms, resulting in an unpaired electron in one of the these atoms that is not forming a bond. Now this bond can be completed by an electron from one of the adjacent bonds, moving the dangling bond to another location. Since this process does not require much energy it is possible to get a net movement of the incomplete bond through the crystal at room temperature. This process is a hole transport through the valence band of the crystal, i.e. the semiconductor becomes hole-conducting upon introduction of p-dopants.

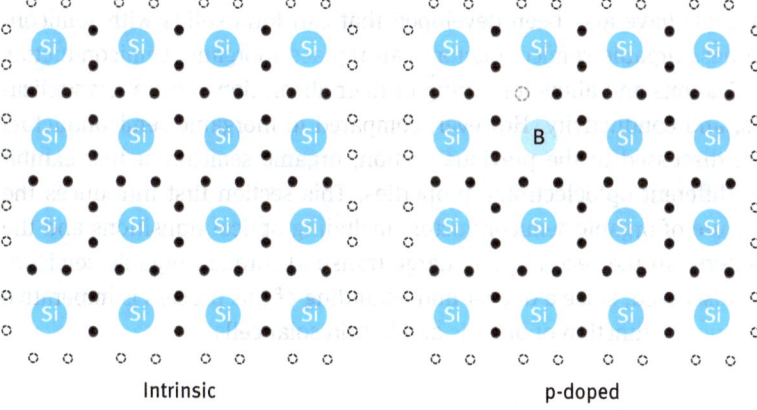

Fig. 2.16: p-type doping of crystalline silicon. Analogous to n-type doping single Si atoms can be substituted by group III atoms like B. These atoms carry only three valence electrons, resulting in one incomplete crystal bond. When an electron from a neighboring bond completes this bond, charge can be transported. Since this process requires only little energy the crystal becomes conductive and the current is carried by holes.

Similar to the case of n-doping this situation can be illustrated in a band diagram as introduction of additional electronic states compared to the intrinsic semiconductor (Figure 2.14). For p-doping, these states are empty electron states associated with the incomplete bonds due to the lower valence of dopants, which are located close to the valence band. Accordingly, the Fermi level in a p-type semiconductor is located between the valence band and these states. At room temperature the Fermi–Dirac distribution typically is broadened to an extent that thermal excitation of electrons from the valence band to these states is possible so that a significant fraction of these states are occupied. Although no band transport of electrons in the dopant states is possible (analogous to n-doping), the holes remaining in the valence band can carry an electric current.

Note that due to the different valency of dopant atoms compared to the crystal atoms, doped semiconductors are electrically neutral, although the densities of mobile electrons and holes in the conduction and valence bands are changed compared to the intrinsic case. This change in charge density is canceled by charges occupying the electron donor (n-doping) or acceptor (p-doping) states introduced by the dopant atoms.

2.2 Introduction to Organic Semiconductors

Organic semiconductors exhibit properties like high absorption coefficients and good processability, which makes them attractive as photoactive materials in solar cells. Most commonly, organic semiconductors are polymeric, even though small molecular materials have also been developed that can form solids with semiconducting properties. Organic semiconductors can replace inorganic semiconductors in various applications and allow for tuning of their absorption spectra, crystallization properties, and conductivity. However, compared to inorganic semiconductors that have been discussed in the previous section, organic semiconductors exhibit fundamentally different optoelectronic properties. This section first introduces the electronic structure of organic semiconductors including optical transitions and the concept of excitons. In the second part charge transport through organic semiconductors will be discussed, since a deeper understanding of this process is imperative when investigating the function of organic and hybrid solar cells.

2.2.1 Carbon Hybridization and Organic Molecules

2.2.1.1 Carbon Orbitals and Orbital Hybridization
When dealing with organic semiconductors the term *organic* means that the materials are carbon based. The chemistry of these materials is organic, where C atoms and their bonds are the basis. However, organic materials also feature other atoms, mainly hydrogen and light nonmetals like N, O, P, and S. In some instances single hydrogen atoms can be substituted by halogens, mostly by F or Cl. Generally, the classification of molecules as organic or inorganic is arbitrary to a certain extent, even though the most commonly used attribute to characterize a compound as organic is that the molecular structure is predominantly determined by C–C bonds.

Since organic molecules mainly feature C and light atoms their molecular structure can be understood when considering the octet rule for all atoms except hydrogen and the duet rule for hydrogen. The octet rule states that for light elements with atomic number < 20, atoms form (covalent) bonds in a way so that each atom has 8 or 0 valence electrons (where electrons in covalent bonds count for both atoms participating in that bond), i.e. the atom forms bonds until its electron configuration acquires the configuration of the nearest noble gas. The duet rule for hydrogen states that hydrogens always participate in exactly one covalent bond, thus having two valence electrons. Even though these rules are only rules of thumb, they hold for the vast majority of organic molecules and can be very successfully employed in order to predict possible molecular structures for a given molecular formula (e.g. for methane with molecular formula CH_4, the only possible molecular structure is the C atom in the center having four single bonds, one to each of the four H atoms).

Let us now have a closer look at the electron configuration of C. C is element number 6 in the periodic table, located in group IV and in the second period, which gives it an electron configuration of $1s^2 2s^2 2p^2$. Since $1s^2$ is the electron configuration of the first noble gas, He, this can also be written as $[He]2s^2 2p^2$. This notation implies that the four electrons in the second shell are valence electrons since the rest of the electrons are in a noble gas configuration. According to the octet rule C participates in bonds in a way that it reaches the electron configuration of the next noble gas, Ne, which is $1s^2 2s^2 2p^6$. This configuration is energetically favorable since all orbitals in the second shell are occupied so the shell is complete.

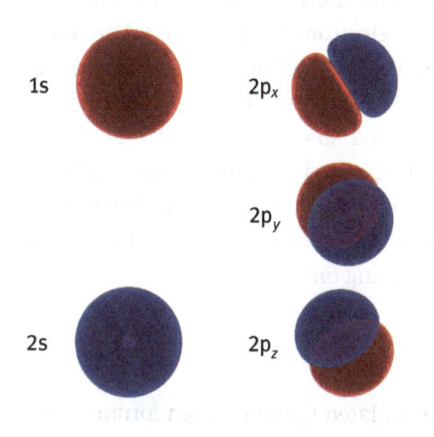

Fig. 2.17: Atomic orbitals of the first and second electron shell of C. The colors indicate the phase of the electron wave function with blue being "+" and red being "−".

The orbitals of a C atom are shown in Figure 2.17, where each orbital can be occupied by two electrons with opposing spin (each orbital is uniquely defined by a set of quantum numbers so according to the Pauli principle each orbital can accommodate two electrons that differ by their spin quantum number). The figure illustrates the electron wave functions $\Psi(x,y,z)$, where the positive or negative phase of the wave function is indicated as blue or red color, respectively. The probability density shows a similar spatial orientation but changes in shape since it is given as $\Psi(x,y,z)^2$. As shown in the figure, the 1s and 2s orbitals are centrosymmetric, while the three p-orbitals p_x, p_y, and p_z are symmetric with respect to one plane in an appropriately defined three-dimensional Cartesian coordinate system.

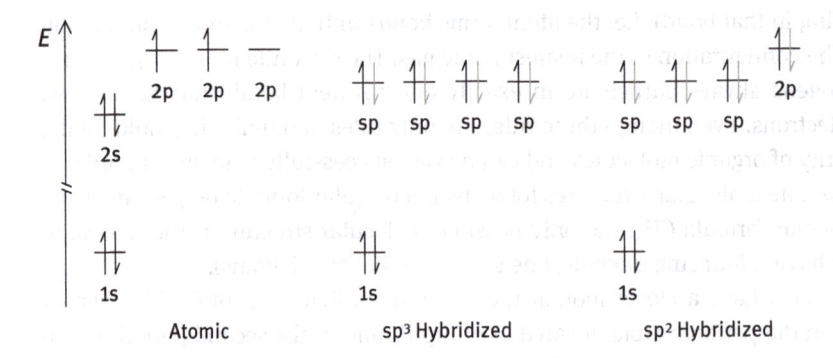

Fig. 2.18: Energy diagram of the atomic orbitals of carbon and their occupation by electrons. The figure illustrates the situation for atomic carbon, for a C atom involved in four single bonds (sp³ hybridized) and for a C atom participating in two single and one double bond (sp² hybridized). Four electrons are drawn in gray to indicate them stemming from the atoms the C atom is forming bonds to.

A schematic energy diagram of the C orbitals and their occupation in atomic carbon is shown in Figure 2.18. The inner shell electrons in the 1s orbital occupy energetically deep states and are strongly bound. Valence electrons in the 2s orbital are energetically located below electrons in the 2p orbitals with the three orbitals p_x, p_y, and p_z having exactly the same energy. The occupation of these orbitals by the six electrons of C can be understood when considering the so-called Aufbau rule, which states that orbitals are occupied on the order of their energies. Thus, the two s-orbitals 1s and 2s are occupied by two electrons each. Due to their energetic degeneracy, however, two of the three p-orbitals are occupied by only one electron since this is energetically more favorable than two electrons sharing one orbital.

2.2.1.2 Single, Double, and Triple Bonds

This, however, only reflects the situation for an isolated C atom. When forming chemical bonds with other atoms the C orbitals are altered, which change their energetic position and accordingly the occupation by electrons. This is called *hybridization* of orbitals – two or more atomic orbitals are combined to form new equi-energetic *hybrid* orbitals. In the case a C atom forms bonds to four different atoms, all orbitals of the second shell combine and form four sp³ hybrid orbitals as shown in Figure 2.18. These orbitals arrange in a tetrahedron around the C atom and are occupied half by the four valence electrons of the C atom and half by electrons of the bonded atoms. sp³ hybridization as found, e.g. in methane (four single bonds to neighboring H atoms) or diamond (four single bonds to neighboring C atoms), is a very common configuration for carbon and the energetically most stable form of bound carbon.

Besides single bonds a C atom can also form double and triple bonds, resulting in different orbital hybridizations. In the case of two single bonds and one double bond the C atom is sp² hybridized as shown in Figure 2.18. Two of the three sp²-orbitals form single bonds while the third forms one of the two bonds in the double bonds. The second bond in the double bond is formed by electrons in the remaining p-orbital.

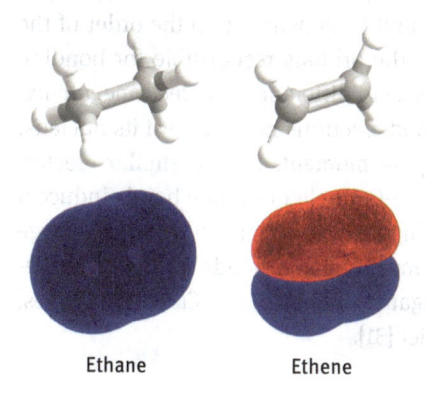

Fig. 2.19: Orbitals of a C–C σ-bond and a C–C π-bond. Orbitals are shown exemplarily for ethane, where each C atom has four single bonds and ethene, where the C atoms each have two single bonds with H atoms and one σ- and one π-bond with each other.

Bonds that are mediated via electrons in s- or sp-hybrid orbitals are typically referred to as σ-bonds, while electrons in p-orbitals form π-bonds. Orbitals forming a σ-bond and a π-bond between two C atoms are illustrated in Figure 2.19 exemplarily for ethane and ethene. Ethane is the simplest molecule with two sp³ hybridized C atoms, each of which forms four single bonds arranged as tetrahedra. One of these bonds is a C–C σ-bond while the other three are σ-bonds to H atoms. In the case of ethene both C atoms are sp² hybridized and form two σ-bonds with H atoms each. Additionally, there is one σ- and one π-bond between the C atoms.

It is also possible to form triple bonds between C atoms as in the simplest case of ethyne (C_2H_2). In this case, there is one σ-bond and two π-bonds between the C atoms, with the first π-bond orbital being similar to the one in ethene and another π-bond orbital being rotated by 90° with respect to the axis connecting the two C atoms. Molecules with triple bonded carbon are chemically very reactive since the third bond can be easily split up. This is because of the high repulsion between the six electrons forming the triple bond, which are all located mainly between the two C atoms. Due to this repulsion it is further not possible to form a quadruple bond between C atoms.

For the analysis of three-dimensional molecular structures it is important to note that molecular moieties can easily rotate around σ-bonds due to the symmetry of the bond orbital. Depending on the other binding partners of the C atoms in a C–C σ-bond, there is a certain lowest energy molecular configuration, but energy differences between different rotation angles around the σ-bond are typically small. On the contrary, since a π-bond orbital is not rotation-symmetric with respect to the bond axis there is no rotational degree of freedom for double and triple bonds. These bonds introduce stiff connections within the molecule and thus have a strong effect on the three-dimensional structure.

2.2.1.3 Structural Formulas

As mentioned before organic molecules do not only consist of carbon and hydrogen but can also contain other atoms like N, O, P, and S. Depending on the order of the bond (single, double, or triple) and the atom, the orbitals responsible for bonding can look different from orbitals in C–C bonds due to differences in electronegativity. The more electronegative atom will pull the bond electrons more toward its nucleus, thus possibly inducing an intramolecular dipole moment. Due to similar electronegativities C–H bonds are nonpolar, whereas C–O single or double bonds induce a polarity in the molecule with a significantly higher portion of the negative charge being located close to the O atom. However, for details on bonds in organic molecules the reader should refer to literature on organic chemistry, e.g. Clayden, Greeves, and Warren [30] or Solomons, Fryhle, and Snyder [31].

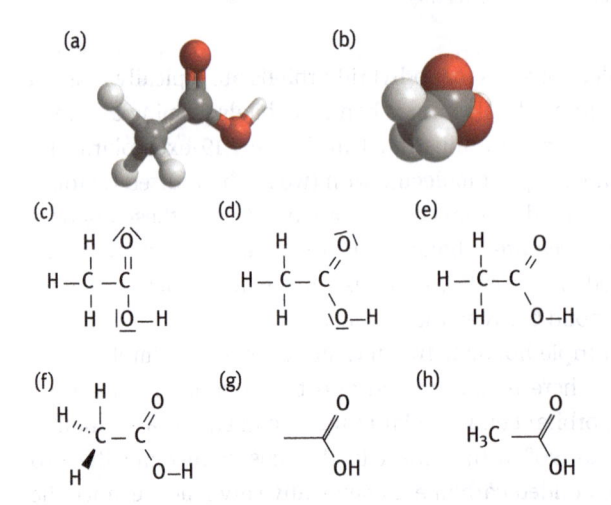

Fig. 2.20: **Different representations of the molecular structure of acetic acid.** (a) Ball-and-stick model, (b) space filling model, (c) and (d) Lewis structure, (e) simplified Lewis structure, (f) Natta projection, (g) skeletal formula without showing C and H explicitly, and (h) skeletal formula with all end groups.

We now take a closer look at how to draw and read structural formulas of organic molecules. Due to their complexity it is often not enough to know the chemical formula since there are different possible structural configurations. For example, there are two different propanol molecules, namely 1-propanol and 2-propanol, both of which have the same chemical formula C_3H_8O but differ in the position of the OH group on the C–C–C backbone. It is therefore useful to draw a schematic of the molecule in order to visualize the spatial arrangement of atoms in the molecule. Different representations of the molecular structure of the simple organic compound acetic acid are summarized in Figure 2.20:

(a) In order to illustrate the three-dimensional structure of a molecule it is often represented as ball-and-stick model, where atoms have a unique color coding (in this case white for H, gray for C, and red for O) and are drawn as balls while bonds are represented as one or more parallel sticks for single or multiple bonds, respectively. However, since atoms overlap their electron shells when forming molecules the ball-and-stick representation gives the wrong impression of the spacing between atoms.

(b) To address this issue another common representation is the space filling model, which shows approximately correct atom sizes (i.e. the size of the electron shell) and distances between atoms. Analogous to the ball-and-stick model, atoms are drawn in different colors but bonds are generally not shown, which makes differentiation between single and multiple bonds impossible.

(c) Besides these two three-dimensional structural formulae there are different simplified notations using element symbols and lines, which allow the structure of a molecule to be easily drawn on paper. In a Lewis structure atom symbols are arranged in a simplified two-dimensional projection of the molecular structure and electron pairs are drawn as lines, either between atoms to indicate bonds or around atoms in the case of valence electrons. Unpaired electrons occurring in radicals are represented as dots. The Lewis structure is useful when constructing a molecule according to the duet and octet rule since all valence electrons are drawn and can be easily counted. Angles between molecules are often simplified and shown as either 90° or 180°.

(d) For some molecules it can also be useful to draw bonds at different angles in the Lewis structure in order to match the real three-dimensional structure a bit closer.

(e) Very commonly Lewis structures are simplified so that only electrons in bonds are drawn but other valence electrons are omitted. This makes the notation easier to read and draw but electron octets are not apparent anymore, so the valency of the atoms has to be checked for each atom individually.

(f) It can further be useful to include some additional information about the three-dimensional arrangement of atoms in the Lewis structure. This is done in the Natta projection, where bonds sticking out of the paper are drawn as bold triangles and bonds retreating into the paper are represented as striped triangles.

(g) Especially for more complex molecules Lewis structures can be confusing because of many atom symbols. Therefore, in organic chemistry the most common notation is the skeletal formula, where the molecule is represented only by its carbon skeleton (including information about the order of the bonds) but C atoms are not explicitly shown. Furthermore, H atoms are implicit, i.e. if less than four bonds are shown for a certain C atom, it can be inferred that the remaining bonds are σ-bonds to hydrogen. For atoms other than implicit hydrogens and carbons the atom symbols are shown and moieties like OH groups are placed in a way indicating the binding situation (for instance, if an OH group is placed on the left-hand side of the molecule, it is written as HO–) but bonds are not explicitly shown as in a Lewis structure.

(h) For clarity, C-containing end groups like CH_3 are also sometimes explicitly shown. Furthermore, for simplicity it is possible to write easy-to-understand end groups using only atom symbols instead of showing the actual structure, e.g. the carboxyl group of acetic acid can be represented as COOH instead of showing the double bond to the O atom and the single bond to the OH group.

Another important portion of information about structural formulas is the notation of polymers. Polymers are large molecules that feature identical repeat units, which allow a condensed notation of their structural formula since it suffices to draw the repeat units of the polymer and give the number of repetitions instead of showing the complete molecule. An example of one of the most common synthetic polymers used in the plastic industry is shown in Figure 2.21, namely polystyrene. A fraction of its skeletal formula is shown in Figure 2.21(a), which clearly illustrates the impracticality of this notation, especially for long polymers with several hundreds of monomer repeat units. A more elegant notation is shown in Figure 2.21(b), where a polystyrene assembled from n styrene monomers is shown. The apparent drawback of this notation is the missing information about the end groups of the polymer. This information is often given in the context and not explicitly included in the structural formula. Depending on the route of synthesis it is also possible that there is a variety of possible end groups and only statistical information about the occurrence of different end groups is known.

A similar notation is also used for copolymers where two or more different repeat units are combined to form the polymer. In many instances, the exact sequence of the monomers is not known but only their stoichiometry, which is then included in the structural formula as indicated in Figure 2.21(c). Another example is the biopolymer deoxyribonucleic acid (DNA), which is a long sequence of only four different nucleotides. This example also illustrates the limitations of a purely stoichiometric notation since the exact sequence of a DNA molecule is carrying the information of interest rather than only the fractions of guanine, adenine, thymine, and cytosine nucleotides.

At this point it is also important to note that there are the so-called block copolymers, which feature a sequence of identical repeat units of one type chemically bonded to a sequence of *m* identical repeat units of another type. These polymers are fundamentally different from statistical copolymers as discussed in the previous paragraph, but are often drawn in the same way as shown in Figure 2.21(c). In the case of a block copolymer the structural formula contains information not only about the repeat units and their stoichiometry but also about the exact structure of the polymer due to its simplicity. Whether or not a polymer is a statistical or a block copolymer is not necessarily apparent from the structural formula and has to be clarified by the context of its presentation.

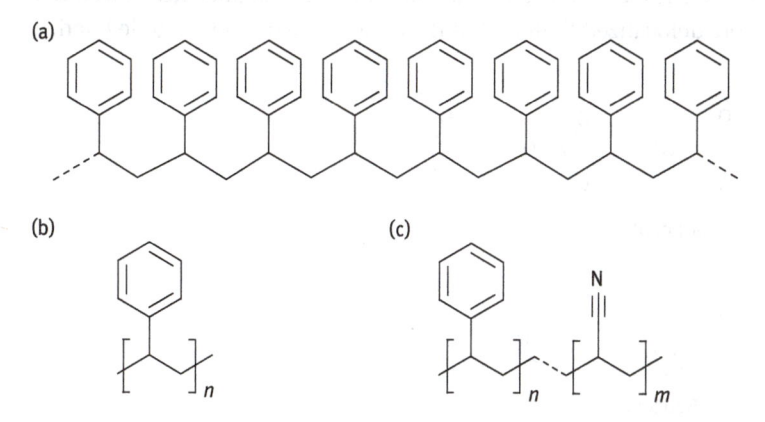

Fig. 2.21: Polystyrene and styrene acrylonitrile resin. (a) A fraction of the skeletal formula of polystyrene. (b) Polymer notation for polystyrene. (c) Copolymer notation for a copolymer featuring an $n:m$ ratio of styrene and acrylonitrile.

2.2.1.4 Mesomerism

When reading structural formulas it is essential to know the concept of mesomerism since there are many instances where a structural formula does not represent the real bonding situation within the molecule. In the case of single bonds the electrons forming this bond are highly localized and can be clearly assigned to the two atoms involved in the σ-bond of question. However, in the case of double and triple bonds it is possible that the electrons are delocalized over a molecular moiety greater than one bond length.

Mesomerism can be understood as a situation where one Lewis formula does not suffice to express the occurrence of single and double bonds in the molecule. In such a case, there are different equivalent Lewis structures, all of which illustrate a possible bond configuration. This is illustrated for two different molecules in Figure 2.22, namely a carbonate ion (a) and benzene (b). We first take a closer look at the carbonate. The three possible configurations are to be understood when the rotation of the molecule (which obviously would transfer the first structure into the second and third) would be inhibited and the position of the atoms kept fixed and only the bond-forming electrons are exchanged. In the case that carbonate looked exactly like the first Lewis formula, there would be two different bond lengths present in the molecule since atoms in a single bond are slightly further apart from each other than in a double bond due to the increased interaction between the atoms in the case of a σ- and π-bond. However, experimentally it is found that all three bonds between C and O are of identical length in a carbonate ion. The real electronic structure is therefore an intermediate between the three Lewis formulas shown in Figure 2.22(a). Roughly speaking, there is one σ-bond and a 1/3 π-bond between the C atom and each O atom, i.e. there are two π-electrons shared among three bonds and these electrons are accordingly more delocalized than that in the case of one discrete double bond.

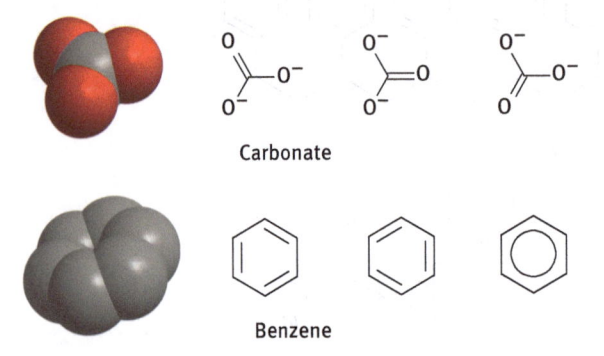

Fig. 2.22: Illustration of mesomerism. Equivalent structural formulas for (a) a carbonate ion and (b) benzene.

Another important example for mesomerism is the benzene molecule. The six-carbon ring of benzene is assembled from alternating single and double bonds and so apparently there are two equivalent Lewis formulas as shown in Figure 2.22. Overall, benzene features six π-electrons, which are completely delocalized along the carbon ring. To illustrate this benzene is sometimes drawn as a hexagon of single bonds with a ring in the center representing these delocalized electrons. This representation is more accurate than either of the two possible Lewis formulas since the latter implicate a localization of the π-electrons. However, it is not possible to count the number of π-electrons in this representation and so the hybridization state of each C atom has to be known in order to accurately read the structural formula.

2.2.2 Electronic States in Organic Semiconductors

2.2.2.1 Conjugated Molecules
Mesomerism is an important concept for understanding the properties of organic semiconductors since their electronic structure is largely determined by the presence

of delocalized π-electrons. Molecules with extended systems of delocalized π-electrons are referred to as conjugated molecules and the π-electron system is called a conjugated electron system. Conjugated organic molecules are compounds with alternating single and multiple bonds (most commonly double bonds). This alternating nature of bonds results in a certain degree of mesomerism for π-electrons in these bonds, since it is often possible to shift all double bonds by one C atom.

One of the most popular and best-understood conjugated molecules is the polymer polyacetylene, which is shown in Figure 2.23. This polymer has served as a model system for an organic semiconductor due to its simplicity – it basically is *only* a conjugated backbone (plus a single hydrogen bounded to each C atom). The semiconducting nature of acetylene was first described and later extensively studied by Shirakawa, Heeger, and MacDiarmid in the late 1970s [5, 32, 33]. In the year 2000 these three scientists were awarded with the Nobel Prize in Chemistry for "the discovery and development of semiconducting polymers," emphasizing the importance of this material class for the scientific community as well as the semiconductor industry.

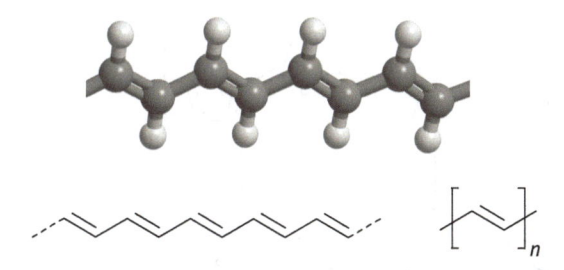

Fig. 2.23: Stick-and-ball representation and structural formula of polyacetylene. Due to alternating single and double bonds along the backbone of the molecule, polyacetylene is a conjugated polymer with semiconducting properties.

The remarkable optoelectronic properties of polyacetylene can be directly attributed to the presence of mobile π-electrons in a stable C atom network constituted by the σ-bond C atom chain. In a similar fashion this is the case in all conjugated molecules. Due to their delocalization conjugated π-electrons can move relatively freely through the molecule and can be excited to higher energy states by relatively low-energy optical transitions, i.e. by visible or near-IR photons. Furthermore, the delocalization of π-electrons is not necessarily constrained to one molecule but electrons can be delocalized over several adjacent molecules in molecular crystals. The result is that not only can optical transitions of comparably low energies be observed in conjugated molecules but these materials are also semiconducting to a certain extent since charges can travel from one molecular site to the other or even between molecules due to pronounced overlap of available quantum states owing to the delocalization of π-electrons.

Besides the delocalization of π-electrons along a system of alternating single and double carbon bonds, conjugated systems can also exist between different types of atoms as mesomerism is not limited to C–C bonds (see, e.g. the carbonate ion discussed earlier). One of the most prominent example for this is polythiophene, a semiconducting conjugated polymer, which is widely used in organic electronics due to its favorable optical properties and high charge carrier mobilities. The structural formula of polythiophene is shown in Figure 2.24 along with the two extreme contributing structures, one with two double bonds per thiophene and the other with one double bond in the thiophene and one between thiophenes. Even though both structures are conceivable when adhering to the octet rule, the upper structure is energetically more favorable than the lower and thus the real density of π-electrons in polythiophene is higher on the thiophene rings than on the bridging bonds. This example illustrates that knowing the structural formula of a conjugated molecule alone does not suffice to understand its electronic properties due to a system of conjugated π-electrons. In most cases it is necessary to find an estimate for the electron distribution on the molecule using computer simulations.

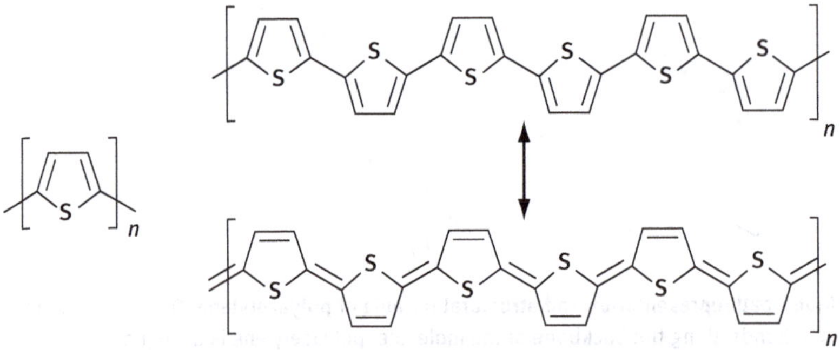

Fig. 2.24: Chemical structure and contributing structures of polythiophene. Since polythiophene contains S atoms in the conjugated system, the two mesomeric structures are energetically not equal and the charge distribution inside the molecule is closer to the extreme localized case shown on top than to the other extreme shown at the bottom.

Another important class of conjugated molecules are metal complex compounds, where the conjugated system forms a charge transfer complex with a centered metal ion. Two prominent examples of such molecules are shown in Figure 2.25. A charge transfer interaction acts as an attractive bond between the conjugated system and the metal ion as a fraction of the electronic charge is transferred between the two, leading to an electrostatic potential between the charged domains of the molecule. Such metal complex dyes have large π-electron systems and can absorb visible light very efficiently, which is reflected in outstandingly high extinction coefficients like $70,000\ M^{-1}\ cm^{-1}$ in the case of copper phthalocyanine. Many metal complex dyes are also chemically very stable and therefore find application in inks, coatings, and different plastics.

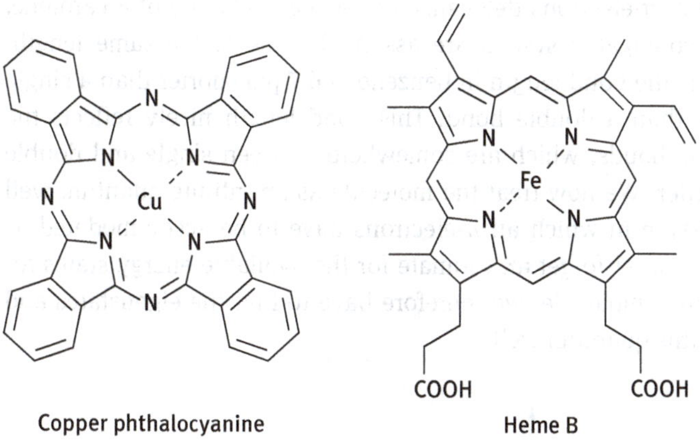

Copper phthalocyanine   **Heme B**

Fig. 2.25: Chemical structures of copper phthalocyanine and heme B as model systems for metal complex dyes with conjugated electron systems. Copper phthalocyanine is a strong blue dye very commonly used for many applications and the pigment produced in highest volume to date. Heme B is a widely found cofactor of hemoproteins like hemoglobin.

2.2.2.2 Molecular Orbitals

So far we talked about the representation of the molecular structure of organic molecules including covalent bonds between atoms. For organic semiconductors π-bonds are of particular interest, since delocalization of π-electrons as in conjugated molecules leads to interesting optoelectronic properties in these materials. We now take a closer look at the implications of this delocalization of electrons and in particular which electronic states are present in conjugated molecules. The nature of these electronic states, namely being separated by forbidden energy regions, renders conjugated molecules applicable for solar cells since photon absorption can lead to excitation across these forbidden regions, similar to excitation across the bandgap in inorganic semiconductors.

Analogous to single atoms, where electrons can occupy energetically located states in the atom shell that are separated by forbidden regions, electrons occupy quantum mechanical states in molecules. These states are therefore referred to as *molecular orbitals* (MO). As discussed earlier, electrons in σ-bonds are strongly bound and thus located in orbitals at low energies. In contrast, due to their delocalization, π-electrons are typically less strongly bound and occupy higher energy states in the molecule.

Particle in a Box Approximation. To get an idea of the energies of states in the conjugated system and their separation we first discuss a simplified model, the quantum mechanical particle in a box. Consider β-carotene, a natural red-orange pigment with a conjugated π-electron system stretched over 21 alternating carbon single and double bonds (see Figure 2.26 for the chemical structure of β-carotene).

Thus, overall there are 22 π-electrons delocalized over the backbone of β-carotene. All C–C bonds in the conjugated system are assumed to be of the same length, namely 140 pm, which is the bond length in benzene and 7 pm shorter than a single bond but 5 pm greater than a double bond. This bond length nicely reflects the mesomeric nature of the bonds, which are somewhere between single and double bonds as discussed earlier. We now treat the molecule as an infinite quantum well with a length a of 2940 pm in which all π-electrons have to be accommodated as shown in Figure 2.26. In order to get an estimate for the available energy states for π-electrons in a β-carotene molecule, we therefore have to find the eigenstates and energy eigenvalues for this quantum well.

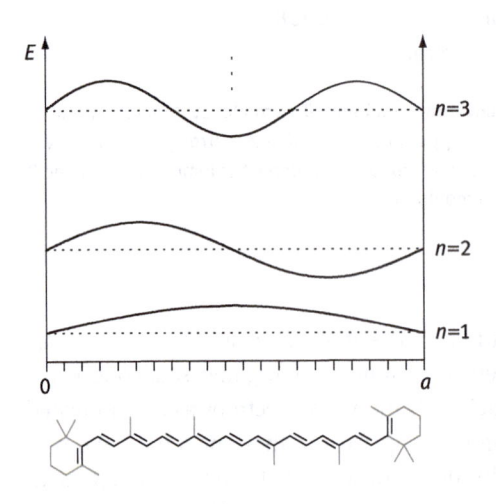

Fig. 2.26: Quantum mechanical particle in a box model for β-carotene. All π-electrons are treated as being confined in an infinite quantum well with a length a of 21 carbon bonds of 140 pm bond length each. The molecular structure of the molecule is shown on the bottom with the conjugated system indicated in black.

First, we write down Schrödinger's equation in one dimension:

$$\left(-\frac{\hbar^2}{2m}\frac{d^2}{dx^2} + V(x)\right)\Psi(x) = E\Psi(x), \tag{2.32}$$

Since the potential is infinite outside the box ($V(x) \to \infty$ for $x < 0$ and $x > a$), the probability to find the particle outside the box is 0, which means for the probability distribution $P(x)$:

$$P(x) = \Psi^*(x)\,\Psi(x) = 0 \Rightarrow \Psi(x) = 0 \tag{2.33}$$

for $x < 0$ and $x > a$. Inside the box we set the potential $V(0 \leq x \leq a) = 0$ and can directly use the Schrödinger equation to find a differential equation for $\Psi(x)$ that reads

$$\frac{d^2}{dx^2}\Psi(x) = -\frac{2mE}{\hbar^2}\Psi(x). \tag{2.34}$$

With a harmonic ansatz for $\Psi(x)$

$$\Psi(x) = A \sin(\alpha x) + B \cos(\alpha x), \quad (2.35)$$

we directly see that

$$\frac{d^2}{dx^2}\Psi(x) = -\alpha^2 \Psi(x), \quad (2.36)$$

which leads to

$$-\alpha^2 \Psi(x) = -\frac{2mE}{\hbar^2}\Psi(x) \Rightarrow E = \frac{\alpha^2 \hbar^2}{2m}. \quad (2.37)$$

In order to find the correct energy eigenvalues E we now have to apply certain boundary conditions since so far there is no restriction for values of α. Continuity at the edges of the box leads to $\Psi(0) = \Psi(a) = 0$ since $\Psi(x) = 0$ outside the box as discussed earlier. Considering this at $x = 0$ we directly see that $B = 0$ and find

$$\Psi(x) = A \sin(\alpha x). \quad (2.38)$$

Furthermore, from $\Psi(a) = A \sin(\alpha a)$ we see that

$$\alpha a = n\pi \Rightarrow \alpha = \frac{n\pi}{a} \quad (2.39)$$

with integer n. We can now use this integer to number the eigenstates

$$\Psi_n(x) = A \sin\left(\frac{n\pi}{a}x\right), \quad n = 1, 2, 3, \ldots \quad (2.40)$$

and use the electron mass m_0 to get the energy eigenvalues

$$E_n = \frac{n^2 \hbar^2 \pi^2}{2m_0 a^2}, \quad n = 1, 2, 3, \ldots \quad (2.41)$$

Note that we do not allow a state with $E = 0$, i.e. the particle is not allowed to stand still.

We now have to find the n of the highest eigenstate that is occupied in our system, since we can then calculate the energy necessary to excite an electron from this state into the next highest state, which gives us the optical energy gap of the molecule. The assumption is that no lower energy states are left empty before higher energy states are occupied, which is a reasonable assumption for realistic systems, since excited states relax to the lowest possible energy state typically within short times. In our particle in a box approximation the eigenstates Ψ_n are the molecular orbitals for π-electrons. We therefore refer to the state with the highest n that still contains electrons as the *highest occupied molecular orbital* (HOMO) and call the next higher state the *lowest unoccupied molecular orbital* (LUMO). Considering Pauli's exclusion principle we can now have two electrons in each orbital due to the spin quantum number, which allows occupation of orbitals by two electrons with opposing spins of $+1/2$ and $-1/2$. For the

22 π-electrons in the conjugated system of β-carotene we get $n_{HOMO} = 11$ and accordingly $n_{LUMO} = 12$. From these numbers we can calculate the energy gap as

$$E_{HOMO-LUMO} = \frac{(n_{LUMO}^2 - n_{HOMO}^2)\hbar^2\pi^2}{2m_0 a^2}, \quad (2.42)$$

which is 1.0 eV. This value is only a very rough estimate, which becomes apparent when comparing it to the real energy gap of β-carotene. The pigment shows a strong absorption around 450 nm, corresponding to an energy of 2.75 eV, meaning that in this instance the particle in a box approach is off by roughly a factor of 3. Nevertheless, even though being very simplistic, it reproduces not only the correct order of magnitude for the energy gap but also the commonly observed trend of decreasing energy gaps for increasing sizes of the conjugated system. This is exemplarily shown in the chart in Table 2.1, which summarizes energy gaps and the corresponding wavelength for a series of molecules with increasing number of phenyl rings in a stick-like configuration (benzene to pentacene) and for two dyes with four and five phenyl rings in a disk-like configuration (pyrene and perylene).

Tab. 2.1: Chart of common conjugated molecules with increasing numbers of phenyl rings summarizing their molecular structure and energy gaps. The first series shows the reduction of the energy gap when putting one (benzene) to five (pentacene) phenyl rings in a stick-like configuration, whereas the last two examples show disk-like arrangement of four and five phenyl rings in pyrene and perylene, respectively.

Name	Structure	λ_{Gap} (nm)	E_{Gap} (eV)
Benzene		254	4.9
Naphtalene		311	4.0
Anthracene		375	3.3
Tetracene		471	2.6
Pentacene		582	2.1
Pyrene		352	3.5

Tab. 2.1 (continued)

Name	Structure	λ_{Gap} (nm)	E_{Gap} (eV)
Perylene		434	2.9

It is important to note that already for these simple molecules the energy gap is determined not only by the number of π-electrons in the conjugated system but also by their spatial arrangement inside the molecule. In parts this is already predicted by the particle in a box model through the parameter a, the length of the box. However, the actual shape of the molecule is of course not taken into account. For pentacene, Equation (2.42) yields an estimate of 6.0 eV for $E_{HOMO-LUMO}$ when considering the molecule length (1.2 nm) as a and a value of 1.00 eV when completely "unfolding" the molecule to a chain of 22 carbons with a length of 2.9 nm (identical to the conjugated system of the particle in a box model for β-carotene). Both values are far from the experimentally observed value of 2.1 eV, which outlines the limitation of the particle in a box approximation.

Perimeter Free Electron Orbital Theory. For benzene and polycyclic aromatic hydrocarbons like naphthalene and anthracene often another simplified approach is chosen, the so-called *perimeter free electron orbital theory*. This theory takes the ring-like structure of the conjugated system into account rather than unfolding it to a straight chain. The model takes the angular moment of the π-electrons into account, which are confined to a circular movement along the perimeter of the molecule, while σ-bonds and electrons in lower orbitals are considered to screen the ionic potential resulting in a constant-effective potential throughout the molecule. Under these assumptions only the kinetic energy of π-electrons has to be taken into account and kinetic energy differences directly translate to energy gaps between molecular orbitals.

Consider benzene as the simplest case of a conjugated ring molecule. Each C atom brings in one electron to the conjugated system, so there are six π-electrons moving along the circular edge of the molecule. From quantum mechanics we know that the angular momentum **L** is quantized:

$$|\mathbf{L}| = n \cdot \hbar, \quad n = 0, \pm 1, \pm 2, \ldots \tag{2.43}$$

We can now write $|\mathbf{L}| = n\hbar = r \cdot p$, where p is the absolute value of the momentum and r the radius of the circular movement, i.e. the radius of the benzene molecule, and use the expression $E_{kin} = p^2/(2m_0)$ for the kinetic energy of the electron to find

$$E_{kin} = \frac{\hbar^2}{2m_0 \cdot r^2} \cdot n^2. \tag{2.44}$$

With this, the kinetic energy of the electron only depends on the angular momentum quantum number n, and since we state that the potential energy is constant throughout the molecule as discussed earlier, the kinetic energy is the only magnitude of interest for our energy-level considerations. In benzene, we now have to accommodate six π-electrons, which requires three different values of n since due to the spin quantum number two electrons can occupy one n-state, analogous to the particle in a box model. The state with $n = 0$ is at a lower energy than the $n = -1$ and the $n = +1$ states. These two, however, have equal energies so that the HOMO is given by $n = \pm 1$ and has the energy $E_{HOMO} = \hbar^2/(2m_0 \cdot r^2)$. We also see that the next lowest energy state, which in our model is the LUMO, is given by $n = \pm 2$ with energy $E_{LUMO} = 2\hbar^2/(m_0 \cdot r^2)$. In order to find the energy gap of benzene we now simply have to calculate the HOMO–LUMO difference, which is

$$E_{HOMO-LUMO} = E_{LUMO} - E_{HOMO} = \frac{3\hbar^2}{2m_0 \cdot r^2}. \tag{2.45}$$

To find the radius r we approximate the benzene molecule with a circle and set the perimeter to be six times the C–C bond length (140 pm in benzene) since there are six C atoms in the benzene molecule. This leads to a radius r of 133.7 pm and we find a HOMO–LUMO gap of 6.4 eV, which is not too far from the correct value of 4.9 eV but overestimates it. For the next polycyclic aromatic hydrocarbons, naphthalene and anthracene, the estimate of $E_{HOMO-LUMO}$ gives 3.8 and 2.7 eV, respectively, which now underestimates the real values of 4.0 and 3.3 eV, respectively. This systematic deviation between estimated and experimentally observed energy gaps among these three molecules illustrates the limitations of the perimeter free electron orbital theory, outlining that more complex approaches are necessary in order to predict the correct energy levels of the molecular orbitals in conjugated molecules.

Molecular Orbital Theories. As discussed in the previous two paragraphs, neither the particle in a box model nor the perimeter free electron orbital theory is successful in predicting molecular orbital energies and energy gaps that are close to the real values, even though both approaches yield energies of the correct order of magnitude and describe in many cases the trend correctly when comparing different molecules. Furthermore, none of the models gives realistic information about the shape of the molecular orbitals, i.e. the electron wave functions on the molecule. However, for many problems it is important to have an impression of the wave functions of the HOMO and LUMO since they also determine the charge density on the molecule in the ground state and first excited state. Designing a molecule and thus controlling its charge density in the HOMO and LUMO is an important strategy for improving the efficiency of optoelectronic devices like organic light-emitting diodes or photovoltaics, since it allows control of the charge density at optoelectronically active interfaces for a given molecular distribution and orientation.

The most commonly used methods to calculate precise predictions of molecular orbitals are based on linear combinations of atomic orbitals (LCAO). We briefly outline the fundamental idea. Consider a molecule with a number of atoms, each of which has atomic orbitals. The first LCAO assumption is that for a total number of m atomic orbitals there are also m molecular orbitals since the former combine to form the latter. The jth molecular orbital Φ_j can then be found simply as a linear combination of all m atomic orbitals χ_i:

$$\Phi_j = \sum_{i=1}^{m} a_{ij}\chi_i \qquad (2.46)$$

with linear coefficient a_{ij}. These a_{ij} determine how much a certain atomic orbital χ_i is contributing to the jth molecular orbital. An important property of the Φ_j in LCAO methods is that they are linear combinations of wave functions that are centered around the nuclei of the molecular atoms. Thus, the molecular structure is implicitly taken into account. Furthermore, the delocalization of a certain molecular orbital is given by the linear coefficients. For localized molecular orbitals only a few a_{ij} are significant while others vanish, in contrast to extended molecular orbitals, which combine a larger number of individual atomic orbitals.

Even though the LCAO approach appears to be a good strategy, we still have no clear idea what the molecular orbitals look like without knowing more about the linear coefficients. Ultimately we are interested in the electron wave functions inside the molecule and the respective eigenenergies so we always have to solve Schrödinger's equation. We know, however, that even for single atoms this is only possible using certain approximations. Hydrogen, the simplest atom, which consists only of a single proton and a single electron, is the only atom allowing an analytical solution, whereas numerical methods or simplifications are required when dealing with atoms with more than one electron, where electron–electron interactions have to be taken into account. A first simplification for conjugated molecules is to separate σ- and π-electrons, since the former are highly localized between bonded C atoms. It is then possible to separate the wave function

$$\Psi = \Psi_\sigma \cdot \Psi_\pi, \qquad (2.47)$$

which leads to a simpler Schrödinger equation

$$\mathbf{H}\Psi_\pi = E\Psi_\pi \qquad (2.48)$$

if only the π-electrons of the molecule are taken into account.

For an n-atomic conjugated molecule, this Schrödinger equation is a $6n$-dimensional set of partial differential equations since the problem is three dimensional in both position $\mathbf{x}$ and momentum $\mathbf{p}$ of each π-electron. Even after separating the wave function there is no analytical solution to the problem. Furthermore, since numerical approaches are computationally intensive, approximations are helpful in order to get to a solution.

An often used method to calculate molecular orbitals in conjugated molecules is the Hückel method, introduced in the 1930s, which is mathematically simple but yields surprisingly good results compared to more precise but computationally intensive ab initio calculations. The method is limited to conjugated hydrocarbons but very elegant in its approach since it is based on the assumption that the molecular structure is solely given by strongly localized σ-bonds, which in turn do not influence the molecular orbitals occupied by delocalized π-electrons. It is therefore sufficient to treat only π-electrons. The molecular orbitals are then found as linear combinations of atomic p-orbitals. For our n-atomic molecule the $6n$-dimensional problem $\mathbf{H}\Psi = E\Psi$ can then be reduced to $\mathbf{H}'\mathbf{x} = E\mathbf{x}$, where $\mathbf{H}'$ is an $n \times n$ matrix and $\mathbf{x}$ the vector of the coefficients of the linear combination.

In the 1960s the method was further developed by the theoretical chemist and 1981 Nobel Prize winner Roald Hoffmann to the so-called extended Hückel theory [34]. In contrast to the original theory it includes not only π-electrons but additionally also σ-electrons. The extended Hückel method is one of the most commonly used LCAO methods for determining molecular orbitals and is implemented in common chemistry software. It is a fast method for relatively precise prediction of energies and shapes of molecular orbitals and therefore very popular. Extended Hückel results are often further used as starting points when using more complex but also more precise methods like ab initio quantum chemistry approaches or the CNDO/2 (*Complete Neglect of Differential Overlap*) method.

Using an appropriate numerical method for molecular orbital calculations it is then possible to calculate the eigenenergies of different states and the shape of the respective electron wave functions in a molecule. In order to illustrate the complex structure of molecular orbitals we investigate the orbitals around the HOMO–LUMO gap in pentacene, which are shown in Figure 2.27. It is clear that the wave function changes its shape as well as the number of knots along the fourth highest occupied molecular orbital (HOMO–3) to the fourth lowest unoccupied molecular orbital (LUMO+3). Nevertheless, the structure of the different orbitals reflects the molecular structure. This is a hint that the extended Hückel approach is fundamentally valid, even if it might lead to slightly wrong numbers.

As will be discussed in the next section it is imperative to know the structure of the molecular orbitals since their quantum mechanical overlap determines the probability of optical transitions between them.

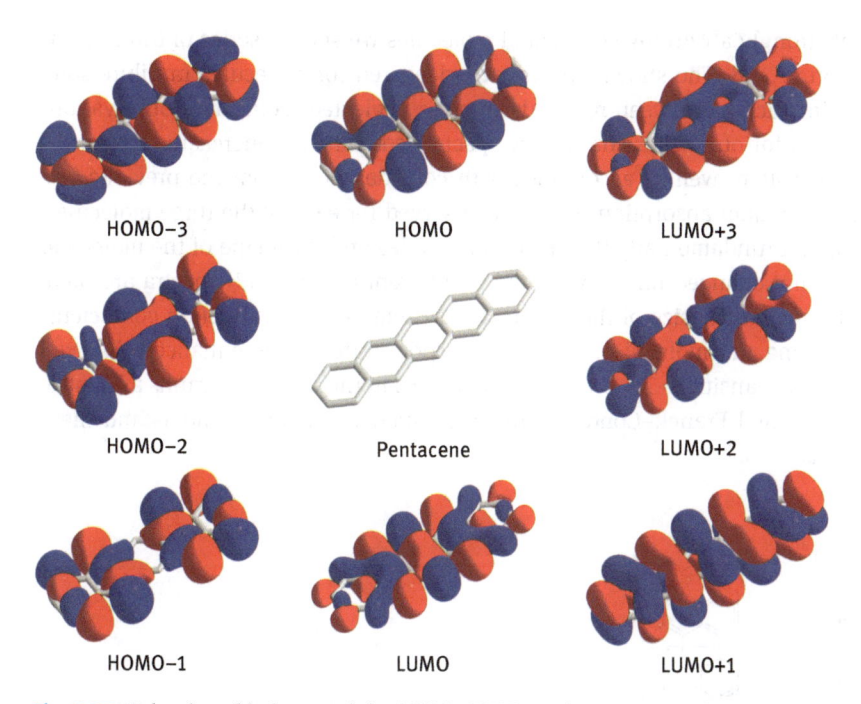

Fig. 2.27: Molecular orbitals around the HOMO–LUMO gap in pentacene. The molecular structure is shown in the center along with the four highest occupied and the four lowest unoccupied molecular orbitals. Blue and red indicate positive and negative phases of the wave function.

2.2.2.3 Optical Transitions

Now that we understand which energy levels are available in a conjugated molecule, namely its molecular orbitals, the next step toward photovoltaic operation is to excite molecules with photons in order to harvest sunlight with organic semiconductors. The subsequent step, the dissociation of excited states into charge carriers, will be discussed later.

After we had a look at different (poly)cyclic aromatic hydrocarbons like the benzene to pentacene series and how relatively simple quantum mechanical considerations can predict at least the correct order of magnitude for HOMO–LUMO gaps as well as the trend among molecules, we now have to know what the real absorption spectra look like. From both the particle in a box model and the perimeter free electron orbital theory the prediction is that allowed energies for electrons in the conjugated systems are sharp lines, so that the absorption spectra would look like the spectral lines of atomic orbitals. An optical transition would then be observable exactly at this energy, i.e. at one specific wavelength as summarized in Table 2.1.

However, numbers given in this table are only peak values of the observed absorption features, which are spectrally broadened. As an example for absorption spectra of strongly absorbing conjugated molecules we show a series of three rylene dyes in Figure 2.28. The shown spectra are measured for molecules in dilute solutions, i.e. interaction between molecules can be neglected. Four main observations are of interest for our discussion: (1) the spectra red-shift with increasing size of the conjugated system, well in accordance with considerations from the previous sections; (2) three main absorption peaks are observed for each of the three molecules, indicating that fundamentally the absorption modes and the shape of the molecular orbitals are not changed but only shifted to different energies; (3) spectra are broadened with increasing size of the conjugated system; and (4) extinction coefficients increase among the molecules, i.e. the absorption strength is enhanced. We will discuss optical transitions in conjugated molecules in the next paragraphs and introduce the so-called Franck–Condon principle, which will help us understand these observations.

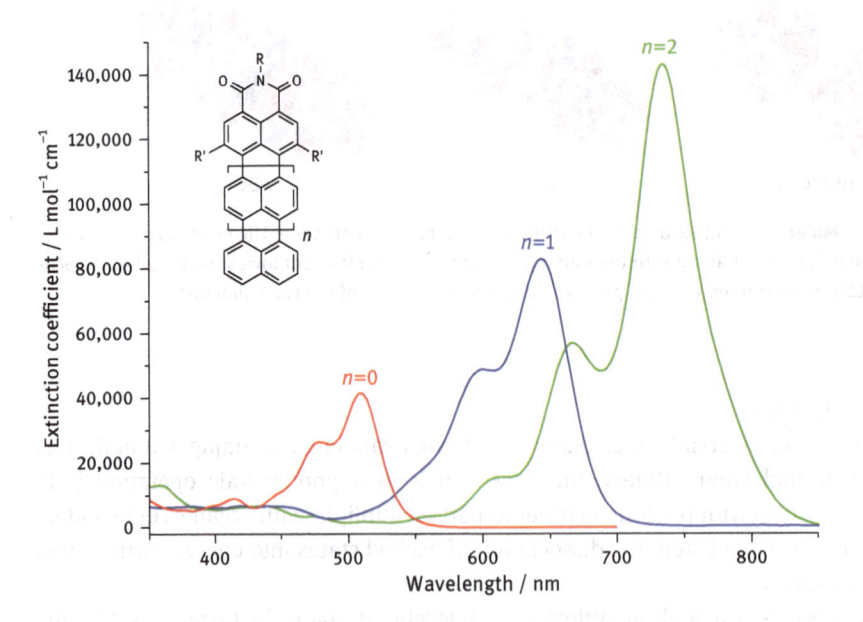

Fig. 2.28: Absorption spectra of three rylene dyes with increasing conjugation along with the chemical structures of the molecules. Absorption features red-shift while the spectrum maintains the two most prominent absorption peaks. Figure after [35] with friendly permission from Chen Li.

Another interesting effect that is generally observed in optical spectra of conjugated molecules is illustrated in Figure 2.29, where absorption and photoluminescence emission spectra of a thin polymer film are shown at different temperatures. When changing the temperature of the sample two effects are apparent: (1) spectral features are broadened and less distinct upon increasing temperatures and (2) spectral features are shifted in energy. We will first discuss the basic mechanisms of absorption and emission of photons and then look at optical transitions in more detail.

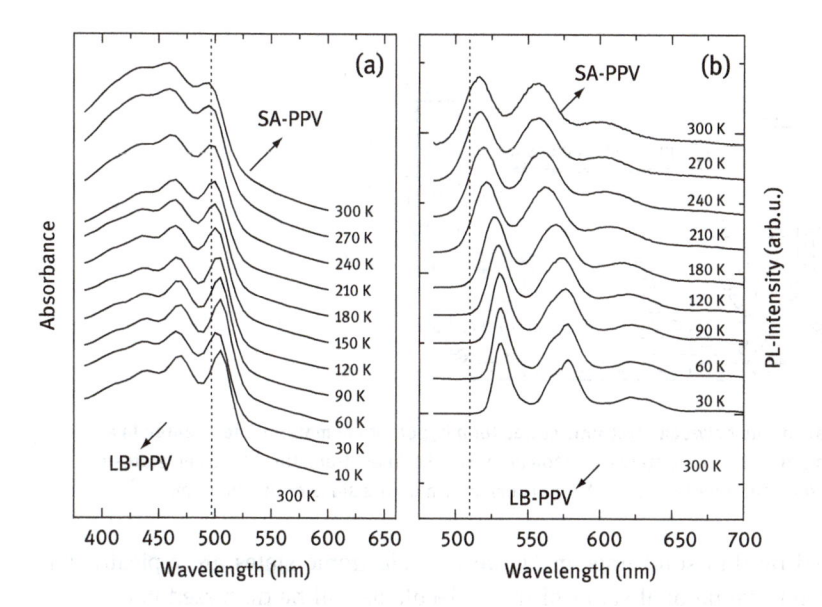

Fig. 2.29: Temperature-dependent absorption and emission spectra of poly(*p*-phenylene vinylene). Absorption and photoluminescence emission measurements are performed on thin films of the polymer on glass substrates. Figure from [36]. Licensed under a Creative Commons Attribution License.

Absorption and Emission. Both absorption and emission of photons in semiconducting polymers are transitions between electronic states as discussed earlier. In organic molecules, these transitions are typically transitions of an electron in the HOMO of the molecule, which can be excited to a higher state upon absorption of a photon. When falling back from an excited state this transition can be accompanied by the emission of a photon. The emission is referred to as photoluminescence if it follows optical excitation, whereas it is also possible to occupy excited states thermally or via application of electric fields. For instance, in a light-emitting diode charges are injected from external electrodes into higher states in the emitting materials and luminescence is observed upon reoccupation of ground states.

In order to illustrate optical transitions a Jablonski diagram is often used. The Jablonski diagram of a hypothetical conjugated molecule is shown in Figure 2.30. The lowest energy state shown is the S_0 state, which is the HOMO of the molecule. It is a spin 0 state as it is occupied by two electrons of opposing spin. Upon photon absorption one electron can be excited to a higher energy level as schematically shown for S_1 and S_2, the two lowest excited singlet states. Optical transitions between singlet states including the absorption of photons are spin-allowed since the spin of the two HOMO electrons remains 0, which is why these transitions are likely and occur on fs timescales.

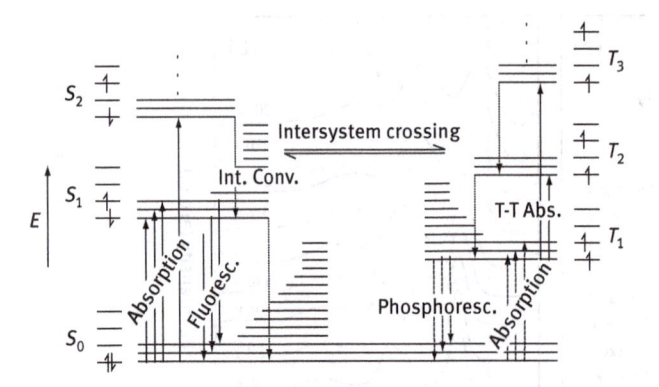

Fig. 2.30: **Transitions between electronic states for a hypothetical material represented in a Jablonski diagram.** Electronic states are indicated by solid lines, transitions between states by arrows. Electron configurations are schematically shown on the sides of the diagram.

As indicated by thin solid lines in Figure 2.30 electronic states are typically split due to different vibrational states of the molecule as will be discussed later. States in Jablonski diagrams are therefore often referred to as *vibronic* states in order to express that they contain both electronic and vibrational contributions. When a photon is absorbed by a conjugated molecule and an electron is lifted to a higher vibronic state, there is always a relaxation down to the lowest vibrational state of the respective electronic energy level. This principle is known as *Kasha's rule*. This relaxation occurs on a short timescale and is typically completed after 10 ps [37]. In the case that there are vibrational states available over a broad energy range the lower electronic state can be reached via internal conversion, which is a non-radiative process. For larger forbidden energy regions, however, the relaxation from a higher electronic state to a lower electronic state, e.g. a $S_1 \rightarrow S_0$ transition, is accompanied by the emission of a photon. Such a transition between the lowest vibrational excited state and one vibronic state in the ground state is referred to as fluorescence and is also a fast process since it is spin-allowed.

Due to a spin-flip process like the electron spin interaction with its orbital momentum it is also possible to get transitions between singlet and triplet states, the so-called *intersystem crossing*. Since it involves a change in spin state intersystem crossing is quantum mechanically unlikely and can therefore be slow but depends strongly on the material's properties, which determine the strength of spin-orbit coupling. For instance, intersystem crossing rates of 10^5 to 10^8 s^{-1} are observed in organic dye molecules, whereas in cytosine intersystem crossing times of only hundreds of fs are found [37, 38]. After intersystem crossing from an excited singlet state the system occupies a triplet state with total spin 1. Similar to singlet states S_0, S_1, S_2, ..., there are different triplet states T_1, T_2, T_3, ..., but the lowest triplet state is always higher than the S_0 (in our case the HOMO) since it has to be the state where the two electrons are in different orbitals.

Analogous to the mechanisms for singlet excitations there is the possibility for non-radiative internal conversion between triplet states, and excited triplet states tend to relax down to the lowest vibrational state. After occupation of the T_1 state via intersystem crossing and vibrational relaxation there can be again intersystem crossing and the system can thus relax down to the S_0 state. This mechanism is called phosphorescence and is typically a slow process, occurring on timescales between tens of ns up to tens of s [37]. On the other hand, there is the possibility of triplet–triplet absorption, which can occur if a photon is absorbed by a triplet state which then induces a transition to a higher triplet state. Such a transition does not involve a spin flip and is therefore quantum mechanically allowed. In a similar manner there can be emission of photons for a conversion from a higher to a lower triplet state.

In solar cells based on organic materials mainly singlet states are of interest, since the occupation of triplet states is typically accompanied by a loss of excitation energy. Therefore, care is taken that the excitation is split into free charge carriers before there is intersystem crossing. There are also approaches to take advantage of triplet states in organic and hybrid solar cells using materials that allow the so-called *singlet fission* [39]. This is possible in materials where the T_1 energy is less than half the S_1 energy, since upon decay of the singlet state two triplet states can be populated. Accordingly, it is possible to generate up to two electrons per incident photon, i.e. achieve external quantum efficiencies exceeding 100% [40].

Now we will look at optical transitions in more detail. Before we discuss transitions between different vibronic states and their probability we will introduce the Born–Oppenheimer approximation, which is useful in order to simplify the quantum mechanical problem that has to be solved.

Born–Oppenheimer Approximation. During our discussions of molecular orbitals we saw already how important approximations are when trying to solve the Schrödinger equation for complex many-electron systems. For optical transitions between molecular orbitals we are mainly interested in the two electrons in the HOMO that can get elevated to higher states upon photon absorption and occupy excited singlet or triplet states or, on the other hand, fall back down to the HOMO accompanied by emission of a photon. However, a conjugated molecule has many more electrons and all the atom's nuclei, all of which contribute to the total kinetic and potential energy of the system, the latter especially due to Coulomb interactions as both electrons and nuclei carry charges.

An important approximation in the context of the theory of optical transitions is the *Born–Oppenheimer approximation* since its idea is used as a basic assumption in the Franck–Condon approach, which will be discussed in the next section [41]. Fundamentally, the Born–Oppenheimer to be calculated approximation allows the wave function of a complicated system to be calculated in two consecutive steps as it separates the total wave function into contributions of electrons and nuclei, i.e. the total wave function can be written as

$$\Psi = \Psi_E \cdot \Psi_N. \tag{2.49}$$

We will now look at the different contributions to the Hamiltonian of the Schrödinger equation, which we have to solve for our conjugated molecule. First of all, there are the kinetic energy operators for electrons and nuclei. For electrons, this operator reads

$$\mathbf{T_E} = -\sum_{i=1}^{N_E} \frac{\hbar^2}{2m_e} \nabla_i^2 = -\sum_{i=1}^{N_E} \frac{\hbar^2}{2m_e} \left(\frac{\partial^2}{\partial r_{i1}^{E\,2}} + \frac{\partial^2}{\partial r_{i2}^{E\,2}} + \frac{\partial^2}{\partial r_{i3}^{E\,2}} \right), \tag{2.50}$$

where N_E is the number of electrons, m_e the electron mass, $\mathbf{r}_i^E = (r_{i1}^E, r_{i2}^E, r_{i3}^E)$ the position of electron number i, and ∇_i^2 the second derivative with respect to the position of the ith electron.

Analogous, we can write down the kinetic energy operator for the nuclei as

$$\mathbf{T_N} = -\sum_{i=1}^{N_N} \frac{\hbar^2}{2m_i^N} \nabla_i^{N\,2} = -\sum_{i=1}^{N_N} \frac{\hbar^2}{2m_i^N} \left(\frac{\partial^2}{\partial r_{i1}^{N\,2}} + \frac{\partial^2}{\partial r_{i2}^{N\,2}} + \frac{\partial^2}{\partial r_{i3}^{N\,2}} \right), \tag{2.51}$$

where N_N is the number of nuclei, m_i^N the mass of the ith nucleus, and $\nabla_i^{N\,2}$ the second derivative with respect to the position of the ith nucleus.

Furthermore, there are three terms of potential energy, namely the repulsion between electrons, the repulsion between nuclei, and the attractive interaction between nuclei and electrons. The electron–electron repulsion can be written as

$$V_{EE} = \frac{1}{2} \sum_{i=1}^{N_E} \sum_{\substack{j=1 \\ j \neq i}}^{N_E} \frac{q^2}{4\pi\varepsilon_0} \cdot \frac{1}{r_{ij}^{EE}}, \tag{2.52}$$

where q is the elementary charge, ε_0 the vacuum permittivity, and $r_{ij}^{EE} = |\mathbf{r}_i^E - \mathbf{r}_j^E|$ the distance between ith and jth electrons. As both sums run over all electrons, the 1/2 in front corrects for the fact that each electron pair is counted twice.

Analogous, for the nucleus–nucleus repulsion, we find

$$V_{NN} = \frac{1}{2} \sum_{i=1}^{N_N} \sum_{\substack{j=1 \\ j \neq i}}^{N_N} \frac{Z_i Z_j q^2}{4\pi\varepsilon_0} \cdot \frac{1}{r_{ij}^{NN}}, \tag{2.53}$$

where r_{ij}^{NN} is the distance between the ith and jth nucleus and Z_i is the number of protons in the nucleus with number i.

Finally, the attractive potential between electrons and nuclei can be written as

$$V_{EN} = -\sum_{i=1}^{N_E}\sum_{j=1}^{N_N} \frac{Z_j q^2}{4\pi\varepsilon_0} \cdot \frac{1}{r_{ij}^{EN}}, \qquad (2.54)$$

where r_{ij}^{EN} is the distance between electron number i and nucleus number j.

Summing this up we see that finding a solution to the problem of our conjugated molecule involves solving the Schrödinger equation

$$(T_E + T_N + V_{EE} + V_{NN} + V_{EN})\Psi = E\Psi, \qquad (2.55)$$

which is a high-dimensional problem since Ψ is a function of all electron and nuclear positions. Besides, we did not treat the spin states of the individual particles involved in the problem yet. The Born–Oppenheimer approximation now tries to solve this problem by taking advantage of the important assumption that the positions of the nuclei are not as uncertain as the positions of the electrons, or rather not uncertain at all, owing to their much higher mass. This means that the kinetic energy term of the nuclei, T_N, is considered to be 0, i.e. the Schrödinger equation is solved only for electrons in a fixed potential landscape given by the nuclei. The wave function we are interested in still depends on the positions of both electrons and nuclei but the latter are taken as fixed and the equation only contains derivatives with respect to the former, which is why we write this wave function as Ψ_E in order to express that it is the electronic portion of the molecule we try to describe rather than the nuclei. Since there is no uncertainty in the (fixed) positions of the nuclei, Ψ_E^2 is then the probability distribution of the electrons in our conjugated molecule.

The Schrödinger equation for electrons then reads

$$(T_E + V_{EE} + V_{NN} + V_{EN})\Psi = E\Psi \qquad (2.56)$$

and by defining an eigenenergy for the electron wave function $E_E = E - V_{NN}$ we can remove V_{NN} from the equation as it is only a constant:

$$(T_E + V_{EE} + + V_{EN})\Psi = E_E\Psi. \qquad (2.57)$$

It is now possible to find the electronic ground state energy $E_{E,0}$ of the system for a given set of nuclei coordinates, and use this to solve the Schrödinger equation for nuclei in a consecutive step. This yields a ground state nuclei wave function $\Psi_{N,0}$, which one can find from

$$(T_N + V_{NN} + E_{E,0})\Psi_{N,0} = E_N \Psi_{N,0}, \qquad (2.58)$$

where $\Psi_{N,0}$ is now a function of all nuclei coordinates, which are no longer fixed. The important point about the Born–Oppenheimer approximation is that the electron eigenenergies $E_{E,0}$, $E_{E,1}$, $E_{E,2}$, ..., in the second step, where the nuclei Schrödinger equation is solved, is not a function of the electron coordinates anymore, as it was computed beforehand for a given set of nuclear coordinates. It is thus possible to calculate a number of different electronic eigenenergies for different sets of nuclear coordinates and use these (and interpolation between them) in order to solve the nuclei

Schrödinger equation in the next step. Thus, one can obtain solutions not only for the electronic wave function Ψ_E but also for the nuclear wave function Ψ_N and with this a complete (approximate) solution for the Schrödinger equation. Furthermore, the total eigenenergies are approximately found as

$$E = E_E + E_N, \qquad (2.59)$$

After we saw how it is possible to separate the total wave function as $\Psi = \Psi_E \cdot \Psi_N$ it is analogously possible to further separate vibrational, rotational, and electron spin contributions (nuclear spin contributions can be neglected due to their low energies). However, we will not discuss this in detail here as the fundamental ideal of the Born–Oppenheimer approximation is apparent from above discussions. Overall, it is possible to approximate the total wave function as a product of an electronic, a nuclear, a vibrational, a rotational, and a spin wave function

$$\Psi = \Psi_e \cdot \Psi_n \cdot \Psi_v \cdot \Psi_{\text{rot}} \cdot \Psi_s \qquad (2.60)$$

and the total energy as a sum of the respective eigenenergies

$$E = E_e + E_n + E_v + E_{\text{rot}} + E_s \qquad (2.61)$$

(note that $\Psi_e \neq \Psi_E$, $E_e \neq E_E$, $\Psi_n \neq \Psi_N$, and $E_n \neq E_N$). This form of Born–Oppenheimer approximation is especially useful for optical transitions and will therefore be used in the next section where we discuss the Franck–Condon principle.

As we are particularly interested in optical transitions between singlet states and we have already stated that these transitions are fast, we have an additional argument for the validity of the Born–Oppenheimer approximation in the context of our considerations. Due to their much higher mass it is reasonable to assume that nuclei do not significantly change their position during the transition of an electron between two singlet states [42–44]. Thus, in order to quantum mechanically understand such optical transitions it is sufficient to treat the electronic wave function Ψ_e in combination with additional contributions from the spin state (Ψ_s), the vibrational state (Ψ_v), and the rotational state (Ψ_{rot}), but Ψ_N does not have to be computed explicitly.

In the next section we will see how absorption and emission spectra of conjugated molecules can be described by the Franck–Condon principle, which treats transitions between different vibronic states in the molecule according to the Born–Oppenheimer approximation.

Franck–Condon Principle. As discussed along with the Jablonski diagram in Figure 2.30 for a theoretical conjugated molecule, each electronic state is split into many vibrational sublevels. At room temperature typical vibrational energies are on the order of 0.1 to 0.4 eV, which is relatively large compared to the thermal energy $k_B T$ that is around 25 meV. Therefore, at room temperature a molecule in its electronic ground state is typically also in its lowest vibrational state. Especially in solution conjugated molecules can also rotate so that there are rotational modes besides molecular vibrations. However, in contrast to vibrational energies typical rotational energies are in the range of 10 meV so that at room temperature the system can easily switch between different rotational states.

We now combine everything we discussed so far in order to understand the origin of the particular absorption and emission spectra of a conjugated molecule. This leads us directly to the Franck–Condon principle. First of all we adhere to the Born–Oppenheimer approximation since optical transitions between singlet states are fast enough that nuclear motion can be neglected. An optical transition is therefore a straight line upward or downward in a potential landscape as exemplarily shown in Figure 2.31. For a two-atom molecule the nuclear coordinate is the distance between the nuclei, whereas for a many-atom molecule it should be understood as a generalized and simplified coordinate where a change in nuclear coordinate means an overall change in the positions of the individual nuclei.

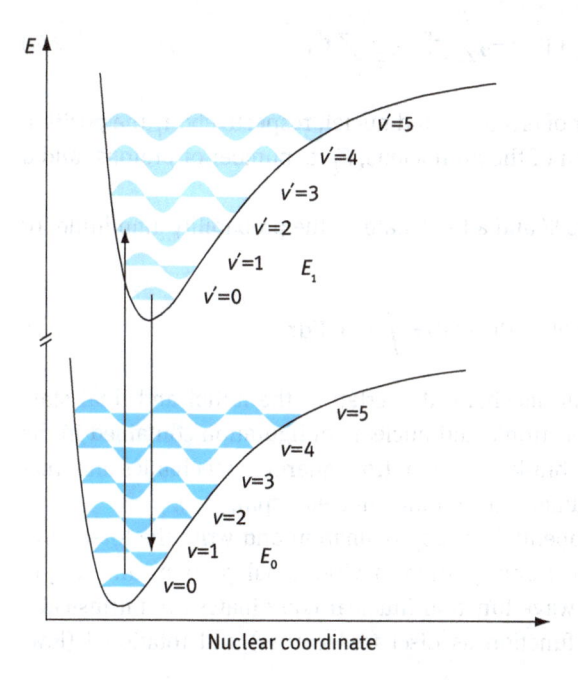

Fig. 2.31: Potential energy landscape for an optical transition between electronic states with vibrational sublevels in a conjugated molecule. According to the Franck–Condon principle transitions are straight lines upward or downward since the nuclear coordinates do not change during the fast transition. The probability for an optical transition is then given by the quantum mechanical overlap of the initial and the final states.

As the figure illustrates the lowest potential energy for a certain electronic state is not necessarily located at the same nuclei configuration of the molecule so that a certain rearrangement of atoms is necessary within the molecule in order to minimize its energy in a given state. Since such a rearrangement is slower than the optical transition, however, the probability for a transition between two vibronic states can be calculated from the electronic portion of the wave function from the Born–Oppenheimer approximation only as all nuclear coordinates can be treated as constant.

We now find an expression for the probability of a certain transition since this will allow us to construct an absorption and emission spectrum. First, it is useful to define a molecular dipole operator μ to quantify the dipole moment of the molecule as a function of the vibronic state. Since a conjugated molecule is an ensemble of electrons on a group of positively charged nuclei, there is a dipole moment associated with the vibronic state as the center of the negative charge (electrons) can be dislocated with respect to the center of the positive charge (nuclei) in the molecule. The dipole operator is simply a sum of electronic and nuclear dipole contributions and can be written as

$$\mu = \mu_e + \mu_n = -q\sum_{i=1}^{N_e}\mathbf{r}_i^e + q\sum_{j=1}^{N_n}Z_j\mathbf{r}_j^n, \qquad (2.62)$$

where N_e and N_n are the number of electrons and nuclei, respectively, $\mathbf{r}_i^e$ the position of the ith electron, $\mathbf{r}_j^n$ the position of the jth nucleus, Z_j its number of protons, and q the elementary charge.

Generally, for an initial state Ψ and a final state Ψ' the probability amplitude for a transition is given as

$$P(\Psi \to \Psi') = \langle \Psi'|\mu|\Psi\rangle = \int \Psi'^* \mu \Psi \, d\mathbf{r} \qquad (2.63)$$

since it depends on the quantum mechanical overlap of the initial and final state with respect to the molecule's electronic and nuclear configuration contained in the dipole operator. Here we use the bra-ket notation from quantum mechanics and integrate over the whole set of coordinates (electronic, nuclear, spin, etc.).

We now recall the Born–Oppenheimer approximation and write the total wave function as a product of an electronic portion, a vibrational portion, and a spin portion, neglecting the nuclear wave function (nuclear coordinates are intrinsically treated in the electronic wave function as discussed earlier) and rotational (low-energy) sublevels. This gives us

$$\Psi = \Psi_e \Psi_v \Psi_s, \qquad (2.64)$$

which we directly plug into Equation (2.63). In combination with the expression from Equation (2.62) for the electronic and nuclear dipole contribution, this leads to

$$\begin{aligned}P(\Psi \to \Psi') &= \langle \Psi'_e \Psi'_v \Psi'_s | \mu_e | \Psi_e \Psi_v \Psi_s\rangle + \langle \Psi'_e \Psi'_v \Psi'_s | \mu_n | \Psi_e \Psi_v \Psi_s\rangle \\ &= \langle \Psi'_e | \mu_e | \Psi_e\rangle\langle \Psi'_v | \Psi_v\rangle\langle \Psi'_s | \Psi_s\rangle + \langle \Psi'_e | \Psi_e\rangle\langle \Psi'_v | \mu_n | \Psi_v\rangle\langle \Psi'_s | \Psi_s\rangle.\end{aligned} \qquad (2.65)$$

The origin of the second equality is not directly apparent and only approximately true. First of all the factorization in Equation (2.64) is only possible since the different wave functions depend on different sets of coordinates: Ψ_v is a function of the nuclear coordinates $\mathbf{r}_n$, and Ψ_s depends on the spin coordinates $\mathbf{r}_s$; Ψ_e generally is a function of both the electronic coordinates $\mathbf{r}_e$ and the nuclear coordinates, but the latter dependency typically is assumed to be a simple and smooth function and can therefore be neglected, according to the so-called *Condon approximation*. This factorization can then be applied to separate the first bra-ket expression since the electronic portion of the dipole operator also depends only on $\mathbf{r}_e$.

A similar argument is used in order to separate the expression for μ_n to yield the final two terms in Equation (2.65). We now take advantage of the fact that electronic eigenstates are orthogonal, so that the second term of the sum is 0 since $\langle \Psi'_e | \Psi_e \rangle = 0$. With this we find the following total expression for the transition probability, where for clarity we write down the bra-kets as integrals once more and sort the terms in different order:

$$P(\Psi \to \Psi') = \langle \Psi'_e | \mu_e | \Psi_e \rangle \langle \Psi'_v | \Psi_v \rangle \langle \Psi'_s | \Psi_s \rangle$$
$$= \int \Psi'^*_e \mu_e \Psi_e d\mathbf{r}_e \cdot \int \Psi'^*_v \Psi_v d\mathbf{r}_n \cdot \int \Psi'^*_s \Psi_s d\mathbf{r}_s. \quad (2.66)$$

As we see in the integral form each of the three terms depends on a different set of coordinates since we directly applied the factorization from Equation (2.64). The last term can be understood as a spin selection rule, i.e. a transition between two vibronic states is only possible if spin-allowed.

The first term is the so-called transition dipole moment and is a measure of how much the electronic states overlap. In the simplified picture of a Franck–Condon transition shown in Figure 2.31 this is determined by the shift between the lower and the higher states since the nuclear coordinates determine the shape of the electronic state as they do not change during the fast transition as discussed earlier.

The second term is referred to as *Franck–Condon factor* and is proportional to the overlap of the vibrational state in the lower and higher electronic state. As there is no change in nuclear coordinates, which means that only straight transitions are allowed in the schematic shown in Figure 2.31, this overlap is fixed for two given electronic states and therefore a material's property. As we will see in the following paragraph the shape and position of absorption and emission peaks directly depend on the overlap of ground state and excited state vibrational levels and thus reflect the Franck–Condon factor. As discussed earlier, relaxations to lower vibrational states are fast and the splitting between vibration levels typically exceeds $k_B T$. Therefore, absorption (i.e. excitation) events always are transitions from the lowest vibrational state ($v = 0$) in the electronic ground state to one of the vibrational states in the excited electronic state ($v = 0, 1, 2, ...$). On the other hand, due to the same arguments emission events always are transitions between the lowest vibrational state in the excited electronic state and one of the vibrational states in the electronic ground state. Exemplarily, Figure 2.31 displays a $0 \to 2$ absorption transmission and a $0 \to 1$ emission transmission.

Since vibrational levels do not (or only marginally) depend on the electronic portion of the wave function, absorption and emission spectra are typically highly symmetric around the 0 → 0 transition, which is the same in absorption and emission. This is shown for a hypothetical conjugated molecule in Figure 2.32. The peak height depends on the Franck–Condon factor, i.e. the quantum mechanical overlap between the respective vibrational levels, whereas the peak position is a function of the energy difference between them.

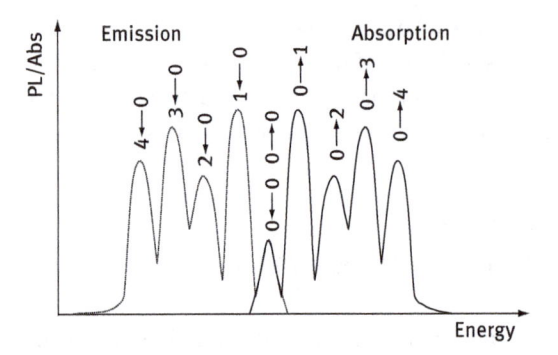

Fig. 2.32: Absorption and emission spectra for hypothetical conjugated molecules showing the spectral influence of the Franck–Condon factor. Absorption is shown as the solid line, emission as the dotted line. The peaks are labeled with the respective vibrational levels in the ground and excited electronic states.

Both absorption and emission features are spectrally broadened for increasing temperature since thermal energy can overcome small energy gaps in the case that the photon energy does not exactly match the transition energy. For small vibrational splitting it is even possible that the lowest vibrational level is not always the one occupied. Furthermore, rotational levels lead to another small broadening and play a larger role if there is more thermal energy available.

The Franck–Condon principle also applies for conjugated polymers, although typically the whole polymer cannot be considered as one extended π-electron system but rather as a series of chemically connected individual conjugated sections. This is the case because polymers are typically not straight and fully extended but folded in a complicated fashion since the backbone is not completely stiff. In this context the term *conjugation length* is used, which refers to the number of monomer repeat units in the polymer that form one conjugated system. Such conjugated systems are typically separated by kinks in the polymer backbone, which disrupt the overlap of the π-electron system.

Examples of absorption and emission spectra for six different conjugated polymers are shown in Figure 2.33. Spectra are recorded at room temperature and in dilute solution, i.e. the individual polymer molecules do not interact strongly. All polymers show the absorption/emission symmetry predicted by the Franck–Condon principle, but the individual vibronic features are more broadened than that shown in Figure 2.32 due to the measurement being carried out at room temperature and also because of interactions between polymer and solvent molecules.

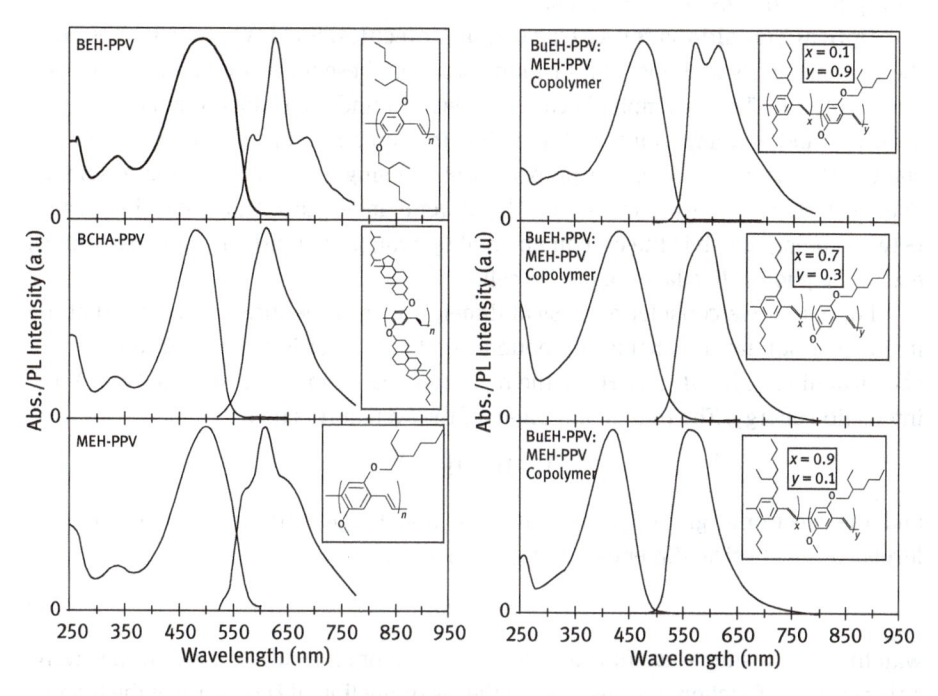

Fig. 2.33: Room temperature absorption (thick lines) and emission (thin lines) spectra for six different conjugated polymers. Reprinted with permission from [45]. Copyright 1997 American Chemical Society.

Although the Franck–Condon principle applies for isolated molecules, i.e. molecules in the gas phase, it can also be used to describe molecules in solution. The nuclear coordinate in Figure 2.31 is then to be understood as a solvation coordinate. Upon excitation of the molecule from the ground state to a certain vibrational level in an excited electronic state, there is a rapid rearrangement of solvent molecules, which assemble into a different configuration in order to minimize their energy with respect to the new vibronic configuration of the molecule. Since this rearrangement can be equally fast as the transition between vibrational states and significantly faster than the lifetime of an excited singlet state, the energy of the molecule relaxes to a lower value not due to a change in its vibrational state but rather due to the rearranged environment. For these processes the characteristic timescales strongly depend on the solvent, e.g. small solvent molecules like methanol allow a quick rearrangement, whereas the singlet lifetime might exceed the solvent rearrangement time in viscous solutions.

The net result for a molecule in solution, similar to molecules in the gas phase, is that emission energies will be lower than or at most equal to the excitation energies. This is the commonly observed *Stokes shift*, i.e. a shift in energies between the absorption and emission maximum.

Optical Transitions for Aggregates of Molecules. So far we were interested in the vibronic levels and optical transitions between these levels for an individual isolated molecule. This is simpler because the system under consideration is smaller. It is also a valid and important starting point since an ensemble of molecules will still show a behavior reflecting the properties of the single molecule in the ensemble. The next step, however, is to consider several molecules and allow interaction between them, which is the case in any real system, particularly in condensed films, e.g. in the photoactive layer of a solar cell.

For a start we consider a physical dimer, a pair of identical molecules that are interacting but are not chemically bound (the term dimer is often used to refer to a chemical dimer). Let $\mathbf{H}_1$ and $\mathbf{H}_2$ be the hamiltonians for molecules 1 and 2 and $\mathbf{V}$ the interaction energy. The total hamiltonian of the system $\mathbf{H}$ is then

$$\mathbf{H} = \mathbf{H}_1 + \mathbf{H}_2 + \mathbf{V} \tag{2.67}$$

and the electronic ground state wave function (neglecting vibrational and spin levels) of the physical dimer can be approximated as

$$\Psi_0 = \Psi_{1,0} \cdot \Psi_{2,0} \tag{2.68}$$

with the ground state wave functions $\Psi_{1,0}$ and $\Psi_{2,0}$ for molecules 1 and 2, respectively. This new wave function, however, is not the eigenfunction of $\mathbf{H}$ because of the interaction energy contribution to the total hamiltonian. In order to improve the approximation for Ψ_0 a contribution by the mixing of the two ground state wave functions $\Psi_{1,0}$ and $\Psi_{2,0}$ has to be added. With this the ground state energy can be approximated as

$$E_0 = E_{1,0} + E_{2,0} + \langle \Psi_{1,0}\Psi_{2,0} | \mathbf{V} | \Psi_{1,0}\Psi_{2,0} \rangle, \tag{2.69}$$

where $E_{1,0}$ and $E_{2,0}$ denote the ground state energies of molecules 1 and 2, respectively. The term $W_0 = \langle \Psi_{1,0}\Psi_{2,0} | \mathbf{V} | \Psi_{1,0}\Psi_{2,0} \rangle$ is the coulombic interaction energy between the ground states of the two molecules due to their electrical dipole moments but is generally not in very good agreement with real values since other interactions like spin–spin coupling are not taken into account.

We now consider the case of the first electronic excitation of the dimer. The wave functions for the first excited state of molecules 1 and 2 are denoted as $\Psi_{1,1}$ and $\Psi_{2,1}$, respectively. If one molecule is in its excited state, the excitation can jump between the two molecules but the energies for the combined states $\Psi_{1,1}\Psi_{2,0}$ and $\Psi_{1,0}\Psi_{2,1}$ are not identical due to the interaction $\mathbf{V}$. Thus, the system oscillates between these two states, which can be expressed when writing the combined excited states $\Psi_1^{\pm}$ as a linear combination

$$\Psi_1^\pm = c_1\Psi_{1,1}\Psi_{2,0} + c_2\Psi_{1,0}\Psi_{2,1} = \frac{1}{\sqrt{2}}\Psi_{1,1}\Psi_{2,0} \pm \frac{1}{\sqrt{2}}\Psi_{1,0}\Psi_{2,1}. \quad (2.70)$$

The second equality holds for identical molecules only. Apparently, there are two different wave functions, corresponding to two different energy states E^+ and E^-. With the interaction energy for the first excited state ΔE_1 between the two different "pure" excited states $\Psi_{1,1}\Psi_{2,0}$ and $\Psi_{1,0}\Psi_{2,1}$, the energy states can be written as

$$E^\pm = E_{1,1} + E_{2,0} + W_1 \pm \Delta E_1 \quad (2.71)$$

since $E_{1,1} = E_{2,1}$. Therefore, analogous to W the coulombic interaction for each of the two excited states is

$$W_1 = \langle \Psi_{1,1}\Psi_{2,0} | \mathbf{V} | \Psi_{1,1}\Psi_{2,0} \rangle = \langle \Psi_{1,0}\Psi_{2,1} | \mathbf{V} | \Psi_{1,0}\Psi_{2,1} \rangle \quad (2.72)$$

and the energy splitting is

$$\Delta E_1 = \langle \Psi_{1,1}\Psi_{2,0} | \mathbf{V} | \Psi_{1,0}\Psi_{2,1} \rangle, \quad (2.73)$$

i.e. the interaction energy between the two pure excited states.

These considerations can be followed in an analogous fashion for a larger ensemble of molecules, i.e. a physical oligomer, as shown in Figure 2.34. For n original molecular levels each excited state then splits into n states, while the ground state is shifted to a lower energy. Thus for systems with several molecules the ΔE_1, ΔE_2, ..., become small so that the original levels are shifted by W_1, W_2, ..., and effectively broadened.

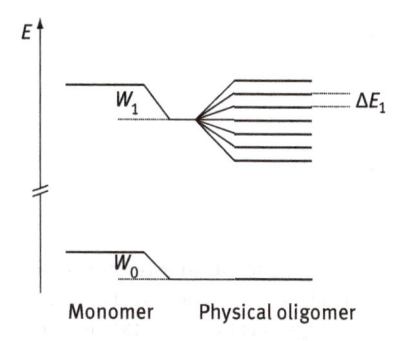

Fig. 2.34: Splitting of energy levels for an ensemble of molecules. For an n-molecular crystal the ground-state energy is shifted while the excited state of the monomer splits into n new energy levels with the energy splitting dependent on the interaction.

The picture of physical oligomers is commonly used to understand the optical behavior of molecular crystals. A molecular crystal can be considered as being built up from unit cells, similar to an inorganic crystal. Each unit cell consists of several molecules that electronically interact (a physical oligomer) so that there is an energy-level splitting as discussed earlier. Additionally there are interactions between the unit cells so that eventually there are regions of allowed and forbidden energies, somewhat similar to energy bands in inorganic solids. The splitting between these energy bands is called *Davydov splitting*, which was predicted by the Russian-Ukrainian physicist Alexander Davydov in 1948.

In order to understand whether or not an optical transition between the ground state and one of the excited states in a dimer (or more generally in a molecular crystal) is allowed, it is necessary to consider the involved dipole moments. In a physical dimer depending on the alignment of molecules with respect to each other the overall dipole moment is different for the state $\Psi_{1,1}\Psi_{2,0}$ than that for $\Psi_{1,0}\Psi_{2,1}$ and so are the overall energies. The two most extreme cases, which are often referred to as *H-* and *J-aggregate*, are depicted in Figure 2.35. An H-aggregate is a parallel alignment of molecules, which results in parallel ground-state dipole moments. In an H-aggregate the transition to the lower lying excited state is dipole-forbidden, so that the absorption is blue-shifted with respect to the monomer since only an excitation to the higher excited state is possible. In contrast for head-to-tail alignment in a J-aggregate the transition to the higher energy level is dipole-forbidden, so that the absorption is red-shifted. For a tilted alignment that is neither completely head-to-tail nor parallel there is a spectral contribution from both states with intensities depending on selection rules and quantum mechanical overlaps. In such cases it is possible to measure the energy splitting between Davydov energy levels since both features can be observed simultaneously.

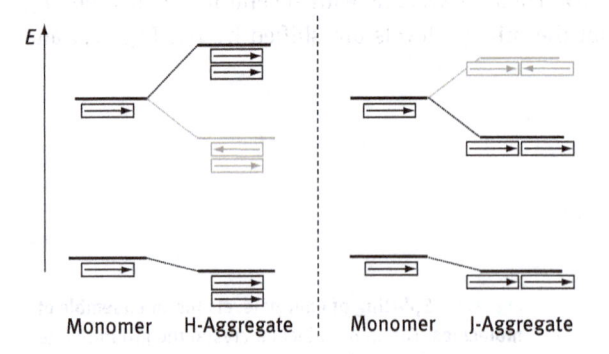

Fig. 2.35: Energetic situation for physical dimers with parallel and head-to-tail arrangement of the molecules, leading to formation of H- and J-aggregates, respectively. In an H-aggregate the transition to the lower excited state is dipole-forbidden, whereas in a J-aggregate this is the case for the higher energy transition. Absorption spectra are therefore blue-shifted and red-shifted compared to the monomer, respectively.

2.2.2.4 Excitons and Exciton Diffusion

A concept in the context of optically excited states in molecules, molecular crystals, and organic semiconductors that is of particular interest for photovoltaic devices is the *exciton*. Excitons can be understood as strongly bound electron-hole pairs, which form upon excitation of a ground-state electron. For instance, if an electron is lifted from the HOMO to the LUMO in an organic semiconductor it basically leaves a hole behind in the HOMO, to which it is coulombically bound. If the molecule is not isolated but interacting with other molecules the excitation induces a reorganization of the surrounding molecules. This collective response is referred to as *exciton*, a term that is typically used only for timescales where the excited state has relaxed to its lowest vibrational state. Due to electron and hole being very close to each other and not free, in the sense that there is a coulombic interaction holding them together, excitons are non-charged quasiparticles. It is this particle nature that makes the exciton picture useful in order to understand the behavior of photoactive solids since excitons can move in the material, i.e. the excitation can be transported to certain regions in the material or to interfaces between different semiconductors. Furthermore, excitons can be split into an electron polaron and a hole polaron, which subsequently have to be collected at external electrodes in order to guarantee the function of a solar cell device.

Frenkel exciton   Wannier exciton   Charge transfer exciton

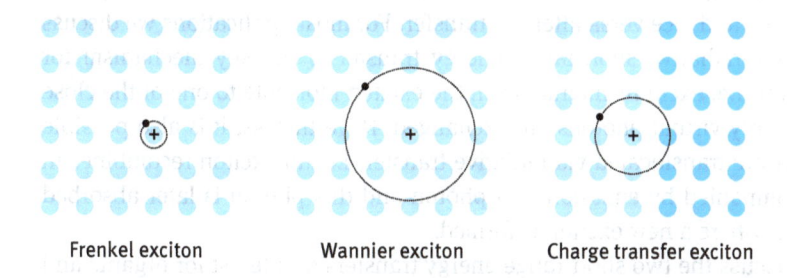

Fig. 2.36: **Schematic representation of the three commonly used exciton classifications.** Frenkel excitons are highly located and strongly bound. They are found in organic semiconductors. Wannier excitons show a weaker binding and delocalization over several lattice points, which is typical for inorganic semiconductors with high relative permittivity. Charge transfer excitons show a delocalization on the order of nearest neighbors and occur at interfaces between different materials.

Most commonly, excitons are classified into three different types as summarized in Figure 2.36, namely *Frenkel excitons*, *Wannier excitons*, and *charge transfer excitons*. Frenkel excitons are found in organic semiconductors due to their molecular nature and low relative permittivity. They are strongly bound (binding energies on the order of 1 eV) and highly localized, typically on one molecule. In contrast, Wannier excitons show a much stronger delocalization and are typically found in inorganic semiconductors, where high relative permittivities lead to low binding energies on the order of tens of meV. The electron and hole can reside on different lattice points and the total size of the exciton spans several unit cells. Depending on the exact binding energy and the material properties, Wannier excitons can also spontaneously split into free electron and hole polarons since the binding can be overcome by thermal energy. The third important class are charge transfer excitons, which represent the situation where electron and hole reside on neighboring molecules or crystal sites. Binding energies of charge transfer excitons are typically between the binding energies of Frenkel and Wannier excitons due to their moderate delocalization. Most commonly charge transfer excitons are found at interfaces between two different semiconducting materials, usually at the interface of organic semiconductors or organic–inorganic junctions. They play a crucial role in organic and hybrid photovoltaics and are often referred to as *charge transfer states*.

Analogous to excited states in isolated molecules excitons can be singlets and triplets, where the latter cannot be directly excited by a photon but is generated only upon formation of a singlet exciton and subsequent intersystem crossing.

We have already mentioned earlier that one of the most interesting properties of excitons is that they can migrate since this is of eminent importance for the function of organic optoelectronic devices. Since excitons are uncharged excited states the underlying mechanism of exciton diffusion is an energy transfer. The net process is that the exciton and its energy are transferred from one molecule (the donor) to another (the acceptor). Before the transfer the donor is excited while the acceptor is in its ground state and vice versa after the transfer. For most applications we discuss in this book non-radiative short-range energy transfer is the key mechanism for exciton transport, i.e. excitons migrate from the excited molecule to one of the close neighbors and only virtual photons are exchanged. Nevertheless, it is also possible that excitations are translocated via radiative transfer, i.e. an exciton recombines in the donor accompanied by emission of a photon and this photon is later absorbed by the acceptor, where a new exciton is formed.

Here we discuss the two short-range energy transfers of interest for organic and hybrid solar cells, namely Förster resonance energy transfer and Dexter energy transfer.

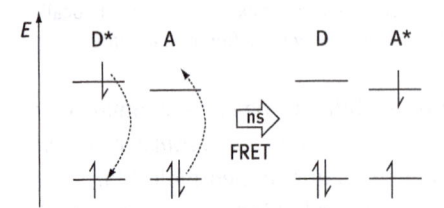

Fig. 2.37: Schematic of the mechanism of Förster resonance energy transfer (FRET). Non-radiative dipole–dipole coupling leads to the excitation of the acceptor molecule upon relaxation of the excited state in the donor.

Förster Resonance Energy Transfer (FRET). The mechanism of FRET is shown in Figure 2.37: upon relaxation of the excited state in the donor the acceptor is excited so that the exciton changes its location [46]. During the energy transfer a virtual photon is exchanged and the energy is transferred via dipole–dipole coupling over distances much smaller than the wavelength corresponding to the excitation. The emitted virtual photon only exists in the near-field region since this dipole–dipole coupling requires direct interaction of donor and acceptor. FRET essentially is an emission–absorption coupling (the donor emits and the acceptor absorbs the energy) and therefore connected with a loss in excitation energy since emission spectra are red-shifted compared to absorption spectra (e.g. Figure 2.33). FRET relies on spin-allowed absorption and emission events.

Two main factors determine the efficiency of FRET: (1) the distance between the donor and the acceptor and (2) the overlap between the emission spectrum of the donor and the absorption spectrum of the acceptor. The efficiency of FRET is an inverse 6th power law dependence on the distance r between the donor and the acceptor and can be written as

$$\eta_{\text{FRET}} = \frac{1}{1 + (r/R_0)^6} \tag{2.74}$$

with the so-called Förster radius R_0, which is given by

$$R_0^6 = \frac{9000 Q_0 \kappa^2 J \ln 10}{128 \pi^5 n^4 N_A} \tag{2.75}$$

with the donor fluorescence quantum yield Q_0, the refractive index of the material n, Avogadro's constant N_A, and the so-called dipole orientation factor κ^2, which depends on the relative orientation of the dipole moments of donor and acceptor. κ^2 is a function of the solid angles describing the spacial arrangement of the dipole moments and ranges from 0 for the case of perpendicular dipoles to 4 for exactly parallel dipoles [47]. J is the overlap integral of the donor's emission spectrum and the acceptor's absorption spectrum and is given as

$$J = \int f_D(\lambda) \varepsilon_A(\lambda) \lambda^4 d\lambda \tag{2.76}$$

with the normalized emission spectrum of the donor f_D and the molar extinction coefficient of the acceptor ε_A.

Due to the 6th power law the efficiency of FRET dramatically drops for donor–acceptor distances exceeding the Förster radius. This strong dependency is owing to the necessary dipole–dipole interaction, which is weak for larger distances between donor and acceptor. Due to that FRET is a very useful mechanism for distance measurements on the nm scale. For instance, FRET can be used to determine the distance between protein moieties via labeling of two sites with donor and acceptor dyes. Upon excitation of the donor the light emitted by the acceptor, which is red-shifted, can be probed, with the emission intensity depending strongly on the distance between the dyes [48].

As is apparent from Equation (2.75), the Förster radius also depends on κ^2, i.e. on the angle between the two molecular dipole moments (R_0 gets larger for more parallel configuration) and inversely on the refractive index. The latter is the case since higher refractive indices lead to stronger screening of the electronic communication between the molecules, leading to a weaker coupling.

Furthermore, R_0 is vitally dependent on the spectral overlap of donor emission and acceptor absorption since the energy transfer is resonant. The virtual photon exchange takes place only if the energy emitted by the donor is the same as the one necessary for excitation of the acceptor, i.e. there has to be an overlap between vibronic transition energies.

Dexter Energy Transfer (DET). The second mechanism we discuss is Dexter energy transfer (DET) [49]. In contrast to FRET this type of energy transfer is also possible for triplet states and is not necessarily accompanied by a loss in excitation energy due to vibrational relaxations. However, it also depends very strongly on the distance between the donor and the acceptor and therefore is a short-range process.

The fundamental mechanisms of singlet and triplet DET are depicted in Figure 2.38. DET involves a rapid transfer of an electron from the excited state in the donor to the excited state in the acceptor, accompanied by the transfer of a ground-state electron in the opposite direction. This means that both an electron and a hole move from the donor to the acceptor, i.e. the exciton changes its position.

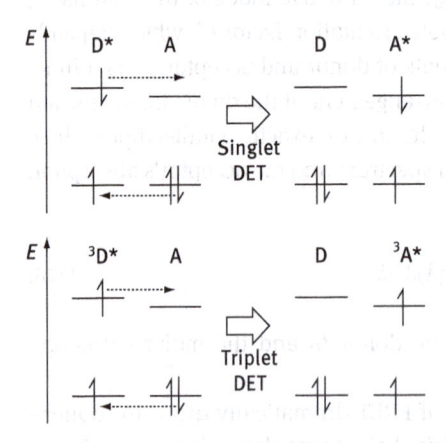

Fig. 2.38: Schematic representation of the mechanisms of singlet (top) and triplet (bottom) Dexter energy transfer. Essentially, Dexter energy transfer is a relocation of the excitation from one site to the other, where the excited electron moves from the donor to the acceptor while a ground-state electron moves in the opposite direction (i.e. a hole is also transferred from the donor to the acceptor).

DET is not a dipole–dipole coupling mechanism involving a virtual photon but depends on the quantum mechanical overlap of the molecular orbitals of the donor and the acceptor. Therefore, DET is an exchange interaction since the close proximity of the donor and the acceptor allows a quantum mechanic mixing of states so that the excited state can hop between the involved molecules. Macroscopically this dependence on the available energy states in the donor and the acceptor and their interaction is reflected in the dependence of the DET rate on the spectral overlap integral J from Equation (2.76). However, DET is possible for both optically allowed and forbidden transitions since it is a direct transfer of the quantum state between the donor and the acceptor, which is why triplet excitons also can migrate via DET.

The strong dependency of the DET efficiency on the distance between the donor and the acceptor is reflected in the exponential behavior of the transfer rate k_{DET}:

$$k_{DET} \propto J \exp\left(\frac{-2r}{L}\right). \tag{2.77}$$

As mentioned earlier, J is the spectral overlap integral for the donor and the acceptor and r is the distance between the two molecules. L is the parameter with which the exponential decay scales. It is the sum of the van der Waals radii of the donor and the acceptor, i.e. a measure for how close the two materials can come. From this formula it is apparent that DET is efficient only for distances on the order of a few nm.

DET plays an important role for exciton migration in molecular crystals and is the only mechanism possible for the motion of triplet excitons. Due to their close proximity DET is possible between neighboring molecules so that the short-range limitation does not play a role as, e.g. in dilute solutions. The spectral overlap integral J has to be interpreted carefully for ensembles of molecules, since the real spectra of the individual molecules are not known. Due to energetic disorder owing to complicated molecular interactions in the bulk the vibronic states that are found in isolated molecules are modified. There is a Gaussian distribution of excited states so that the spectral overlap integral cannot be understood in terms of a Franck–Condon formalism but rather reflects the complicated quantum mechanical situation of the donor and the acceptor in interaction with the surroundings, i.e. a collective behavior of all involved molecules. For vibrationally relaxed excitations that migrate between identical molecules or molecular moieties, DET is an efficient mechanism for the virtually loss-free transfer of the excitation energy as discussed in the next paragraph.

Exciton Diffusion. Since excitons in organic and hybrid solar cells are commonly Frenkel excitons they do not easily split into charge carriers but diffuse through the material via a sequence of energy transfers between molecules. This exciton diffusion is crucial for the function of these types of solar cells since it is the mechanism that transports excitons to interfaces of the material, where energy differences between HOMO and LUMO levels of donor and acceptor materials can provide the energy necessary to separate excitons.

As long as the distribution of excitonic states is isotropic, the diffusion of excitons via several energy transfers can be described by a simple diffusion equation

$$\frac{\partial n}{\partial t} = D\nabla^2 n - \frac{n}{\tau}, \tag{2.78}$$

where n is the exciton density, D the diffusion coefficient, and τ the exciton lifetime. This is valid for both singlet and triplet excitons since both can effectively migrate via energy transfer hopping from one molecular site to another. From the diffusion equation it is also possible to define an effective exciton diffusion length

$$L_D = \sqrt{D\tau}, \tag{2.79}$$

which is the average displacement of an exciton from its initial position when it recombines, i.e. when the excited electron falls back to its ground state. The equation for L_D illustrates that efficient exciton diffusion over long distances requires a long lifetime of excited states on the one hand and an efficient mechanism for their translocation on the other, which determines the diffusion coefficient D.

Exciton diffusion through extended molecular crystals is governed by the energetic disorder in the system due to various interactions between molecules and their disorder. Disorder plays an important role especially in amorphous and polycrystalline films as typically found in the active layers of organic and hybrid solar cells. In disordered systems of molecules the energy splitting is not symmetric any more as in the case of physical oligomers discussed earlier, but leads to a Gaussian distribution of HOMO–LUMO gaps for large numbers of molecules involved. During diffusion an exciton moves through this Gaussian density of exciton states before it recombines.

The energetic picture of exciton diffusion is shown in Figure 2.39. Upon absorption of a photon in the organic semiconductor a high-energy exciton is generated, which resides in a certain state in the molecule or section of the polymer where the photon was absorbed. At low temperatures the exciton energy quickly relaxes to its final equilibrium value E_{ex} due to a series of lossy energy transfers to neighboring exciton states. E_{ex} is the average exciton energy in the material, somewhat similar to the Fermi level for electrons in semiconductors. At low temperatures there is virtually no exciton–phonon coupling so that only steps to lower energies occur during exciton diffusion. Therefore, E_{ex} is located at the lower end of the Gaussian density of exciton states, resulting in a low exciton diffusion coefficient due to the low density of exciton states, i.e. a large spatial separation between states, which lowers the probability of exciton hopping.

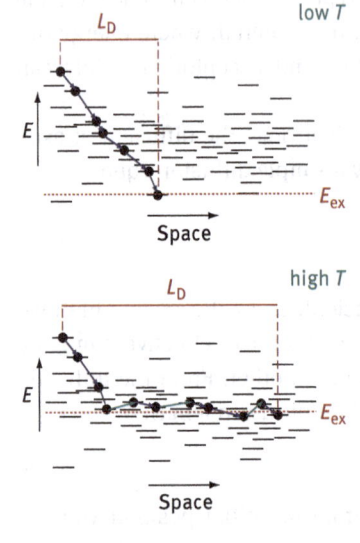

Fig. 2.39: Exciton diffusion at low and high temperatures. After formation upon photon absorption an exciton moves through a Gaussian density of exciton states. At low temperature this movement occurs mainly toward states of lower energies and does not involve interactions with phonons (blue arrows). Eventually, the exciton reaches states at the lower end of the Gaussian distribution. At higher temperatures excitons can move to higher energy states due to interaction with phonons (turquoise arrows).

In contrast, at higher temperatures there is a pronounced contribution of phonons in the diffusion process, which enables the occupation of higher energy states after the exciton translocates due to a phonon absorption process. This results in a higher average exciton energy E_{ex}. Furthermore, the movement in an energy region with a higher density of exciton states enables a higher exciton diffusion length due to a higher diffusion coefficient. This leads to two subsequent steps of exciton diffusion, namely a first sequence of exciton hoppings to lower energy states and a second sequence of hoppings between states of similar energies due to interactions with phonons.

The density of exciton states, the energy transfer rates, and the exciton lifetime are strongly dependent on the properties of the organic semiconductor and differ massively among different materials. For instance, typical exciton diffusion lengths in conjugated polymers as employed in organic and hybrid photovoltaics are on the order of 5–10 nm due to typically high disorder in polymeric films [50]. For strongly crystalline films of small conjugated molecules longer exciton diffusion lengths can be observed. This is attributed to the higher order in the molecular crystal. Due to their forbidden ground-state relaxation and corresponding long lifetime, triplet excitons can show outstandingly high diffusion lengths exceeding 1 μm as shown in the case of highly crystalline films of rubrene [51].

2.2.3 Charge Transport in Organic Semiconductors

The active material of a solar cell has to play a dual role since it has to be a good light absorber and possible exciton transporter on one hand and also a good conductor for photogenerated charge carriers on the other. The mechanism of charge generation in conventional inorganic solar cells as well as in excitonic solar cells will be discussed in detail in Chapter 3. The main difference is the role of excitons, which in many inorganic semiconductors can be separated thermally at room temperature, whereas an energy step at a donor–acceptor interface is required to overcome their binding energy in organic semiconductors.

Here we discuss the transport of charges in organic semiconductors. For a very comprehensive view on charge transport in organic semiconductors see the book or the review article by Bässler and Köhler [25, 52].

2.2.3.1 Charge Hopping in Disordered Systems

The nature of charge transport in solids can substantially differ depending on the material's properties. One extreme case, which is found in highly crystalline inorganic semiconductors, is a strongly undisturbed charge transport in a broadband, where the carrier can travel as a highly delocalized plane wave. These waves are scattered at impurities or when interacting with phonons but high charge mobilities exceeding $1,000\,\text{cm}^2\,\text{V}^{-1}\,\text{s}^{-1}$ can be observed in doped semiconductors [53]. In contrast, in highly disordered systems charge transport is completely dominated by scattering, and carriers effectively travel via a hopping process from one site to the other. Such hopping transport is present in many organic semiconductors, especially when integrated into real devices, where the processing often induces a low crystallinity, e.g. due to blending of donor and acceptor materials and formation of mixed films. In these situations charge mobilities are below $1\,\text{cm}^2\,\text{V}^{-1}\,\text{s}^{-1}$ and in many cases only on the order of $10^{-4}\,\text{cm}^2\,\text{V}^{-1}\,\text{s}^{-1}$ [54, 55].

The Hamiltonian for Charge Transport. We now take a closer look at the general description of charge transport in a solid in terms of a one-electron Hamiltonian formalism, which applies for low charge densities and covers both hopping and band transport. This paragraph is based on the discussion of the Hamiltonian model for charge transport in *Pope and Swenberg* (see there for further details) [23]. The overall Hamiltonian can be written as a sum of several contributions:

$$H = H_e + H_p + H_J + H_{e-p}^{\text{diag}} + H_{e-p}^{\text{n-diag}} + H_{\text{dis}}^e + H_{\text{dis}}^J + H_{e-p}^{\text{diag}'} + H_{e-p}^{\text{n-diag}'}. \tag{2.80}$$

We will now discuss the individual contributions to H, which are to be understood in terms of the second quantization formalism, where creation and annihilation operators are used. All sites in the crystal are numbered with integer n or m and sums are to be computed over all sites even though we do not specifically write down their total number. The first term is the electronic state contribution to the total energy of the system, which is given by the energy of an excited state ε (we assume that all excited states *ab initio* have the same energy since the LUMO is the same for all molecules in our molecular crystal) and the creation and destruction operators, a_n^+ and a_n, of an excited electron at site n in an orbital of this energy, respectively:

$$H_e = \sum_n \varepsilon a_n^+ a_n. \tag{2.81}$$

The second term is a similar contribution not from the electronic states but the vibrational states with energy $\hbar\omega_\mathbf{k}$, where $\mathbf{k}$ is the wave vector of the vibrational mode. With the creation and destruction operators b_n^+ and b_n, respectively, this phonon contribution can be written as

$$H_p = \sum_\mathbf{k} \hbar\omega_\mathbf{k} \left(b_\mathbf{k}^+ b_\mathbf{k} + \frac{1}{2}\right). \tag{2.82}$$

The sum $H_e + H_p$ denotes the total energy in the system if the electronic and vibrational states are taken into account but there is no coupling between the molecules and the lattice. Such interactions are added to the system with the other contributions to the Hamiltonian. First, there is the term H_J, which describes the interaction of molecules due to possible energy transfer between different sites n and m. This interaction is characterized by the interaction energy J_{nm} and is given by

$$H_J = \sum_n \sum_{\substack{m \\ m \neq n}} J_{nm}^{(0)} a_n^+ a_m. \tag{2.83}$$

Additionally, the electron can be influenced by lattice vibrations, i.e. there is an electron–phonon coupling. This coupling is a consequence of disorder in the system, as discussed in the introduction of this section: the electron is scattered at phonons due to disorder in the system, which can change its electronic state or the probability for an electronic transition. The former is represented as diagonal interaction with $n = m$, while the latter comes into play as off-diagonal element $n \neq m$ in the Hamiltonian.

The diagonal disorder term, which results in a change in site energy ε, can be written as

$$H_{e-p}^{\text{diag}} = \sum_{\mathbf{k}} \sum_n g_{n,\mathbf{k}}^2 \hbar \omega_{\mathbf{k}} a_n^+ a_n (b_{\mathbf{k}} + b_{-\mathbf{k}}^+), \tag{2.84}$$

where $g_{n,\mathbf{k}}$ is a dimensionless constant describing the strength of linear coupling to lattice vibrations. Since $H_{\text{dis}}^{\text{diag}}$ is linear in the creation and destruction operators a_n^+ and a_n, respectively, this term is the linear, dynamic, diagonal disorder.

The off-diagonal disorder term in Equation (2.80) represents the change in the probability amplitude for a transition between sites due to linear, dynamic coupling to lattice vibrations. It is given by

$$H_{e-p}^{n-\text{diag}} = \sum_n \sum_{\substack{m \\ m \neq n}} \sum_{\mathbf{k}} f_{nm,\mathbf{k}}^2 \hbar \omega_{\mathbf{k}} a_n^+ a_m (b_{\mathbf{k}} + b_{-\mathbf{k}}^+) \tag{2.85}$$

with dimensionless coupling constant $f_{nm,\mathbf{k}}$.

Besides the dynamic disorder contributions there are effects due to static local variations in site energy and in general the local environment of each site. This can be written as a site-dependent differential energy $\delta\varepsilon_n$ and a site-dependent differential electronic coupling constant δJ_{nm}. Analogous to the terms H_e and H_J these differential contributions due to disorder lead to the terms

$$H_{\text{dis}}^e = \sum_n \delta\varepsilon_n a_n^+ a_n \tag{2.86}$$

and

$$H_{\text{dis}}^J = \sum_n \sum_{\substack{m \\ m \neq n}} \delta J_{nm} a_n^+ a_m. \tag{2.87}$$

Furthermore, contributions due to dynamic electron–phonon interactions can be added in second order if necessary, i.e. quadratic in the lattice vibration creation and destruction operators $b_{\mathbf{k}}^+$ and $b_{\mathbf{k}}$, respectively. This is done analogously to the terms H_{e-p}^{diag} and $H_{e-p}^{n-\text{diag}}$ as

$$H_{e-p}^{\text{diag}'} = \sum_{\mathbf{k}} \sum_{n} \delta g_{n,\mathbf{k}}^2 \hbar \omega_{\mathbf{k}} a_n^+ a_n (b_{\mathbf{k}} + b_{-\mathbf{k}}^+)^2 \qquad (2.88)$$

with nonlinear electron–phonon coupling constant $\delta g_{n,\mathbf{k}}$ and as

$$H_{e-p}^{n-\text{diag}'} = \sum_{n} \sum_{\substack{m \\ m \neq n}} \sum_{\mathbf{k}} \delta f_{nm,\mathbf{k}}^2 \hbar \omega_{\mathbf{k}} a_n^+ a_m (b_{\mathbf{k}} + b_{-\mathbf{k}}^+)^2 \qquad (2.89)$$

with coupling constant $\delta f_{nm,\mathbf{k}}$.

For our further discussion, however, we will only consider contributions linear in the phonon operators and neglect the last two terms in Equation (2.80).

Considering the total Hamiltonian we now see that the nature of charge transport in our molecular solid is defined by the relative magnitudes of the coupling constants J_{nm}, $g_{n,\mathbf{k}}$, and $f_{nm,\mathbf{k}}$, and the static disorder energies and static disorder coupling $\delta \varepsilon_n$ and δJ_{nm}, respectively. In accordance with *Pope and Swenberg* [23] we discuss four different situations as follows:

Case I   The first case is present for strong nearest-neighbor overlap energies, i.e. $J_{n,n\pm1}$ is large compared to all other coupling constants and energies. For this case the total Hamiltonian can be approximated as $H_e + H_p + H_J$, which yields Bloch waves as eigenfunctions. Electrons are highly delocalized over several molecules, resulting in band transport with single scattering events due to weak disorder contributions. This situation is applicable to metals, where overlap energies are on the order of 1 eV and also on highly crystalline organic films at low temperature.

Case II   Dynamic disorder is large compared to nearest-neighbor overlap energies and static disorder is neglected, i.e. $g_{n,\mathbf{k}}$ and $f_{nm,\mathbf{k}}$ play a major role compared to $J_{n,n\pm1}$ and both $\delta \varepsilon_n$ and δJ_{nm} are considered to be 0. In this case the eigenstates are not independent of the lattice, i.e. the electronic states couple to the surrounding. Such a situation is present when charge carriers in the medium have to be understood as polarons, which commonly transport via uncorrelated hopping from site to site.

Case III   Static disorder is the dominant mechanism. This is the case for large $\delta \varepsilon_n$ and δJ_{nm}, which effectively extinguishes the effect of electronic coupling. In a situation where J_{nm} alone would induce a delocalization of the electron over neighboring sites, a large δJ_{nm} induces a localization due to the local environment of the site. In a similar fashion large values of $\delta \varepsilon_n$ induce deviations from the excited-state energy ε and site n, which do not allow the carrier to be delocalized owing to insufficient energy. Such a situation is present in many amorphous polymeric semiconductors where disorder in the molecular arrangement plays a major role. Intermolecular interaction is then strong, which leads to significant changes in the energies of vibrational relaxation when comparing the solid material with its vapor phase. Charge mobilities in this case are low (typically $<10^{-6} \text{cm}^2 \text{V}^{-1}\text{s}^{-1}$) and charge transport is a series of hopping events including trapping events in low-energy states at some sites that slow down charge transport.

Case IV Nearest-neighbor interactions $J_{n,n\pm 1}$ are of comparable magnitude to dynamic disorder coupling constants $g_{n,\mathbf{k}}$ and $f_{nm,\mathbf{k}}$. This is the case for many molecular crystals, like anthracene, which are dominated by the van der Waals force.

In many organic semiconductors employed in organic and hybrid solar cells a situation somewhere between Case II and Case IV is present. Charge transport is to be understood as a series of polaron hoppings, which, however, are not completely uncorrelated but show a contribution of electronic nearest-neighbor coupling.

Marcus Theory. An important approach in the context of charge transport in organic semiconductors, where charge localization plays a role rather than pure band transport and polarons moving via hopping processes, is the Marcus theory. It was developed in the late 1950s by the Canadian chemist Rudolph Marcus and describes the rates for electron transfer processes between a donor and an acceptor, e.g. during a chemical redox reaction. It can be applied as a semi-classical theory for charge transport and works under the assumption that $k_B T > \hbar \omega_\mathbf{k}$, where vibrational dynamics are not governed by quantum mechanics.

In Marcus theory an electron jump process from a charged molecule A^- to an uncharged molecule B can be pictured as shown in Figure 2.40. The initial state $(A^- + B)$ and the final state $(A + B^-)$ exhibit a parabolic energy landscape on a reaction coordinate. Between the two states there is a difference ΔG_0 in total Gibbs free energy, i.e. $(A + B^-)$ is energetically more favorable. Due to the electronic coupling of the two states J_{AB} there is a certain reduction in activation energy, which the electron has to overcome for the jump from A to B. However, due to a necessary reorganization of the system due to a change in electronic configuration, a reorganization energy λ has to be provided in order to allow for the electron transfer. This reorganization energy is large compared to the electronic coupling, rendering the nature of electron transfer a hopping rather than a band transport since donor A and acceptor B are clearly distinguished.

Since a full derivation of Marcus theory is not in the scope of this chapter we only give the final result of Marcus theory, which is an electron transfer rate k_{Marcus}. This rate depends on the reorganization energy λ, the difference in total Gibbs free energy ΔG_0, and the electronic coupling J_{AB} as follows:

$$k_{\text{Marcus}} = \frac{2\pi}{\hbar} |J_{AB}|^2 \frac{1}{\sqrt{4\pi\lambda k_B T}} \exp\left(-\frac{(\lambda + \Delta G_0)^2}{4\lambda k_B T}\right). \tag{2.90}$$

Due to the quadratic dependency of the exponent on $\lambda + \Delta G_0$ there is a so-called *normal* and an *inverted* Marcus region. In the normal Marcus region an increased driving force ΔG_0 leads to a higher transfer rate, i.e. a larger energy step down of the electron enables a quicker transfer from donor to acceptor. The rate has its maximum for $\Delta G_0 = \lambda$, where the gained free energy exactly matches the reorganization energy. For very large ΔG_0, however, the opposite is the case and the rate is reduced. This situation is referred to as inverted Marcus region and is present if the intersection of the two parabolas in Figure 2.40 is on the left-hand side of the minimum of the parabola of state (A^- + B). (The schematic in Figure 2.40 does not show an example for the inverted region.)

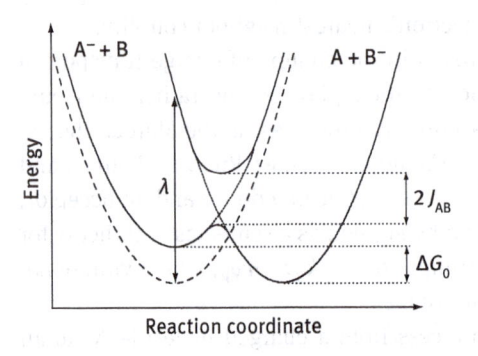

Fig. 2.40: Energy diagram for an electron jump process from molecule A to molecule B according to Marcus theory. ΔG_0 denotes the difference in total Gibbs free energy and J_{AB} the electronic coupling between A and B. λ is the reorganization energy of the system.

Although the inverted Marcus region was experimentally only observed a few decades after its prediction, it can play a role in real processes. As shown by researchers in the lab of Rumbles in 2012 the inverse Marcus region can be observed for charge separation between conjugated donor polymers and different derivatives of the C_{60} fullerene acceptor [56]. The polymer–fullerene pair is the most widely used materials combination in organic solar cells, outlining the importance of their findings. The relative electron transfer efficiency between polymer and fullerene as a function of the free Gibbs energy are shown in Figure 2.41 for combinations of three different donor polymers with 12 different acceptor fullerenes. In all cases a maximum in transfer efficiency is found for a certain ΔG_0 where too high or too low values of ΔG_0 reduce the efficiency.

Miller-Abrahams Hopping. For real extended systems a mathematical treatment of the complete Hamiltonian from Equation (2.80) is not practical due to the large number of involved molecules and the complexity of the formulation. In comparison, Marcus theory is simpler but still requires knowledge over (1) the individual coupling constants J_{nm} between all molecules and (2) the corresponding reorganization energies, which differ in the case of local, static disorder for each site. For many cases it is therefore useful to apply the Gaussian disorder model, which yields good predictions for the hopping transport of charges through a manifold of sites with energetic disorder [52]. The resulting motion of carriers is similar to the exciton diffusion process from Figure 2.39, with hopping generally occurring to states of lower energy and thermally induced hopping to higher energies.

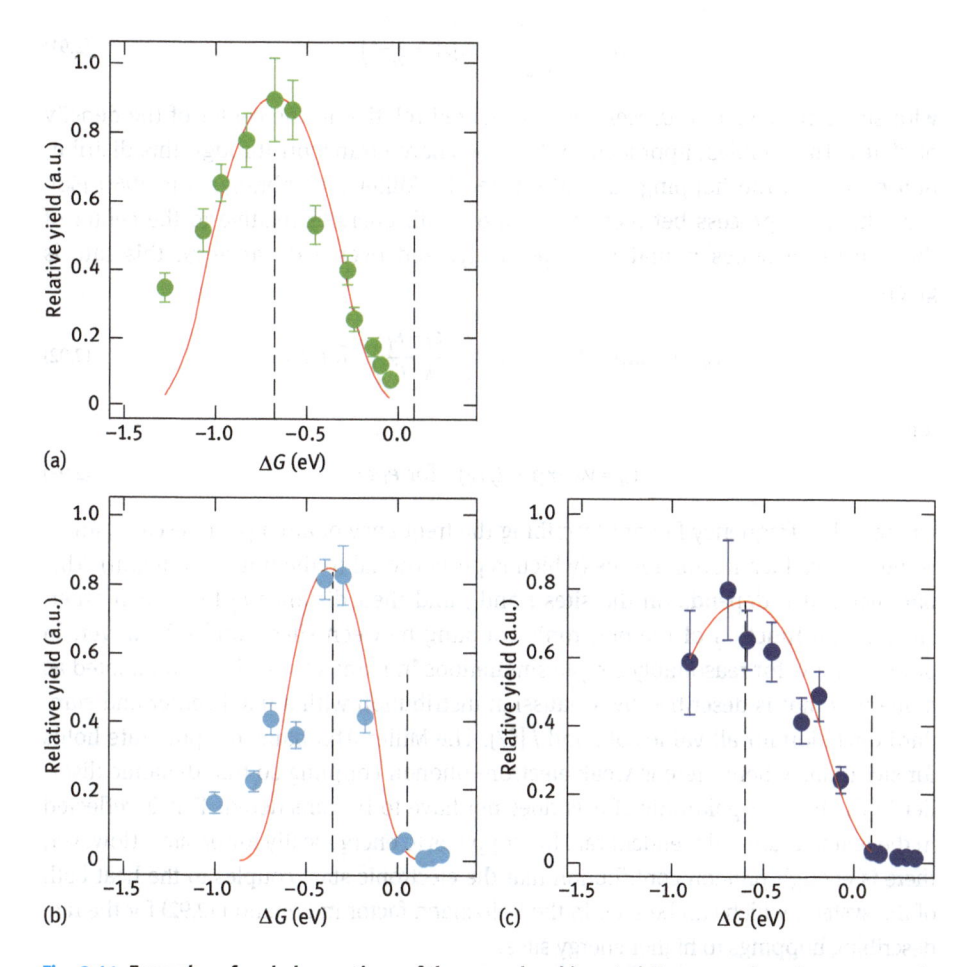

Fig. 2.41: Examples of real observations of the normal and inverted Marcus regions for systems of organic donor and acceptor. Three conjugated donor polymers are each tested in combination with 12 different fullerene acceptors. Reprinted with permission from [56]. Copyright (2012) American Chemical Society.

In the Gaussian disorder model the distribution of energy states in a system of disordered conjugated molecules is simply approximated with a Gaussian distribution [57]. This model is very general as it can be applied for molecular crystals and polymeric materials and works for pristine materials as well as multiple component solids. The real situation in such solids is complicated due to energetic as well as dynamic disorder induced by, e.g. differences in molecular orientation, topological defects in polymer chains, impurities, molecular crystal grain boundaries, and mixed phases in multimolecular films. All this disorder is summarized in one simple distribution of available energies ε

$$\rho(\varepsilon) = \frac{1}{\sqrt{2\pi\sigma^2}} \exp\left(-\frac{\varepsilon^2}{2\sigma^2}\right) \qquad (2.91)$$

with standard deviation σ. Here, ε is measured relative to the center of the density of states. The simplest approach to describe charge transport through this distribution of sites is the hopping rate introduced by Miller and Abrahams in 1960 [58]. For a hopping process between sites i and j with energies relative to the center of the density of states ε_i and ε_j, respectively, and over a distance r_{ij}, this rate is given by

$$v_{ij} = v_0 \exp(-2\gamma r_{ij}) \cdot \exp\left(-\frac{\varepsilon_j - \varepsilon_i}{k_B T}\right) \quad \text{for } \varepsilon_j > \varepsilon_i \qquad (2.92)$$

and

$$v_{ij} = v_0 \exp(-2\gamma r_{ij}) \quad \text{for } \varepsilon_j \leq \varepsilon_i, \qquad (2.93)$$

where v_0 is a frequency factor (describing the frequency of attempts-to-escape) and γ is the inverse localization radius (which is proportional to the transfer integral). This parameter also depends on the sites i and j and their distance r_{ij} from each other, since it is a function of the electronic coupling between sites i and j. However, in order to allow for reasonably simple simulations in many cases γ is approximated as a quantity that is described by a Gaussian distribution with a fixed center and standard deviation for all values of i and j [57]. The Miller-Abrahams hopping rate holds for situations where there is weak electron–phonon coupling so that dynamic disorder in the form of polaronic effects does not have to be considered. This is reflected in the temperature-independent rate for hoppings to energetically lower sites. However, there is enough phonon contribution that the electronic state couples to the heat bath of the system, which can be seen in the Boltzmann factor in Equation (2.92) for the rate describing hoppings to higher energy sites.

The energetic situation for Miller-Abrahams charge transport in an organic semiconductor is shown in Figure 2.42. After a charge generation event, e.g. upon absorption of a photon in a photoactive material, a charge carrier is placed in a high-energy state in the Gaussian density of states with standard deviation σ. Over a series of several hoppings to lower energies and thermally induced hoppings to higher energies, the carrier eventually arrives at an average energy, around which the states that it occupies during hopping are distributed. This energy is referred to as the *transport energy* and is temperature dependent. At temperatures approaching 0 K there is no heat bath that can induce thermal hopping to higher energies so that the carrier ends up at the lower end of the density of states. This is also reflected in the temperature-dependent occupational density of states, which is shifted by $\sigma^2/k_B T$ with respect to the center of the density of states. This occupational density of states can be understood as distribution of energies available for a transported charge carrier due to Miller-Abrahams hopping to lower and higher energy states.

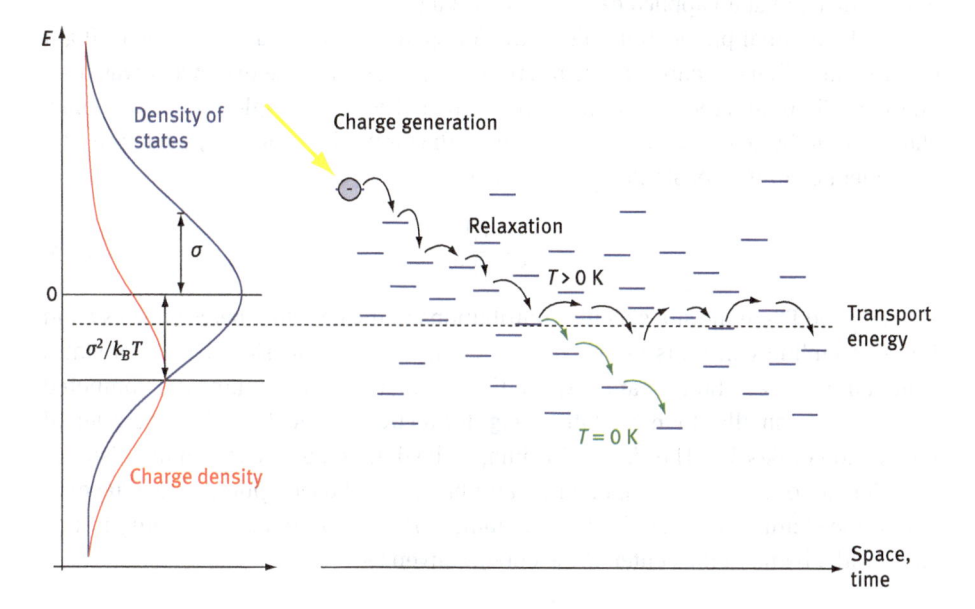

Fig. 2.42: Energetic picture of charge transport in a disordered molecular semiconductor film via Miller-Abrahams hopping. After charge separation, e.g. due to a photoexcitation, the charge hops in a Gaussian density of states with standard deviation to energetically lower sites while diffusing through the material. Due to coupling to the heat bath in the materials, thermally induced hops to energetically higher sites are also possible, leading to a charge distribution that is shifted by $\sigma^2/k_B T$ with respect to the center of the density of states. Figure after [59].

One implication of Equations (2.92) and (2.93) is that the charge mobility decreases over a series of hoppings since there are more thermally induced hoppings required once the charge carrier arrives at the transport energy. Due to the Boltzmann factor in Equation (2.93) the hopping rate between two sites i and j is always lower for $\varepsilon_j > \varepsilon_i$ than for downward hops, so that the first hopping processes occur more quickly, resulting in a higher charge mobility, while later hops are slower due to their temperature dependence, resulting in a decreased mobility. Depending on the thickness of the active film this mobility relaxation can be complete before the carrier reaches an external electrode. This has to be taken into account when performing mobility measurements like time-of-flight or charge extraction experiments [60].

Bässler Model. In 1993 Bässler developed a model based on the Miller-Abrahams hopping rates using Monte Carlo simulations [57]. Due to the Gaussian density of states there are no analytical solutions possible for the problem of a charge carrier moving through Miller-Abrahams hopping. Therefore, Monte Carlo simulations, which essentially are computer-based simulated idealized experiments, are an appropriate approach in order to yield useful results for charge transport in a disordered organic material at an applied external bias voltage.

In the original publication this methodology was used to simulate time-of-flight experiments. There, a cubic lattice of $70 \times 70 \times 70$ sites with site distance a was used and periodic boundary conditions were applied. The site energies ε_i are randomly chosen from the Gaussian density of states with standard deviation σ, while the site-dependent coupling constants y_{ij} are given as

$$y_{ij} = \frac{\Gamma_i \Gamma_j}{2a}, \qquad (2.94)$$

where Γ_i and Γ_j are the site-specific contributions to the coupling between sites i and j. Both coupling constants are chosen randomly from a Gaussian density of states with standard deviation σ_Γ and expectation 0. Charge carrier jumps are computed within a cube smaller than the lattice, e.g. for a cube of size $7 \times 7 \times 7$ in the original publication of Bässler. The size of this jump cube determines the maximum distance a carrier can overcome during one hop and has to be chosen appropriately depending on the simulated material. The probability for a jump from site i to site j in this cube (with i being in the center of the cube) is given as

$$P_{ij} = \frac{v_{ij}}{\sum_{k \neq i} v_{ik}}, \qquad (2.95)$$

i.e. the jump probability depends on the Miller-Abrahams hopping rate and with this on both the involved site energies and their coupling. In order to determine to which site the charge carrier is attempting to jump, a random number is chosen from a uniform distribution and each site is given a length in random number space relative to P_{ij}. Furthermore, the time t_{ij} for the jump to occur is determined using a random number t_{i0} from an exponential distribution as

$$t_{ij} = \frac{t_{i0}}{\sum_{k \neq i} v_{ik}}. \qquad (2.96)$$

With the reduced energetic disorder parameter $\hat{\sigma} = \sigma/k_B T$ as a computational variable the system is simulated over a total time t and the mean position $\langle x \rangle$ as well as the mean energy $\langle \varepsilon \rangle$. From this it is also possible to calculate the charge diffusivity

$$D = \frac{\langle (x - \langle x \rangle)^2 \rangle}{2t}, \qquad (2.97)$$

where x is the position in the lattice.

The temporal evolution of the mean energy $\langle \varepsilon \rangle$ and the energy distribution is shown in Figure 2.43. Analogous to the hopping picture for an individual charge carrier shown in Figure 2.42 the distribution of charge carrier energies also starts at a certain level and relaxes toward a lower distribution over time. Eventually, the mean energy reaches a constant value $\langle \varepsilon_\infty \rangle$ and the distribution is invariant and of Gaussian shape for long times. In this situation charges are transported via thermally induced hopping around the transport energy and the system is in dynamic equilibrium.

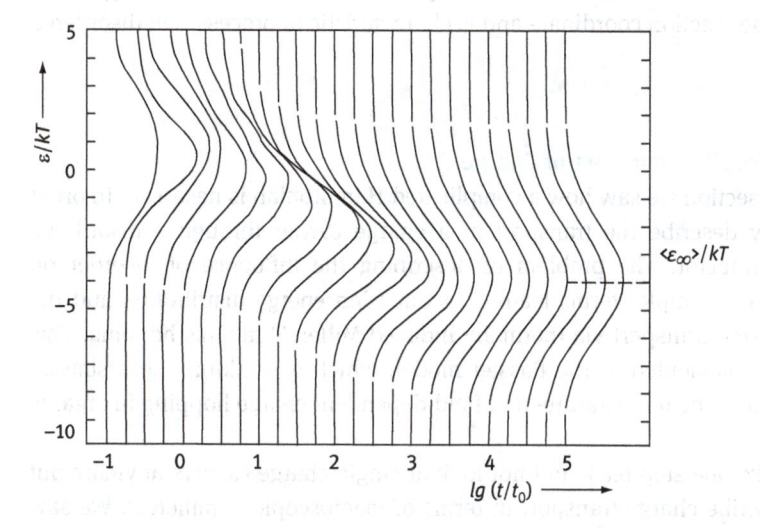

Fig. 2.43: Distribution of charge carrier energies as a function of time as predicted by the Bässler model. Data is shown for a reduced energetic disorder of $\hat{\sigma} = 2$. The figure illustrates how the energy distribution shifts with time and its center reaches the transport energy $\langle \varepsilon_\infty \rangle$ for long times. Figure from [57] with friendly permission by Wiley and Sons (1993).

As shown in the original publication, the Bässler model can be used to predict the dependence of the charge carrier mobility on both the temperature and the applied electric field. In the high field limit this dependence is found to be

$$\mu(\hat{\sigma},\sigma_\Gamma,E) = \mu_0 \exp\left(-\left(\frac{2}{3}\hat{\sigma}\right)^2\right) \exp\left(C(\hat{\sigma}^2 - 4\sigma_\Gamma^2)E^{\frac{1}{2}}\right) \quad \text{for } \sigma_\Gamma \geq 0.75 \quad (2.98)$$

and

$$\mu(\hat{\sigma},\sigma_\Gamma,E) = \mu_0 \exp\left(-\left(\frac{2}{3}\hat{\sigma}\right)^2\right) \exp\left(C(\hat{\sigma}^2 - 2.25)E^{\frac{1}{2}}\right) \quad \text{for } \sigma_\Gamma < 0.75, \quad (2.99)$$

where E is the electric field, μ_0 the mobility at zero field and 0 K, and C a numerical constant, which depends on the lattice parameter a. The temperature dependence is implicit since $\hat{\sigma} = \sigma/k_B T$ is temperature dependent.

The field dependence of the mobility predicted by the Bässler model is in accordance with a Pool-Frenkel-like dependence $\ln \mu \propto \sqrt{E}$, which is experimentally observed for many disordered organic semiconductors [61, 62]. Furthermore, the temperature dependence is found to be stronger than that for an Arrhenius-type reaction, namely $\mu(-\hat{\sigma}) \propto \exp(\hat{\sigma}^2)$ instead of $\mu(-\hat{\sigma}) \propto \exp(\hat{\sigma})$. This kind of dependence has been shown to occur for stochastic processes with Gaussian energy distributions along the reaction coordinate and is characteristic of processes in disordered systems [63].

2.2.3.2 Macroscopic Parameters of Charge Transport

In the previous section we saw how a complicated Hamiltonian is necessary in order to appropriately describe the transport of a charge carrier through a disordered organic semiconductor. The problem of describing the influence of disorder on charge transport in simpler terms leads to a Gaussian energy distribution and the modeling of charge transport via thermally induced Miller-Abrahams hopping. This description is implemented in the Bässler model, which uses Monte Carlo simulations to understand the temperature- and field-dependent charge hopping in organic semiconductors.

Now we take one step back and not look at single charge carriers anymore but at a way to describe charge transport in terms of macroscopic parameters. We saw these parameters already in the previous section and now bring them together. Ultimately, in any electronic device, e.g. in a solar cell, we are interested in the current density through the device as a function of the internal charge carrier density and the external electric field $\mathbf{E}$. The carrier density further depends on generation and recombination mechanisms, e.g. due to absorption of photons and subsequent separation of excitons, and annihilation of electrons and holes, respectively. This is described by two macroscopic rates, an electron-hole generation rate G and a recombination rate R. This current density has two contributions, namely a drift contribution due to the external electric field and a diffusion contribution due to local differences in the charge carrier density.

Since an electrical current in a semiconductor can be carried by both electrons and holes there are two sets of equations describing the drift and diffusion currents. Each set contains one equation describing the drift-diffusion current density and one describing the time dependence of the charge carrier density. For electrons and holes these equations read

$$\mathbf{j}_n = qD_n \nabla n + qn\mu_n \mathbf{E}, \tag{2.100}$$

$$\frac{\partial n}{\partial t} = \frac{1}{q}\nabla \mathbf{j}_n + G - R \qquad (2.101)$$

and

$$\mathbf{j}_p = -qD_p\nabla p - qp\mu_p \mathbf{E}, \qquad (2.102)$$

$$\frac{\partial p}{\partial t} = -\frac{1}{q}\nabla \mathbf{j}_p + G - R, \qquad (2.103)$$

where n and p denote the density, $\mathbf{j}_n$ and $\mathbf{j}_p$ the current density, and D_n and D_p the diffusion coefficient for electrons and holes, respectively. Equations (2.100) and (2.102) are called *drift-diffusion equations* since the first term describes diffusion transport and the second drift transport of charges.

Furthermore, the diffusion coefficients D_n and D_p are connected to the mobilities via the *electrical mobility equation*, a special case of the *Einstein relation* for charged particles, which reads

$$D_n = \frac{\mu_n k_B T}{q} \qquad (2.104)$$

for electrons and

$$D_p = \frac{\mu_p k_B T}{q} \qquad (2.105)$$

for holes.

The drift-diffusion picture is useful for many applications since all information about static and dynamic disorder is contained in the charge mobilities, the charge densities, and the generation and recombination rates. The macroscopic parameters used to describe the field-dependent motion of charges in a solid can be understood as average values and are observables, i.e. they can be directly or indirectly measured in experiments.

2.2.3.3 Doping of Organic Semiconductors

Analogous to inorganic semiconductors it is possible to dope organic semiconductors, which can be useful for many applications where the conductivity σ of the material has to be increased. Doping increases the density of either electrons or holes, which in turn increases the conductivity, depending on both the charge carrier mobility and density as

$$\sigma = q(n\mu_n + p\mu_p). \qquad (2.106)$$

For organic semiconductors there are also two types of doping, namely n-doping and p-doping. The mechanisms for both doping mechanisms are exemplarily shown in Figure 2.44. p-Doping is realized by introducing dopant molecules with a very low

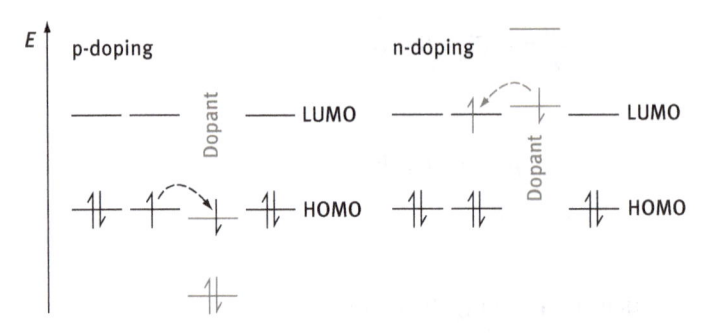

Fig. 2.44: Doping of organic semiconductors. p-Doping can be realized by introducing a few molecules with very low LUMO into the molecular crystal. An electron can then be transferred from the HOMO of the material into the LUMO of the dopant, which effectively leaves a hole behind. Analogous, n-doping is possible via incorporation of molecules with very high HOMO. An electron can then hop from the HOMO of the dopant into a LUMO in the doped material.

LUMO level that has to be energetically below the HOMO of the doped material. Electrons can then jump from the HOMO of the doped material into the LUMO of the dopant, leaving a hole behind. This hole can travel as a hole polaron through the material and the hole density is increased.

Similarly, n-doping is possible via the introduction of molecules with high HOMO levels. If the HOMO of the dopant is higher than the LUMO of the doped material an electron can jump into the LUMO and the electron density in the material is increased.

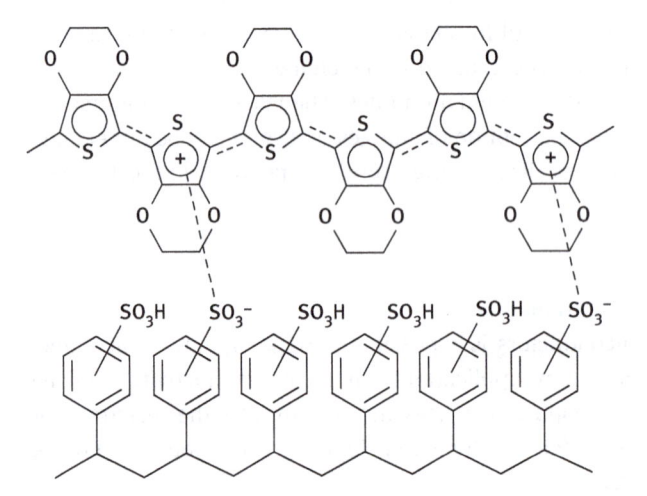

Fig. 2.45: Structural formula and doping mechanism of poly(3,4-ethylenedioxythiophene) polystyrene sulfonate (PEDOT:PSS). Some of the sulfonate groups of PSS are deprotonated and carry a negative charge, while single monomers in the PEDOT are positively charged. These positive charges can travel as hole polarons through the PEDOT and lead to the high conductivity of PEDOT:PSS.

Figure 2.45 shows the structural formula and the mechanism of doping for one of the most famous examples of a doped semiconducting polymer, poly(3,4-ethylenedioxythiophene) polystyrene sulfonate (PEDOT:PSS). PEDOT:PSS is a salt-like mixture of the two polymers PEDOT and PSS, which is p-doped due to positively charged EDOT monomers. These stem from deprotonation of sulfonate groups of the PSS. PEDOT:PSS is a widely used highly conducting polymer combination, which is employed as a contact material in solar cells, in light-emitting diodes, and as an antistatic agent.

2.3 Junctions

Having discussed the fundamentals of inorganic and organic semiconductors we now take a look at junctions between different types of materials. Every photovoltaic device consists of at least three different materials, namely (1) a transparent front electrode, typically a transparent conducting oxide or a highly doped polymer film, (2) a photoactive semiconducting material, which is the absorber of the solar cell, and (3) a metallic back contact. This very simple metal–semiconductor–metal configuration already contains two junctions, and in most solar cells the photoactive layer consists of at least two different parts so that there is another junction in the active part of the device. Furthermore, especially in organic and hybrid solar cells different interfacial layers are employed in order to render one or both external contacts selective for one carrier type, either electrons or holes, respectively, outlining the important role junctions play for the proper function of a solar cell. We will therefore discuss the properties of different junction types in detail in the following sections.

2.3.1 Inorganic p–n Homojunctions

First we take a look at the *p–n homojunction*, often simply referred to as *p–n junction*, which occurs within an inorganic material if one side is p-doped and the other side n-doped. This type of junction can be found in many inorganic photovoltaic devices, e.g. in conventional Si solar cells. Furthermore, p–n junctions are responsible for the function of most semiconductor devices and can be found in diodes, light-emitting diodes, transistors, and integrated circuits. p–n junctions are fabricated via various synthesis routes, for instance by implantation of two different types of dopant ions or via epitaxial growth of two different doped layers on top of each other.

Mainly two properties of the p–n junctions are responsible for their importance in semiconductor electronics, namely (1) their rectifying behavior, i.e. their property to allow many orders of magnitude higher currents in one bias direction (forward bias) than in the other (reverse bias) and (2) their internal electric field, which enables the spatial separation of photogenerated electrons and holes and enables the fabrication of efficient photovoltaic devices.

This internal field is a consequence of the difference in Fermi levels in the p-doped and n-doped part of the junction, which leads to the formation of a *space charge region* with an equilibrium built-in potential V_{bi}. This space charge region is depleted of free charge carriers and therefore also referred to as a *depletion region*.

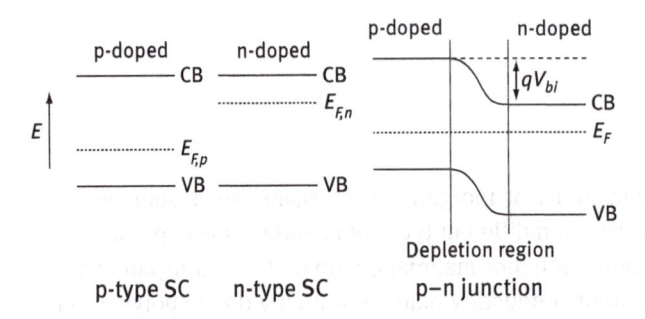

Fig. 2.46: Band bending in a p–n junction. On the left the band structures and Fermi levels are shown for isolated p-type and n-type semiconductors. After contacting them, there is a band bending as the Fermi level is the same throughout the whole crystal.

The band structure of a p–n junction is schematically shown in Figure 2.46. First, we consider two isolated chunks of material, where one is n-doped and one is p-doped. The Fermi levels are $E_{F,n}$ and $E_{F,p}$ in the n-type and p-type material, and the work functions are Φ_n and Φ_p, respectively. When brought into contact the Fermi energies equilibrate and there is a band bending as illustrated on the right-hand side of Figure 2.46. This is possible since free electrons and holes can diffuse across the junction and recombine. Far away from the junction electron and hole densities are at their equilibrium values, as in the respective materials when not in contact. When getting closer to the junctions from either side, however, both densities drop due to the annihilation of electrons and holes that diffuse into the space charge region. The space charge itself is carried by immobilized ions in the crystal lattice so that there are no free charge carriers in the space charge region, i.e. it is depleted of free charge carriers. With the bands there is also a bending in the vacuum level since the work functions far away from the junction are the same as in the noncontacted materials.

The situation in a p–n junction is schematically shown in Figure 2.47 for a junction where the n-type region has a higher dopant concentration than the p-type region. Far away from the junction both electron and hole densities are the same as in the noncontacted materials and are directly given by the density of electron donors N_d in the n-type part and the density of electron acceptors N_a in the p-type part, i.e. $n|_{n\text{-type side}} = N_d$ and $p|_{p\text{-type side}} = N_a$.

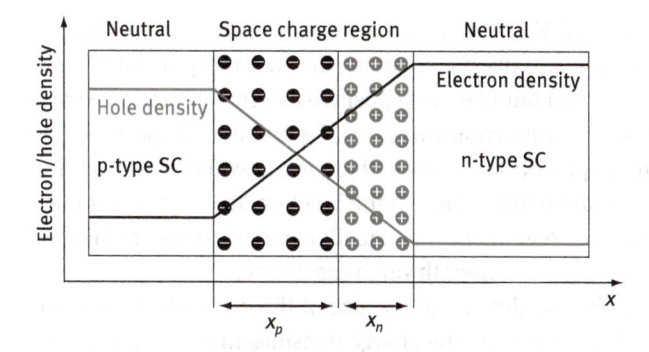

Fig. 2.47: Schematic drawing of the situation inside a p–n junction. When the p-doped and n-doped parts are brought in contact electrons diffuse into the p-type region, leaving positively charged fixed ions behind. Analogously, holes diffuse from the p-type to the n-type region, resulting in negatively charged ions located at the edge of the p-type part of the junction. This diffusion results in the formation of a space charge region, which is depleted of free charge carriers and responsible for the internal electric field.

We further notice that the space charge region is not symmetrical around the junction but extends further into the material of lower charge density (note that a higher doping level leads to a higher charge density since np is constant but $n+p$ is not). We will see later how it is possible to find an analytical expression for the distances x_n and x_p that the depletion region extends into the n-type and p-type regions, respectively.

The space charge at the junction is responsible for the built-in voltage, which is given as the difference of the work functions of the two materials. This difference is equal to the Fermi level shift in the two doped materials with respect to the Fermi level E_i in the intrinsic material at two positions in the n-type and p-type material far away from the junction:

$$V_{bi} = \frac{1}{q}(\Phi_n - \Phi_p) = \frac{1}{q}\left((E_i - E_F)|_{p\text{-type side}} - (E_i - E_F)|_{n\text{-type side}}\right)$$
$$= \frac{1}{q}\left((E_i - E_{F,p}) - (E_i - E_{F,n})\right). \qquad (2.107)$$

With the electron and hole densities $n = N_d$ and $p = N_a$, respectively, and the intrinsic charge carrier density n_i with $n_i^2 = n_0 p_0$ (see Equations (2.27), (2.28) and (2.31)) the built-in voltage can also be expressed in terms of the doping levels (for details see [64]):

$$V_{bi} = \frac{k_B T}{q} \ln\left(\frac{N_d N_a}{n_i^2}\right). \qquad (2.108)$$

Since the built-in voltage has to be cancelled in order to make the p–n junction field-free, the voltage drop across the junction V_j is given as

$$V_j = V_{bi} - V_{ext} \qquad (2.109)$$

for an externally applied potential V_{ext}. Due to the space charge region a p–n junction shows a rectifying behavior, i.e. the conductivity through the p–n junction is low for a negative external potential but high for a positive external potential. Applying a negative potential to the p–n junction increases the amount of space charge and with it the width of the depletion layer since charges are withdrawn from the junction. This results in a low conductivity through the junction. In contrast, for positive external bias the space charge region becomes smaller since charges are injected into the junction, resulting in a higher current through the device.

The Depletion Region. We now determine the size of the depletion region and see how x_n and x_p in Figure 2.47 depend on the charge densities in the n-type and p-type parts of the junction [64]. We first choose an electrostatic potential Φ according to $E_i = -q\Phi + C$, where C is a constant that we set to 0. We then look at a one-dimensional p–n junction as in Figure 2.47 and set $x = 0$ where the n-type region touches the p-type region. We can then write down the Poisson equation in the two materials, which gives

$$\frac{d^2\Phi}{dx^2} = -\frac{q}{\varepsilon} N_d \quad \text{for } x < 0, \tag{2.110}$$

i.e. in the p-type region, and

$$\frac{d^2\Phi}{dx^2} = \frac{q}{\varepsilon} N_a \quad \text{for } x > 0, \tag{2.111}$$

i.e. in the n-type region, with $\varepsilon = \varepsilon_r \varepsilon_0$ being the permittivity of the semiconductor. Since the electrostatic potential can vary only within the space charge region it changes exactly by V_{bi} from the n-type side to the p-type side of the junction, which leads to the boundary conditions

$$\Phi = 0 \quad \text{at } x = -x_p \tag{2.112}$$

and

$$\Phi = V_{\text{bi}} \quad \text{at } x = x_n. \tag{2.113}$$

Via integration of the Poisson equation we find the electric field

$$F = -\frac{d\Phi}{dx} = -\frac{qN_a}{\varepsilon}(x + x_p) \quad \text{for } -x_p < x < 0 \tag{2.114}$$

and

$$F = -\frac{d\Phi}{dx} = \frac{qN_d}{\varepsilon}(x - x_n) \quad \text{for } 0 < x < x_n, \tag{2.115}$$

where we used the fact that the field vanishes at the edges of the space charge region as boundary conditions for the integral. A second integration using the boundary conditions (Equations (2.112) and (2.113)) yields

$$\Phi = \frac{qN_a}{2\varepsilon}(x+x_p)^2 \quad \text{for } -x_p < x < 0 \tag{2.116}$$

and

$$\Phi = -\frac{qN_d}{2\varepsilon}(x-x_n)^2 + V_{bi} \quad \text{for } 0 < x < x_n. \tag{2.117}$$

With these expressions for the electric field and the electrostatic potential, x_p and x_n can be derived using the condition that both F and Φ are continuous at $x = 0$:

$$x_p = \frac{1}{N_a}\left(\frac{2\varepsilon V_{bi}}{q(N_a^{-1}+N_d^{-1})}\right)^{\frac{1}{2}}, \tag{2.118}$$

$$x_n = \frac{1}{N_d}\left(\frac{2\varepsilon V_{bi}}{q(N_a^{-1}+N_d^{-1})}\right)^{\frac{1}{2}}. \tag{2.119}$$

Accordingly, the total width of the space charge region d_{SCR} is

$$d_{SCR} = x_n + x_p = \left(\frac{2\varepsilon}{q}\left(\frac{1}{N_a}+\frac{1}{N_d}\right)V_{bi}\right)^{\frac{1}{2}}. \tag{2.120}$$

The dependency of x_p and x_n on the charge densities $p = N_a$ and $n = N_d$ in the p-doped and n-doped region, respectively, has important implications for devices based on p–n homojunctions. Depending on the application, the distance over which the built-in potential drops can be minimized when using identical doping concentrations and maximized if one material is heavily and the other only lightly doped. The latter is employed in conventional Si solar cells, where the space charge region is almost solely located in a weakly doped bottom layer with a highly doped layer on top in order to maximize the size of the space charge region, where charge carriers are generated.

The Current through the p–n Junction. Since the p–n junction is the functional unit of electronic and optoelectronic devices like transistors, diodes, light-emitting diodes, and solar cells, not only is the electrostatic situation in the junction of interest, the dynamics of charge carriers is as well. In particular, we are interested in the current through a p–n homojunction device as a function of the applied external potential. However, here we do not give a full description of the currents in a p–n junction since the comparably long derivation is outside the scope of this chapter. The interested reader is referred to the very comprehensive discussion of a p–n junction in the context of solar cells in the book of Jenny Nelson [64]. Here we only give a brief overview over the most important parts of this discussion.

Consider a p–n junction with an externally applied bias V. In order to find the total current through the device we have to consider the electron current far away from the junction in the n-type region, the hole current far away from the junction in the p-type region, and the current in the depletion region, which is only a recombination current if the device is in the dark and a generation-recombination current under illumination. The latter will not be considered in our discussion since we first are only interested in the dark current.

Due to the bias the system is not in equilibrium and the Fermi level is split into the so-called *quasi-Fermi levels* $E_{F,n}$ and $E_{F,p}$ for electrons and holes, respectively. The splitting between these two levels is exactly qV and can therefore be directly measured as the external potential of the device. Close to the contacts the quasi-Fermi levels fall together, so that the Fermi level is $E_F = E_{F,n}$ at the n-type end of the device and $E_F = E_{F,p}$ at the p-type side. Using the so-called *depletion approximation* both $E_{F,n}$ and $E_{F,p}$ are assumed to be constant throughout the depletion region and separated by qV so that we can write

$$E_{F,n}(x) = E_{F,n}(x_n) \tag{2.121}$$

$$E_{F,p}(x) = E_{F,p}(-x_p) \tag{2.122}$$

$$E_{F,n}(x) - E_{F,p}(x) = qV \tag{2.123}$$

for $-x_p < x < x_n$. With the electrostatic potential Φ from Equations (2.116) and (2.117) and the intrinsic Fermi level $E_i = q\Phi$, we can also write down the electron and hole densities as a function of the quasi-Fermi levels and the intrinsic charge carrier density n_i (after Equations (2.27) and (2.28)) as

$$n = n_i \exp\left(\frac{E_{F,n} - E_i}{k_B T}\right) \tag{2.124}$$

and

$$p = n_i \exp\left(\frac{E_i - E_{F,p}}{k_B T}\right), \tag{2.125}$$

which yields for the product np inside the space charge region

$$np = n_i^2 \exp\left(\frac{E_{F,n} - E_{F,p}}{k_B T}\right) = n_i^2 \exp\left(\frac{qV}{k_B T}\right). \tag{2.126}$$

In order to get the current in the space charge region we then have to find an expression for the current as a function of the charge densities. Here, we have to take recombination into account since electrons and holes injected into the space charge region can annihilate via different mechanisms like Auger recombination, trap-assisted recombination, or bimolecular recombination. Summarizing these mechanisms separately for electrons and holes in two recombination rates U_n and U_p, respectively, we can write down the relation between current and charge density as

$$\frac{\partial n}{\partial t} = \frac{1}{q}\frac{\partial J_n}{\partial x} - U_n \tag{2.127}$$

and

$$\frac{\partial p}{\partial t} = \frac{1}{q}\frac{\partial J_p}{\partial x} - U_p \tag{2.128}$$

with electron and hole currents J_n and J_p, respectively. It can be shown that integrating these equations over the space charge region $-x_n \leq x \leq x_p$ leads to an approximated overall current in the space charge region J_{SCR} of

$$J_{SCR}(V) = J_{SCR,0}\left(\exp\left(\frac{qV}{2k_B T}\right) - 1\right), \tag{2.129}$$

where $J_{SCR,0}$ is a prefactor given by

$$J_{SCR,0} = \frac{qn_i(x_p + x_n)}{\sqrt{\tau_n \tau_p}} \tag{2.130}$$

with electron and hole lifetimes τ_n and τ_p. This is true for the case where the recombination mechanism is trap-state-mediated Shockley–Read–Hall recombination, which is the dominant recombination mechanism in most p–n homojunctions and will be discussed in particular in the context of organic solar cells in Chapter 3.

In the case of bimolecular, radiative recombination the factor of two in denominator in the exponential vanishes and the recombination current is given as

$$J_{rad}(V) = J_{rad,0}\left(\exp\left(\frac{qV}{k_B T}\right) - 1\right) \tag{2.131}$$

with a material-dependent constant $J_{rad,0}$.

We now have to add the electron and hole currents from the neutral region. There, the interesting part plays the minority carriers, i.e. electrons in the p-type region and holes in the n-type region. In order to illustrate the reason for this we picture a p–n homojunction device under forward bias, where current is easily conducted. In forward bias, the n-type side of the device is connected to the negative terminal and the p-type side to the positive terminal, resulting in injection of electrons as majority carriers into the n-type region and holes as majority carriers into the p-type region. However, in order to allow net current through the device these injected carriers have to be transported across the space charge region and to the opposing terminal, i.e. electrons have to migrate through the p-type region and holes through the n-type region. We are therefore interested in the electron current directly behind the space charge region on the p-type side $J_n(-x_p)$ and the hole current on the opposing side $J_p(x_n)$.

Since the currents can be obtained from the drift-diffusion equations given in Equations (2.100) and (2.102), we have to know the electron and hole densities. With the equilibrium electron and hole concentrations in the p-type and n-type regions $n_{p,0} = n_i^2/N_a$ and $p_{n,0} = n_i^2/N_d$, respectively, the charge densities at the edges of the space charge region are

$$n(x=-x_p) = \frac{n_i^2}{N_a}\left(\exp\left(\frac{qV}{k_BT}\right)-1\right) + n_{p,0} \qquad (2.132)$$

and

$$p(x=x_n) = \frac{n_i^2}{N_d}\left(\exp\left(\frac{qV}{k_BT}\right)-1\right) + p_{n,0}. \qquad (2.133)$$

Since the whole potential drops in the space charge region the drift contribution in the drift-diffusion equation is 0 and charge transport through the neutral regions outside the space charge region is only diffusive, i.e.

$$J_n(x<-x_p) = qD_n\frac{dn}{dt} \qquad (2.134)$$

and

$$J_p(x>x_n) = -qD_p\frac{dp}{dt} \qquad (2.135)$$

with electron and hole diffusion coefficients D_n and D_p, respectively. The solution to these equations is a total diffusion current

$$J_{\text{diff}}(V) = J_n(-x_p) + J_p(x_n) = J_{\text{diff},0}\left(\exp\left(\frac{qV}{k_BT}\right)-1\right) \qquad (2.136)$$

with the prefactor

$$J_{\text{diff},0} = qn_i^2\left(\frac{D_n}{N_aL_n} + \frac{D_p}{N_dL_p}\right), \qquad (2.137)$$

which depends on electron and hole diffusion lengths L_n and L_p, respectively.

The overall current in the p–n homojunction device is then the sum of this diffusion current and the two recombination currents from Equations (2.129) and (2.131):

$$J(V) = J_{\text{diff}}(V) + J_{\text{SCR}}(V) + J_{\text{rad}}(V). \qquad (2.138)$$

Since the exponent is qV/k_BT for both the diffusion current and the radiative recombination current and $qV/2k_BT$ for the trap-assisted recombination current, the overall current is often approximated as

$$J(V) = J_0\left(\exp\left(\frac{qV}{nk_BT}\right)-1\right), \qquad (2.139)$$

where J_0 is a constant (the so-called *reverse saturation current*) and n the so-called *ideality factor* with $1 \leq n \leq 2$. Apparently the ideality factor depends on the recombination mechanism and is 1 in a trap-free diode and 2 in a diode, which is dominated by trap-state recombination. The case $n = 1$ is therefore referred to as *ideal diode*.

Equation (2.139) is the so-called *Shockley equation*, which is named after the American physicist William Shockley, one of the three winners of the Nobel Prize in Physics in the year 1956 for the invention of the transistor. It is often used to describe the current–voltage curves of well-performing diodes and in the simplest case can be extended to describe a solar cell as

$$J(V) = J_0 \left(\exp\left(\frac{qV}{nk_B T}\right) - 1 \right) - J_{\text{ph}}, \qquad (2.140)$$

where the charge generation due to photon absorption results in a voltage-independent photocurrent J_{ph}. In Chapter 3 we will see how the Shockley equation can further be used as the basis to describe the current–voltage characteristics of a solar cell with nonideal parasitic resistances, namely a series and a parallel resistance, both of which are present to a certain extend in all real devices.

2.3.2 Inorganic Heterojunctions

In the previous section we discussed the formation of a p–n junction upon doping the two sides of a chunk of one material differently, a p–n homojunction. In many applications, however, two separate materials are used, which has similar consequences for the formation of a junction between the materials due to formation of a depletion zone but is qualitatively different due to the presence of energy steps. These energy steps are a result of the possibly different bandgaps of different materials. Heterojunctions form between n-doped and p-doped materials as well as between materials of the same doping.

Heterojunctions are commonly categorized depending on the type of energy-level alignment as schematically shown in Figure 2.48. A straddling alignment, where there is a driving force for both electrons and holes to go from one semiconductor to the other is referred to as *type I junction*. For most electronic applications like solar cells, diodes, and transistors a *type II junction* is used, where the level alignment is staggered. In the case of a large difference in energy levels it is possible that a broken gap may be formed, where the valence band of one semiconductor is higher than the conduction band of the other semiconductor. This situation is present in a so-called *type III junction*.

Analogous to the previously discussed p–n homojunction, the properties of a heterojunction depend on the Fermi levels in the two materials, their bandgaps and vacuum energy levels and their charge carrier densities. However, due to the possibly different bandgaps the band bending is not necessarily continuous as in the case of homojunctions and the occurrence of energy barriers at the interface of the two semiconductors is possible. The band bending in a heterojunction can be approximated using *Anderson's rule* (also called the *electron affinity rule*), which gives reasonably good predictions of the band bending even though it uses various simplifications.

For instance, it does not take into account chemical bonding between the two semiconductors and the occurrence of interface and defect states in the junction.

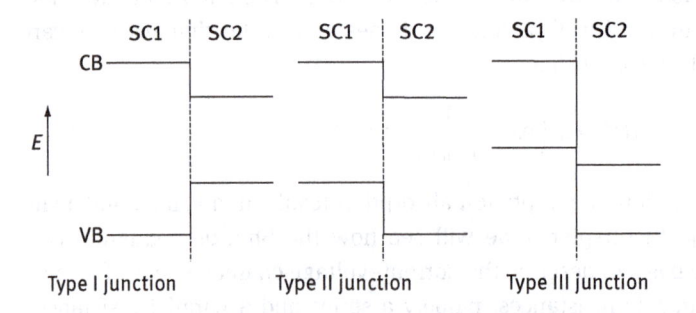

Fig. 2.48: Types of heterojunctions. A straddling energy band alignment is referred to as a type I junction, a staggered alignment as a type II junction. A type III junction is a broken junction, where the valence band of semiconductor 1 is energetically above the conduction band of semiconductor 2.

The energetic situation inside a type II heterojunction between an n-type and a p-type semiconductor is exemplarily shown in Figure 2.49. Quantities like the conduction band energies $E_{C,i}$, the valence band energies $E_{V,i}$, and the electron affinities χ_i are denoted with a subscript $i = n$ and $i = p$ for the n-type and the p-type material, respectively. After contacting, there is a shift in vacuum as for the p–n homojunction, i.e. far away from the junction the electron affinities are unchanged and the vacuum level is bent across the junction.

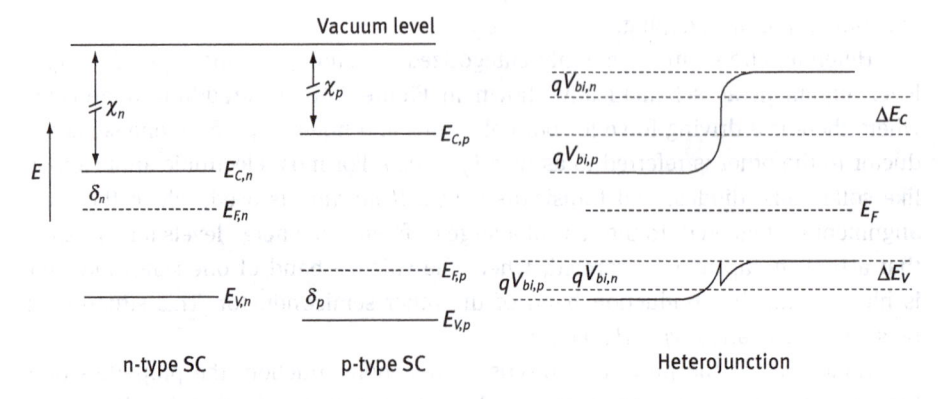

Fig. 2.49: Schematic representation of a heterojunction. The left-hand side shows two different semiconductors with n-type and p-type characters when not contacted. Both have specific Fermi levels, electron affinities, and bandgaps. After contacting, a heterojunction forms and the Fermi level equilibrates. The band bending, the step between the conduction bands and valence bands, and the formation of a depletion region depend on the Fermi levels, charge densities, and vacuum energy levels of the materials.

Across the junction there is a built-in potential V_{bi} analogous to the homojunction, which depends on the original Fermi levels and can be split into a contribution in the n-type and in the p-type semiconductors:

$$V_{bi} = \frac{1}{q}(E_{F,p} - E_{F,n}) = V_{bi,n} + V_{bi,p} \tag{2.141}$$

with

$$\frac{V_{bi,n}}{V_{bi,p}} = \frac{N_a \varepsilon_p}{N_d \varepsilon_n}, \tag{2.142}$$

where ε_n and ε_p are the relative permittivities and N_d and N_a the dopant densities of the n-type and p-type semiconductors, respectively. The latter is found from the Poisson equation.

With $\delta_n = E_{C,n} - E_{F,n}$ and $\delta_p = E_{F,p} - E_{V,p}$ the conduction band energy step is given as

$$\Delta E_C = \delta_n + qV_{bi,n} - (E_{C,p} - E_{V,p} - \delta_p) + qV_{bi,p} = (\chi_p - \chi_n), \tag{2.143}$$

and the valence band energy step as

$$\Delta E_V = (E_{C,n} - E_{V,n}) - (E_{C,p} - E_{V,p}) - (\chi_p - \chi_n). \tag{2.144}$$

As expected this implies that the total energy step $\Delta E_C + \Delta E_V$ equals the difference between the bandgaps of the two materials.

Furthermore, analogous to homojunctions, the distances x_n and x_p that the depletion region extends into the n-type and p-type semiconductors, respectively, are inversely dependent on the dopant densities:

$$\frac{x_n}{x_p} = \frac{N_a}{N_d}. \tag{2.145}$$

As mentioned before, Anderson's rule is only an approximation as it does not consider the different lattice structures of the two contacted semiconductors. In a real device, mismatch of lattice parameters can lead to stress or strain in the junction, causing a change in the density of states. This can happen as new states are introduced, which only exist in the distorted lattice regions close to the interface and not in the undistorted lattice further away from it. Tersoff also outlined the important role of gap states at the interface due to band discontinuities. These states result in an electrical dipole moment that drives the band line-up alignment toward a situation where there is zero dipole. According to Tersoff's study this is the main effect determining the line-up alignment of bands in a heterojunction [65]. For further reading on heterojunctions see also the very comprehensive discussion in Milnes and Feucht [66].

2.3.3 Schottky Junctions

Having looked into the mechanisms of junction formation between semiconducting materials we now come to another junction type, the so-called *Schottky contact* between a semiconductor and a metal. This junction type is important for semiconductor devices and in particular solar cells, since these devices commonly feature metallic external contacts. For instance, in a solar cell one contact is a highly conducting, metal-like transparent conducting layer and the other contact is commonly a metal like aluminum, silver, or gold.

Depending on the energetic offset and the charge density in the semiconductor, Schottky junctions show a behavior very similar to p–n junctions but typically with a smaller built-in potential and a faster switching time. Therefore, Schottky junctions are employed in so-called *Schottky diodes*, which are used for applications where one requires very fast electronic components. Due to their fast response and low potential drop in forward direction Schottky diodes are also used for electronic clamping, which allows the defined addition of a DC voltage signal to an arbitrary AC signal, a procedure that is used for instance to define a common black level when mixing two or more analog video signals.

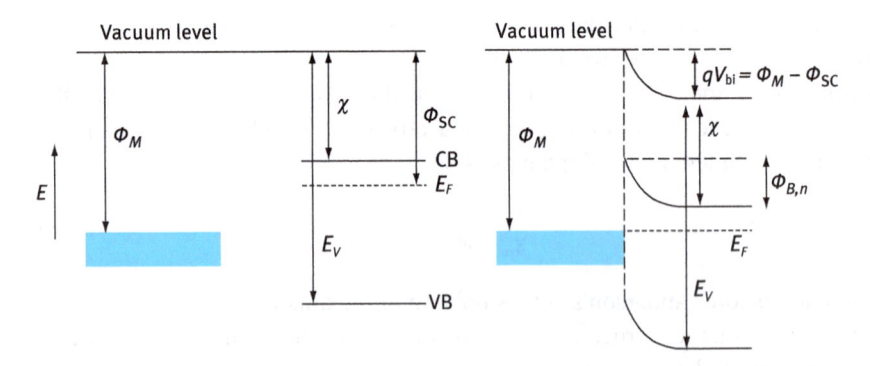

Fig. 2.50: Schematic of the energetic situation in a Schottky junction. A metal is brought in contact with an n-type semiconductor, resulting in a shift in vacuum levels and band bending. Due to the high charge density in the metal the depletion region is formed only in the semiconductor.

Consider a metal and an n-type semiconductor as shown in Figure 2.50. Before contacting, the metal and the semiconductor have work functions Φ_M and Φ_{SC}, respectively. After contacting, these levels are still valid far away from the junction, i.e. the vacuum level is bent. Due to the high charge density in a metal and the absence of a bandgap the Fermi level of the semiconductor is pinned to the highest occupied electron energy in the metal. This results in an overall energy-level shift, which defines the built-in voltage of the Schottky junction as

$$qV_{bi} = \Phi_M - \Phi_{SC}. \tag{2.146}$$

Furthermore, the height of the Schottky barrier, $\Phi_{B,n}$, that the electrons have to overcome when going from the metal into the conduction band of the semiconductor is given as

$$\Phi_{B,n} = \Phi_M - \chi, \tag{2.147}$$

where χ is the electron affinity of the semiconductor, i.e. the energy gap from the conduction band minimum to the vacuum level. This is true for a Schottky contact with an n-type semiconductor, whereas for a Schottky contact with a p-type semiconductor the energy barrier is given as

$$\Phi_{B,p} = (E_C - E_V) + \chi - \Phi_M, \tag{2.148}$$

where E_C and E_V denote the conduction and valence band energy of the semiconductor, respectively.

Since there is a depletion layer forming in the semiconductor close to the junction a Schottky contact shows the characteristic behavior of a diode, with an exponentially rising current in forward direction and low current in reverse direction. For a Schottky contact with an n-type semiconductor the junction is in forward bias if the negative terminal is connected to the semiconductor and the positive terminal to the metal, and vice versa for a Schottky contact with a p-type semiconductor.

In the example we discussed that there is a barrier for electrons between the metal and the semiconductor, which is a consequence of the difference in work functions. In the case this difference is smaller than $k_B T$ electrons can overcome the barrier in both directions and the junction does not show a current rectifying behavior. In this case the semiconductor–metal contact is referred to as *ohmic contact* since it shows the characteristic of a simple resistor. For electrons moving from an n-type semiconductor to a metal an ohmic-like behavior is also present if the band bending points in the opposite direction, which for instance is the case if $|\Phi_{SC}| > |\Phi_M|$. Accordingly, the metals for external contacts in semiconductor devices are typically matched to the work function of the semiconductor.

2.3.4 Organic–Organic Heterojunctions

So far we only looked at junctions between inorganic semiconductors or inorganic semiconductors and metals, where the first step of junction formation is the equilibration of the Fermi levels throughout the junction. However, we are in particular interested in organic semiconductors and therefore will discuss junctions involving organic materials in this and the following section.

Organic–organic junctions are qualitatively different from inorganic junctions due to the very low intrinsic charge densities in organic semiconductors. The Fermi level is therefore located close to the center of the HOMO–LUMO gap in many organic materials. Furthermore, doping of organic semiconductors is not nearly as common as in inorganic semiconductors resulting in low charge carrier densities in devices at room temperature and without applied bias.

The most common case of an organic–organic junction is shown in Figure 2.51. Two materials, often referred to as *donor* (higher energy levels) and *acceptor* (lower energy levels), are brought in contact so that electronic interaction becomes possible. However, as mentioned before, charge densities in the materials are commonly so low that there is no charge diffusion across the interface and accordingly no formation of a depletion region. Therefore, the vacuum level is not bent and the energy landscape of the heterojunction can be simply constructed by using the HOMO and LUMO levels of the materials before contacting. As pointed out by Tang and co-workers this is commonly the case if the Fermi levels in donor and acceptor are located close to the center of the HOMO–LUMO gap [67].

In contrast, if the Fermi levels are close to the HOMO or the LUMO there is a Fermi level alignment effect resulting in a change in vacuum level across the junction. This is accompanied by a bending of molecular orbitals and results in a situation similar to a p–n junction between inorganic semiconductors.

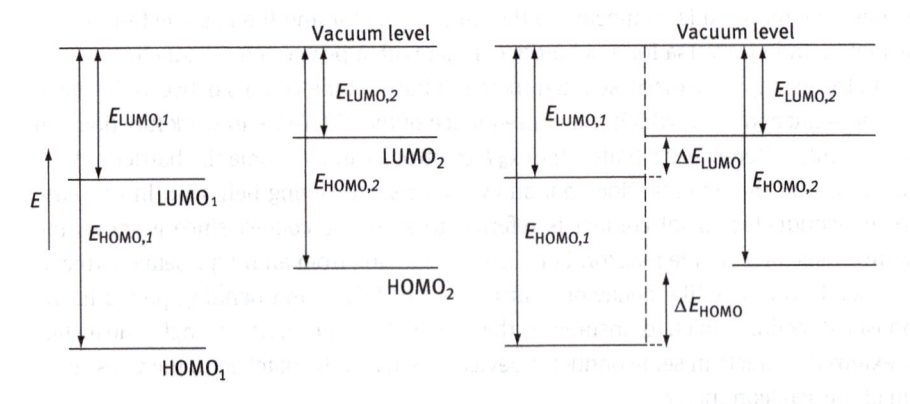

Fig. 2.51: Energy diagram of an organic heterojunction. In many cases organic–organic junctions are governed by vacuum-level alignment. This is the case due to the low charge carrier densities in organic materials that do not allow the formation of a depletion region.

However, another significant difference between inorganic and organic junctions is that organic devices commonly feature majority carriers. If an organic–organic heterojunction diode is biased in forward direction, the current is therefore a recombination current, i.e. holes are transported through the donor and electrons through the acceptor toward the junction where they recombine instead of traveling through the junction to the other side as minority carriers. As we saw in the case of the p–n homojunction in Section 2.3.1, the recombination current has an exponential form similar to the Shockley equation with a factor 1/2 in the exponent in the case of trap-assisted recombination and a factor 1 in the case of bimolecular recombination. Therefore, the Shockley equation is often also used for organic heterojunctions and the ideality factor n is associated with the type of recombination. However, there are limitations to that model as will be discussed in Chapter 3.

2.3.5 Organic–Inorganic Heterojunctions

Finally we are also interested in junctions between organic and inorganic materials. Experimentally it is found that there is a Schottky-type behavior for combinations of organic semiconductors with metals in the case that the work function of the metal is not matched with either the HOMO or the LUMO of the semiconductor. For external contacts of organic semiconductor devices metals or transparent conducting materials are therefore typically chosen in order to allow an ohmic contact between the active material and the electrodes. There are, however, also instances where a metal–organic Schottky diode is employed intentionally, e.g. in the first organic solar cells, where charge separation occurred only at this metal–organic Schottky contact. The properties of these contacts depend strongly on the interface properties and not only on the work functions of the involved materials. For instance, it has been shown that the Schottky barrier height can be tuned by more than 1 eV using self-assembled monolayers with defined dipole moments [68].

A very common junction type is the combination of an organic and an inorganic semiconductor, which is used for instance in hybrid solar cells featuring a metal oxide electrode in combination with conjugated molecules. As shown in Figure 2.52 the energetic situation in such a hybrid junction is given by a vacuum level alignment, i.e. there is no depletion region and the vacuum level is not bent. This is the case for most organic materials due to their low charge densities, analogous to organic–organic junctions. Predictions of the energy landscape are therefore commonly based on experimental measurements of the conduction and valence band energies of the inorganic semiconductor, and the HOMO and LUMO levels of the organic material. These predictions are generally considered valid; however, the real situation is more complicated due to chemical bonding at the interface and electronic coupling between the semiconductor and the molecular orbitals of the organic material.

It is further possible to change the effective work function of the semiconductor via modification with self-assembled monolayers of dipolar molecules. This is a common strategy for tuning the built-in potential of hybrid solar cells or dye-sensitized solar cells, which allows the fabrication of devices with enhanced open circuit voltage [69]. As schematically shown in Figure 2.52 for an inorganic acceptor in combination with an organic donor, a dipole moment pointing toward the inorganic semiconductor lowers its vacuum level and thus increases the energy-level offset, whereas a dipole moment pointing away from the inorganic semiconductor has the opposite effect.

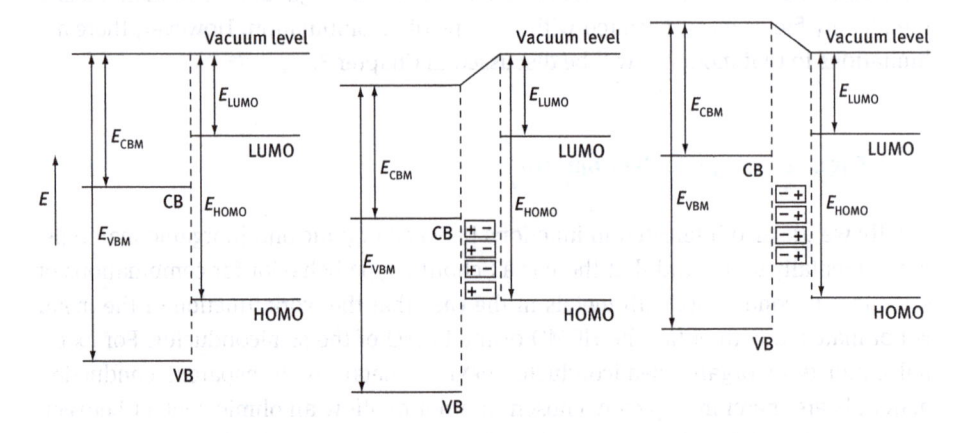

Fig. 2.52: Energy diagram of an organic–inorganic junction without (left) and with (middle and right) interface dipole modification. Similar to organic–organic junctions there is commonly vacuum level alignment between organic and inorganic semiconductors without the formation of a depletion region. The interface properties can be tuned by introduction of self-assembled monolayers of molecules with defined dipole moments, which causes a shift in vacuum levels and results in a different energy-level alignment.

Since virtually all conjugated molecules have a nonvanishing dipole moment this effect has to be taken into account for most hybrid junctions. In the case the organic material has a dipole moment, the relative orientation of organic molecules determines the energetic situation at the junction. Such orientation effects play an important role in organic and hybrid solar cells and will be discussed in later chapters.

3 Working Mechanisms of Organic and Hybrid Solar Cells

Having had a closer look at the fundamental properties of semiconductors and semiconductor junction, the constituting parts of all semiconductor devices, we now move on to the application, which is the focus of this book, namely a photovoltaic device – a solar cell. First we will discuss the basic principles of photovoltaic devices, starting with the fundamental conversion of photon energy into electric energy, before we move on to the photocurrent response and the diode characteristic of a solar cell. With this groundwork, we then look into the working principles of different types of photovoltaic devices, starting with all-inorganic solar cells as a basis and then looking at different organic, hybrid, and perovskite solar cells.

3.1 Basic Principles of Solar Cells

Fundamentally, a solar cell is a light-in–current-out device, where the current can be driven against a certain potential; i.e. the power of incident photons is partly converted into electric power. Since solar cells are operated mainly on the surface of the earth and solely in our solar system, they respond to the incoming sunlight, which strongly depends on pathway the light has taken. E.g. will the spectrum change when the light passes through our atmosphere, where parts of it get absorbed. As solar cells respond to the specific light spectrum, this will have implications for the maximum efficiency a solar cell can achieve and accordingly the ideal design. This will be discussed in Sections 3.1.1 and 3.1.2 before we move on to the spectral photocurrent response and the current–voltage characteristics of a solar cell.

3.1.1 Energy Conversion and the Solar Spectrum

Solar cells employ semiconductors as light absorbers and take advantage of the different time scales for light absorption, thermalization (relaxation to the lower conduction band edge or the lowest vibrational state), and recombination of excited states, which are a consequence of an energy gap that separates the ground state of the material from excited states. This energy gap can be taken advantage of by converting the energy of an excited state into electric energy in the form of a mobile pair of an electron and a hole. Depending on the bandgap and the external electrodes, there is a certain potential difference associated with this charge separation that a photovoltaic device generates a defined photovoltage when operated at open-circuit conditions and a defined photocurrent at short-circuit conditions.

The first solar cell was invented by Alexandre-Edmond Becquerel in 1839, who discovered the photovoltaic effect. In his experiment, he illuminated a silver chloride-coated electrode in an electrolyte solution with a plantinum counter-electrode [70, 71]. Thirty-eight years later, Adams and Day observed a similar effect in a first solid-state device, which combined vitreous selenium with platinum contacts, leading to the generation of a photocurrent and photovoltage under illumination [2]. The photovoltaic effect is related to the photoelectric effect, where an electron can be ejected from a metal by converting the energy of a photon to kinetic energy. However, the photovoltaic effect is generally more efficient since electrons are injected from one material into another material rather than into vacuum, which typically requires high photon energies. Due to the longer lifetime of excited states in materials with bandgaps, the photovoltaic effect is much more efficient in semiconductors.

Today, the photovoltaic effect is useful in various types of solar cells that are integrated into the global electric grid. Most of these devices are operated outside under solar illumination, so we first have to take a closer look at the sunlight available on our planet's surface. The massive amount of power that is incident on the earth is apparent from Figure 3.1, which shows a map of the yearly averaged solar energy input. Across most of the landmasses, there is 1,000–2,500 kWh of energy per square meter available per year. These numbers are tremendous compared to the average global energy consumption per year, which was on the order of 21,000 kWh in 2008. From these numbers it is apparent that only a few ten square meters of moderately efficient solar cell panels would be required per capita in order to provide sufficient electrical power. As will be discussed in Chapter 5, fabrication of these panels for a world population beyond 7 billion people is massively demanding and requires fast and low-cost manufacturing routes.

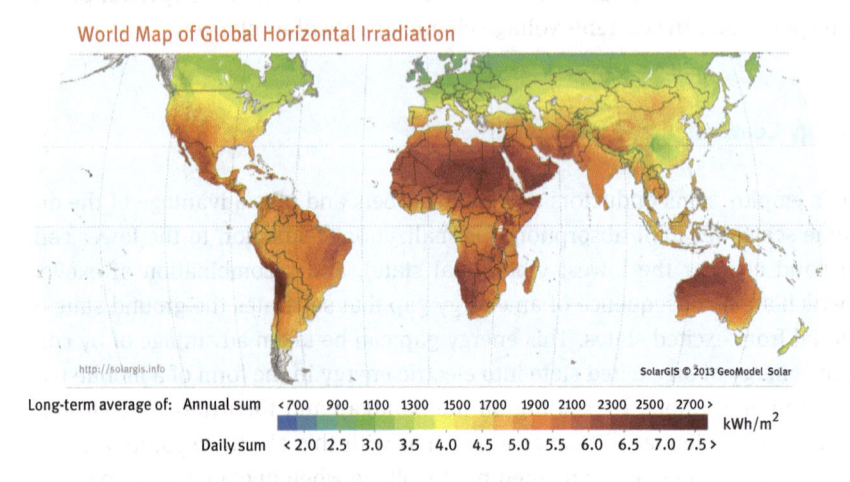

Fig. 3.1: **Annual average solar energy input on the surface of the earth.** Figure based on the *SolarGIS* database. Copyright (2015), SolarGIS, GeoModel Solar [72]

The power conversion efficiency (PCE; often denoted as η) of a photovoltaic cell is defined as the ratio of output power to input power:

$$\text{PCE} = \frac{P_{\text{out}}}{P_{\text{in}}}. \tag{3.1}$$

As we will see later, the PCE typically depends on the input power as P_{out} does not necessarily increase linearly with P_{in}. More importantly, the spectral distribution of the input power plays a crucial role as the solar cell has to be optimized with respect to a certain spectrum of the incident radiation. For instance, under constant illumination with monochromatic photons of a defined energy $h\nu$, the bandgap of the absorber material can be adjusted to match this energy and avoid any thermalization losses. The sun, however, emits a broad spectrum of radiation, which is further altered by absorption processes in the atmosphere, so that the construction of optimized solar cells is more demanding, which will be discussed in Section 3.1.2.

Before we come to this point, we will inspect the availability of solar light on the earth's surface in more detail and look at the specific solar spectrum at which solar cells are operated. The sun carries over 99.8% of the mass of our solar system and consists of approximately three quarters of hydrogen, while the rest consist mostly of helium. About 4.6 billion years ago, the sun formed via the gravitational collapse of a molecular cloud, leading to high density and temperature in its core, which eventually initiated thermonuclear fusion of hydrogen to form helium. This thermonuclear reaction occurs mostly within the innermost 25% of the sun's radius and is responsible for a massive power output in the form of black body radiation. For us, mostly the surface temperature of the sun, which is approximately 5,778 K, is of relevance since it determines the emitted spectrum of electromagnetic radiation.

According to Planck's law of black body radiation, the spectrum emitted by a black body with temperature is given as

$$I(\nu, T) = \frac{2h\nu^3}{c^2} \left(\exp\left(\frac{h\nu}{k_B T}\right) - 1 \right)^{-1}, \tag{3.2}$$

where $I(\nu, T)$ is the energy emitted in normal direction per unit time, unit area, and unit angle, the so-called *spectral radiance*, and ν is the frequency of the electromagnetic radiation.

As illustrated in Figure 3.2, the real spectrum of the sun above our atmosphere is very close to that of a perfect black body at 5,778 K. It is centered between 400 and 700 nm wavelength, and over 40% of solar radiative power are emitted in this region. Approximately 5% of the power lies in the UV region, the rest lies in the near-, mid-, and far-IR regions. When going through the atmosphere, the solar spectrum is significantly modified, as apparent when comparing the cyan and dark blue areas in Figure 3.2. This is mostly due to absorption processes by ozone (O_3) in the UV, O_2, and H_2O in the visible and near-IR regions and CO_2 around 2,000 nm. Obviously, the attenuation due to these absorption processes depends on the distance the sunlight travels through the atmosphere and thus on the angle of incidence, which changes with position on the globe, season, and local time. A common convention is therefore to classify solar spectra in terms of the air mass (AM) the light has gone through, where AM 1.0 refers to the most direct path through the atmosphere to sea level, as schematically shown in Figure 3.3. Additionally, the letters *D* and *G* are often added to the spectral information, denoting direct or global spectra, where global spectra take direct and scattered light into account. Since perfectly perpendicular sunlight input is almost never the case, a spectrum conventionally used for solar cell measurements is the AM 1.5G spectrum, i.e. the spectrum for light going through 1.5 atmospheres, which corresponds to a zenith angle of 48.2° (see Figure 3.3). The corresponding spectrum is shown in Figure 3.2, scaled to an overall illumination intensity of 100 mW cm^{-2}. In order to make solar cell efficiencies comparable, this spectrum and intensity are widely used in photovoltaic research and certified efficiencies are measured under these conditions by using a solar simulator. As the solar cell efficiency often depends on the temperature, a standard solar cell measurement is performed at 25°C. The standard measurement conditions are summarized in Table 3.1.

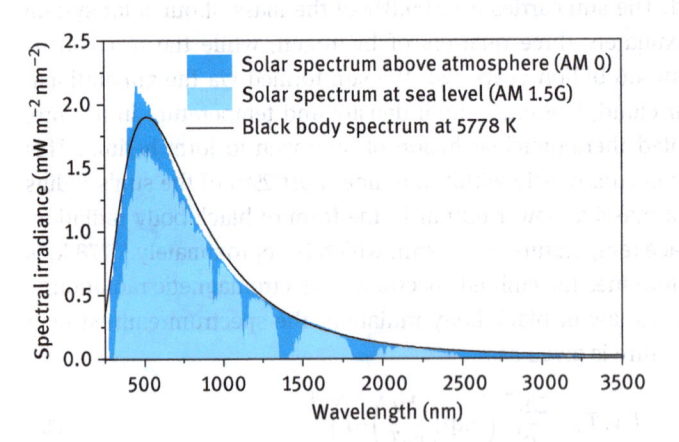

Fig. 3.2: **Comparison of black body radiation at a temperature of 5,778 K, the mean surface temperature of the sun, and the real solar spectra above the atmosphere and at sea level.** The spectrum at sea level exhibits characteristic regions of lower intensity, which can be mainly attributed to photon absorption by O_3 (UV region), O_2 (feature slightly above 750 nm), H_2O (several features), and CO_2 (feature at around 2,000 nm). Data from the web site of the National Renewable Energy Laboratory (NREL) [73]. The black body spectrum is scaled for comparison and given in arbitrary units.

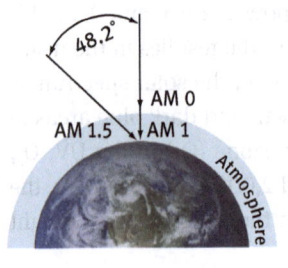

Fig. 3.3: **Schematic demonstration of sunlight passing through the atmosphere leading to AM 0, AM 1, and AM 1.5 spectra.** For testing solar cells the standard spectrum is the AM 1.5G, indicating the global average of sunlight traveling through 1.5 air mass (1.5 atmosphere thickness) at an incident angle of 48.2°.

Tab. 3.1: Standard conditions for solar cell and module efficiency measurements. Through this standardized measurement it is possible to compare efficiency values from different labs. Also certified solar cells are measured under these conditions.

Condition	Value
Spectrum	AM 1.5G (for solar cells for space applications: AM 0)
Intensity	100 mW cm^{-2} (one sun)
Cell temperature	25°C

3.1.2 Shockley–Queisser Limit

Before we take an in-depth view of the power conversion efficiency of photovoltaic devices and how it depends on factors such as absorption efficiency, separation of excited states, and charge–carrier recombination, we will take a brief look at the maximum possible efficiency of a solar cell. This topic has first been addressed in a publication by Shockley and Queisser in 1961 [74]. Shockley and Queisser derived the maximum efficiency of a single semiconductor solar cell under solar illumination, which is today commonly referred to as the *Shockley–Queisser limit*. A more detailed discussion on this and its extension is found, for instance, in the textbook by Jenny Nelson [64].

In short, the Shockley–Queisser limit is a bandgap-dependent maximum efficiency that can be obtained with a solar cell fabricated by a single semiconductor as absorber (for instance, a p–n junction silicon device). The underlying assumption of the limit is that every incident photon impinging on the solar cell with an energy exceeding the bandgap is absorbed and converted into an electron and a hole, while all photons with energies below the bandgap do not contribute. Furthermore, the potential energy of photogenerated carriers is independent of the photon energy; i.e. there is thermalization down to the band edge for all absorption events. The maximum efficiency is then found by considering the solar spectrum and the underlying thermodynamics, which is why the limit is often also referred to as the *detailed balance limit*. What has to be taken into account is that the solar cell is a black body, i.e. there are radiative losses that cannot be avoided. Furthermore, under solar illumination on the earth's surface, there is only a very narrow angle distribution of incident light (sunlight is virtually parallel), while the black body radiation of the device is isotropic. In short, this is the thermodynamic argument for why the device efficiency can be significantly enhanced when using light concentration, which broadens the angle distribution of the excitation light.

The resulting maximum efficiency curve for illumination with AM 1.5G light for a single-junction solar cell is shown in Figure 3.4. High maximum efficiencies exceeding 30% can be obtained for bandgap energies between 1 and 1.5 eV. For higher bandgaps the energy per charge is higher, but a significant amount of incident light cannot be absorbed. On the other hand, for smaller bandgaps most of the sunlight is converted, but thermalization losses reduce the overall power conversion efficiency.

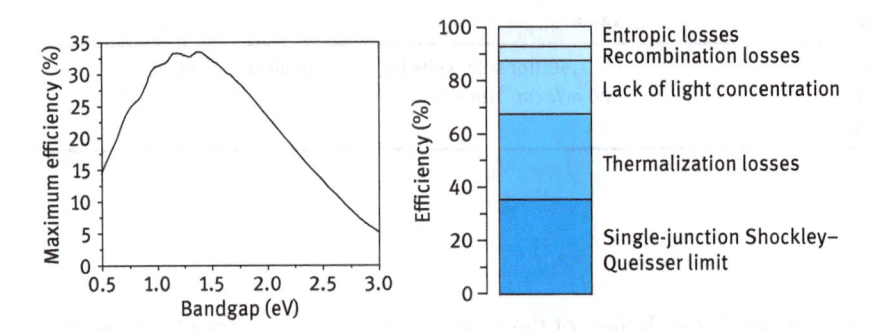

Fig. 3.4: Maximum efficiency for a single-junction solar cell as a function of the absorber bandgap and factors limiting the efficiency for idealized devices. Owing to the shape of the solar spectrum, the maximum efficiency for a single-junction solar cell can be achieved for bandgaps between 1 and 1.5 eV. For a theoretical infinite multijunction device operated under concentrated light, the maximum obtainable efficiency is 0.86.

The conventional Shockley–Queisser limit can be overcome using multijunction devices, which operate based on a stack of semiconductors with different bandgaps. Light passes through the wide-bandgap materials first, where only the high-energy photons are absorbed, while low-energy photons are converted in the rear part of the stack. Such a multijunction device is capable of increasing the achievable power conversion efficiency, since thermalization losses can be minimized. As derived by de Vos in 1980, for an infinite multijunction a maximum efficiency of 68% can be predicted [75]. This value can be pushed even further using light concentration as discussed before, so that ultimately an efficiency of 86% could be realized. This efficiency is lower than that of a theoretical Carnot engine operating between the temperature on earth (~300 K) and the surface temperature of the sun (5,778 K), which is roughly 95% due to the entropy of the photons emitted by the sun.

3.1.3 Photocurrent and Spectral Response

The electrical power output P_{out} of a solar cell operated at a bias voltage V (e.g. owing to a load resistance) is given as

$$P_{out} = V \cdot I, \qquad (3.3)$$

where $I = I(V)$ is the current through the device at the respective voltage. Since a solar cell is a diode, this current is a superposition of the diode dark current and the current that is generated by conversion of incident photons into electrons, the so-called *photocurrent* J_{ph}. We will now have a closer look at this photocurrent, while the diode characteristic of a solar cell will be discussed in Section 3.1.4.

For conventional photovoltaics without upconverters or material that allow multiple exciton generation, the upper limit for the photocurrent is given by the generation of one electron per photon. As discussed in the context of the Shockley–Queisser limit, the bandgap of the material further determines how many photons can be absorbed, since photons of energies below the bandgap of the absorber are not harvested by the device. The maximum photocurrent that can be generated by such a conventional solar cell under AM 1.5G illumination is shown in Figure 3.5 as a function of the threshold wavelength that is absorbed. Assumptions for this calculation are that every photon of an energy above the bandgap of the absorber is converted into one electron, which is then collected at the electrodes of the device and contributes to the photocurrent. The maximum photocurrent thus is an integral over the AM 1.5G spectrum and accordingly reflects intensity dips in several plateaus.

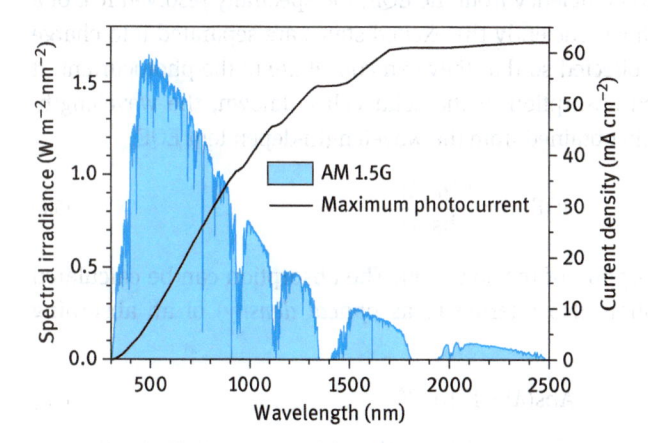

Fig. 3.5: **AM 1.5G spectrum and maximum achievable photocurrent for conversion of all incident photons down to the given wavelength.** The dark blue curve is the AM 1.5G spectrum, and the cyan area is the area under the curve. The maximum photocurrent is calculated via integration of the solar spectrum over all energies greater than the energy corresponding to the respective wavelength and under the assumption that one electron is generated per photon.

As seen in Section 3.1.2, optimization of a solar cell only with respect to the current does not result in the highest efficiencies, since there is a balance between thermalization loss and charge generation so that a bandgap energy corresponding to around 1,000 nm is ideal. Accordingly, high-efficiency solar cells generate photocurrents on the order of 30–40 mA cm^{-2}. Real devices, however, never generate photocurrent at an efficiency of 1, i.e. less than one electron is generated per absorbed photon. Furthermore, the efficiency for conversion of photons with energies below the bandgap is obviously 0. In order to quantify this efficiency and monitor its spectral dependence, one defines the so-called *external quantum efficiency* (EQE) as

$$\text{EQE} = \frac{\text{number of photogenerated electrons per second}}{\text{number of incident photons per second}}. \tag{3.4}$$

The EQE can be understood as a total photocurrent-generation efficiency when integrating over the whole solar spectrum. Typically, however, the EQE is analyzed as a function of photon energy and is wavelength-dependent. Since the EQE contains an electronic efficiency that depends on how loss-free charges are separated and collected at the electrodes and an optical efficiency that depends on how efficiently photons of a specific wavelength are absorbed, it is also useful to define the so-called *internal quantum efficiency* (IQE) as

$$\text{IQE} = \frac{\text{number of photogenerated electrons per second}}{\text{number of absorbed photons per second}}. \tag{3.5}$$

As it separates the absorption efficiency from the EQE, the spectrally resolved IQE of a solar cell is a measure for how efficiently the excited states are separated into charge carriers and subsequently collected so that they can contribute to the photocurrent. If the total spectrally resolved absorption of the solar cell is known, the wavelength-dependent IQE can be readily obtained from the wavelength-dependent EQE:

$$\text{IQE}(\lambda) = \frac{\text{EQE}(\lambda)}{\text{Abs}(\lambda)}, \tag{3.6}$$

where Abs is the total absorption of the solar cell. The absorption can be calculated from the absorbance A (often also referred to as *optical density*) of an absorbing material as

$$\text{Abs}(\lambda) = 1 - 10^{-A(\lambda)}. \tag{3.7}$$

Absorbance is commonly more often used than absorption since it scales linearly with the concentration of an absorbing solution or the thickness of an absorbing film. For a material that transmits an intensity I_1 upon illumination with intensity I_0, the absorbance of the material is defined as

$$A(\lambda) = -\log_{10}\left(\frac{I_1(\lambda)}{I_0(\lambda)}\right), \tag{3.8}$$

Besides EQE and IQE, it is also very common in literature to use the acronym *IPCE*, which stands for *incident photon to current conversion efficiency*, *incident photon to current efficiency*, or *incident photon to converted electron*. As it is ambiguous whether IPCE refers to an internal or an external quantum efficiency, we use the terms EQE and IQE throughout this book.

An example of the absorption, EQE, and IQE spectra of an organic solar cell is shown in Figure 3.6. The figure illustrates one of the most common loss mechanisms in organic solar cells, in which photon harvesting is incomplete. While the IQE is around 90% over the whole spectral range where the polymeric semiconductor is absorbing, the absorption is significantly below 80%, resulting in EQE values of approximately 60%. The drop in IQE below 400 nm and above 725 nm is owing to parasitic absorption in the solar cell due to reflection and absorption in the external contacts. Therefore, sometimes the IQE refers to the absorption only in the active semiconductor material of the solar cell neglecting parasitic losses of contacts, which allows us to judge the performance of the specific semiconductor material used in the solar cell and does not depend anymore on the device architecture, such as the specific substrate or special light management structures of the substrate.

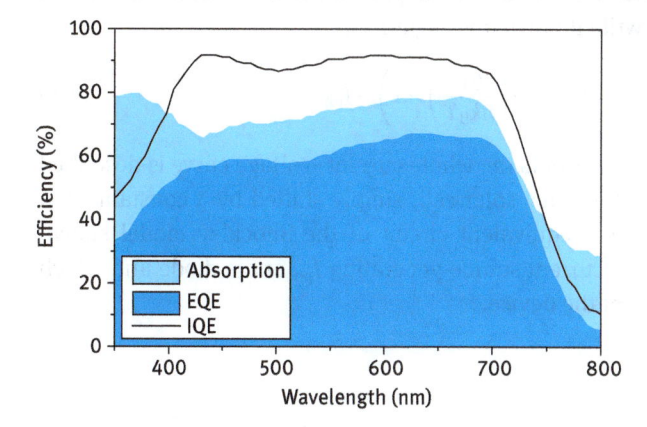

Fig. 3.6. Example of absorption, EQE, and IQE spectra of an organic solar cell. Data are shown for a typical solar cell with a mixed active film consisting of the polymer thieno[3,4-b]thiophene/benzodithiophene and [6,6]-phenyl-C_{71}-butyric acid methyl ester.

This example illustrates the importance of EQE and IQE analysis when optimizing solar cells. A closer look at the IQE of a device allows us to identify spectral regions of lower electronic efficiency. This is the first step toward understanding the underlying loss mechanisms, which has to be overcome to optimize the system.

For a solar cell with known EQE, it is now possible to calculate the photocurrent for operation of the cell under illumination with an arbitrary spectrum $S(\lambda)$ as

$$J_{ph} = q \int \text{EQE}(\lambda) S(\lambda) d\lambda, \tag{3.9}$$

where $S(\lambda)$ is given in units of photons per second. For instance, this integral executed over the AM 1.5G spectrum yields the photocurrent of a solar cell under simulated solar light. However, this calculation is valid only if the solar cell efficiency does not depend on the light illumination intensity. For organic and dye-sensitized solar cells we often observe higher efficiencies at lower illumination intensities. Therefore, the calculation of Equation (3.9) often gives higher values for the J_{ph} than measured under AM 1.5G, as in many cases the EQE is measured without background illumination, leading to a measurement under much lower light intensity than under one sun. More details about EQE measurements are described in Section 4.2.3.

3.1.4 Characteristics of a Solar Cell

Now that we have seen how the photocurrent J_{ph} of a solar cell depends on its EQE and the incident spectrum, we can use J_{ph} to construct the whole current–voltage curve of a solar cell with a Shockley diode behavior by substituting the photocurrent into Equation (2.140). We will now discuss the impact of the different parameters in this equation and therefore write it down once again as:

$$J(V) = J_0 \left(\exp\left(\frac{qV}{nk_BT}\right) - 1 \right) - J_{ph}. \tag{3.10}$$

For voltage-independent photocurrents the whole current–voltage curve is described by the diode behavior of the Shockley solar cell, simply shifted by a constant $-J_{ph}$. This can be illustrated with the equivalent circuit of the Shockley model shown in Figure 3.7, which features a current source-generating J_{ph} and a diode and which accounts for the dark current of the device.

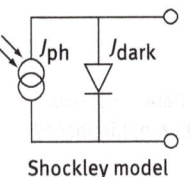

Fig. 3.7: Equivalent circuits for a solar cell as described by the Shockley model. The model contains a diode with current–voltage curve $J_{dark}(V)$ that follows the Shockley diode equation and a current source that generates a voltage-independent photocurrent J_{ph}.

As q and k_B are constants and V is the variable in the equation, there are three parameters of interest left behind, which are the reverse saturation current density J_0, the ideality factor n, and the temperature T. It is important to note that J_0 itself depends on the temperature, but this dependence is a function of the materials used in the device. For an ideal p–n diode with electron and hole diffusion coefficients D_n and D_p, n and p dopant densities N_D and N_A, and electron and hole recombination lifetimes τ_n and τ_p, respectively, the reverse saturation current density is given as

$$J_0 = q \left(\sqrt{\frac{D_n}{\tau_n}} \frac{n_i^2}{N_A} + \sqrt{\frac{D_p}{\tau_p}} \frac{n_i^2}{N_D} \right), \tag{3.11}$$

with the intrinsic charge carrier density

$$n_i = N_c \exp\left(-\frac{E_{gap}}{2k_BT}\right), \tag{3.12}$$

where N_c is the effective density of states and E_{gap} is the bandgap of the material. Thus, the temperature dependency of J_0 is implied in the temperature dependency of the intrinsic charge density.

We will now illustrate how the diode current density–voltage curve changes with the three parameters J_0, n, and T for an ideal Shockley solar cell with a fixed photocurrent density J_{ph}. Figure 3.8 shows the impact of the reverse saturation current on the diode curve. While the curvature of the J–V curve is not affected, the curve is shifted toward higher voltages for decreasing J_0. Accordingly, the area included by the curve in the fourth quadrant is increased and thus the theoretical maximum power output of the solar cell, which is given as

$$P_{max} = \max_V \left(V \cdot J(V) \right). \tag{3.13}$$

For a given material with intrinsic charge density n_i, such an increase in power output might be related to reduced loss of photogenerated charge carriers to recombination, i.e. extended charge carrier lifetimes τ_n and τ_p, which lead to lower J_0 according to Equation (3.11).

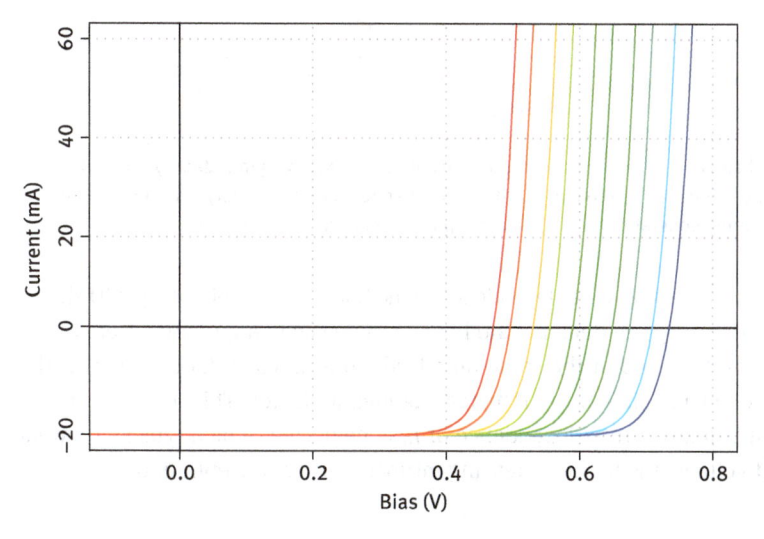

Fig. 3.8: Simulated current–voltage curves for a Shockley diode with varying reverse saturation current. J_0 is increasing from the blue to the red curve while keeping all other parameters constant.

In contrast to the reverse saturation current, which has no effect on the curvature of the J–V curve, the ideality factor determines how spontaneously the diode curve bends. This is shown in Figure 3.9 for values of between 1 and 2, which are the most extreme cases for a Shockley diode, as discussed in Section 2.3.1. The curves are simulated with J_0 being varied so that the open-circuit voltage (the voltage where the diode curve intercepts the voltage axis) is constant. Apparently, an increasing ideality factor results in a less bent J–V curve, which in turn means that P_{max} is reduced. Such an increase in ideality factor can be induced by the presence of trap states that mediate Shockley–Read–Hall recombination as described by the recombination current in Equation (2.129). For p–n junction solar cells the reduction of such trap states is therefore a commonly pursued objective.

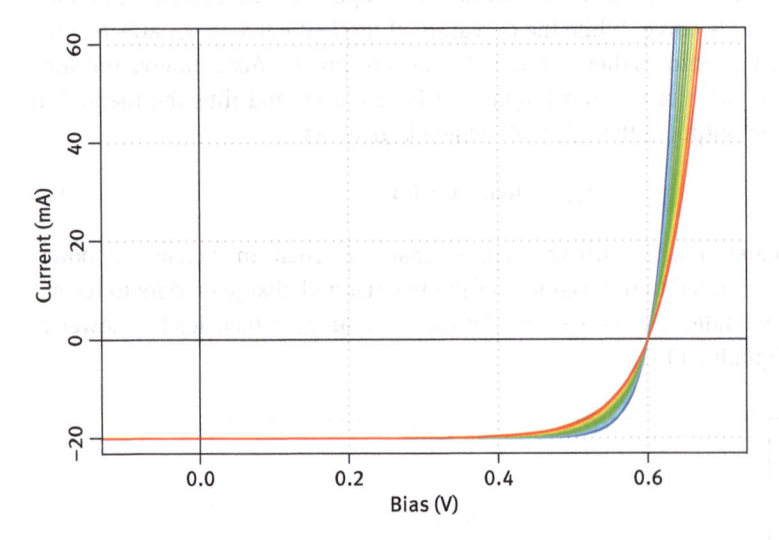

Fig. 3.9: Simulated current–voltage curves for a Shockley diode with varying ideality factor. n is increasing from the blue to the red curve, while J_0 is changed simultaneously in order to keep the open-circuit voltage constant at 0.6 V. All other parameters are kept constant.

Finally, we look at the temperature dependence in the Shockley model. Figure 3.10 shows exemplary J–V curves, which demonstrate how an increasing temperature T enhances the open-circuit voltage and at the same time leads to a less spontaneous rise in the diode current, somewhat similar to an increasing ideality factor. This, however, is the case if the implicit temperature dependence of J_0 is neglected, which typically lowers the open-circuit voltage due to the increasing intrinsic carrier concentration.

3.1.5 Parasitic Resistances and the Generalized Shockley Model

Now that we have seen the impact of parameters like temperature and ideality factor on the J–V curves of solar cells described by the idealized Shockley equation, we want to turn to real devices, which show less ideal J–V curves. Note that *ideal* in this context does not refer to the ideality factor. As we have seen in Section 2.3.1, for a p–n junction diode the ideality factor is determined by the type of recombination in the depletion region and is not directly related to fabrication-related or electronic nonidealities in the device.

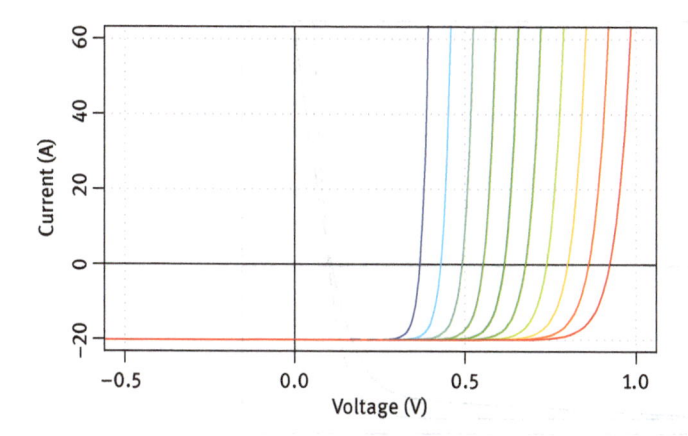

Fig. 3.10: Simulated current–voltage curves for a Shockley diode operated at varying temperatures. T is increasing from the blue to the red curve, while the implicit temperature dependency of J_0 is ignored and J_0 is kept constant together with all other parameters.

All J–V curves shown in Section 3.1.4 have two things in common: (1) a virtually perfect blocking in reverse direction, i.e. the current is constant at J_{ph} and only marginally increases with increasing reverse voltage; and (2) an exponential rise in forward current (even though the slope of the exponential function is different for different ideality factors or temperatures). Real devices typically deviate from this behavior in both forward and reverse directions due to the presence of parasitic resistances. This is exemplarily shown in Figure 3.11, where a Shockley equation J–V curve is compared to a J–V curve as exhibited by a real solar cell. The real device is characterized by a forward current slope that changes from exponential to linear behavior for large bias voltages and a nonperfect blocking in the reverse direction. These two features also affect the shape of the curve in the fourth quadrant, where the power output of a solar cell is positive. The curve is less rectangular in this quadrant, resulting in a lower maximum power.

We will now look at the most important characteristic values that are used to describe the J–V curve of a solar cell. These values are summarized in Figure 3.12. The first two important parameters are the axis intercepts: the *short-circuit current density* J_{SC} is the current that flows at 0 applied bias and can be used as a good approximation of the photogenerated current J_{ph} for well-working solar cells and the *open-circuit voltage* V_{OC} is the bias at which there is no net current flow and is equal to the potential difference between the front and back electrodes of a solar cell in the case they are not in electrical contact. The most important parameter of a solar cell is obviously the *power conversion efficiency* PCE, which is the efficiency at which the device converts radiative energy into electrical energy. At an arbitrary point on the J–V curve, the electrical power of a solar cell is given as $P = V \cdot J(V)$. Conventionally, the diode curve of a solar cell is plotted so that negative voltages correspond to reverse bias, whereas positive voltages are applied in forward direction. Furthermore, the photocurrent has

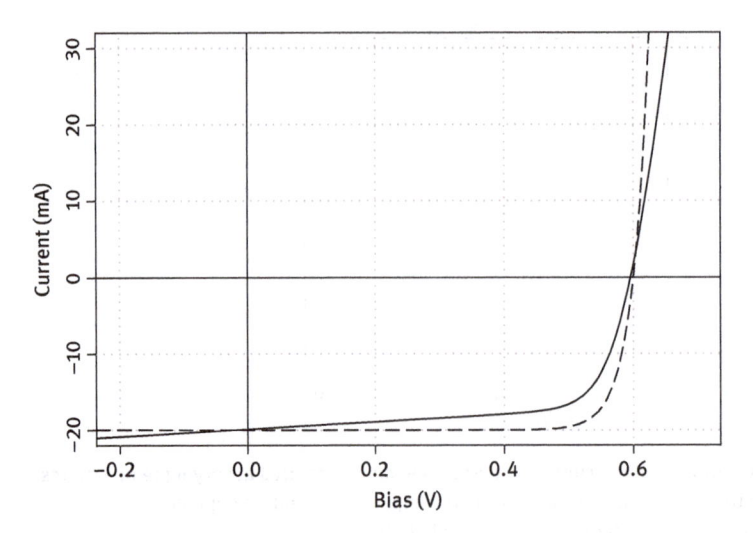

Fig. 3.11: Diode curves for an ideal solar cell following the Shockley equation (dashed line) and a real solar cell (solid line). Real devices are typically characterized by a nonvanishing increase in current with increasing reverse voltage, a sub-exponential current rise in forward direction, and slightly less rectangular diode curves.

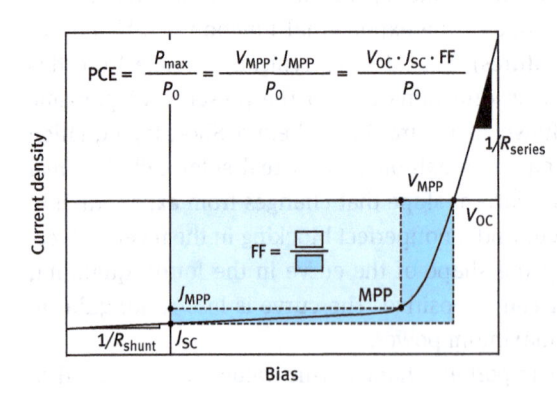

Fig. 3.12: Schematic current density–voltage curve labeled with the most important characteristic values. Per convention, negative bias voltages correspond to the reverse direction of the diode and positive voltages to the forward direction.

a negative sign by convention. Then, the electrical power is negative only in the fourth quadrant, which means that only in that location the solar cell has a net power output, whereas in the first and third quadrants, power is dissipated by the device. In order to find the power conversion efficiency of a solar cell, the point at which the product $V \cdot J(V)$ has a maximum power has to be found, the so-called *maximum power point* with power output P_{max}. This point is defined by the respective current density and voltage values, the maximum power point current density J_{MPP}. The *PCE* is then given as the ratio of P_{max} to the incident light power P_0.

$$PCE = \frac{P_{max}}{P_0} = \frac{V_{MPP} \cdot J_{MPP}}{P_0}. \tag{3.14}$$

Due to the exponential onset in forward direction, P_{max} is always smaller than the product $V_{OC} \cdot J_{SC}$, since the J–V curve is not perfectly rectangular. However, the more spontaneous the rise in current is and the more the solar cell shows a blocking behavior in reverse direction, the more rectangular the curve is, leading to a higher value of P_{max} for given V_{OC} and J_{SC}. This means that the diode quality plays a crucial role for the PCE of a solar cell, where devices with ideality factors of $n = 1$ perform best. An often used measure for how rectangular the J–V curve is in the fourth quadrant is the so-called *fill factor* FF, which is defined as

$$FF = \frac{V_{MPP} \cdot J_{MPP}}{V_{OC} \cdot J_{SC}}. \tag{3.15}$$

When optimizing solar cells it is imperative to maximize that the fill factor once the system generates optimum J_{SC} and V_{OC}. This is the focus of intense research for various photovoltaic systems since a charge generation and recombination have to be well understood in order to allow good control over the fill factor. For a solar cell with known FF, V_{OC}, and J_{SC}, the PCE can also be written as $PCE = V_{OC} \cdot J_{SC} \cdot FF/P_0$ by using Equation (3.14) and Equation (3.15).

J–V curves of real devices, as shown in Figure 3.11, are in many cases sufficiently well described when adding two parasitic resistances to the Shockley solar cell model from Figure 3.7, a series resistance R_s, and a shunt resistance R_{sh} parallel to the diode. This model is referred to as *generalized Shockley model* and the respective equivalent circuit is shown in Figure 3.13. R_s causes a voltage drop of $J(V)R_s$ at all applied bias voltages V, resulting in an effective voltage $V - J(V)R_s$ that replaces the applied voltage in the exponent of the Shockley equation. R_{sh} allows a parasitic current to flow parallel to the ideal diode in the equivalent circuit, which is given as $-(V - J(V)R_s)/R_{sh}$. These considerations lead to the *generalized Shockley equation*

$$J(V) = J_0 \left(\exp\left(\frac{q(V - J(V)R_s)}{nk_B T} \right) - 1 \right) - J_{ph} - \frac{V - J(V)R_s}{R_{sh}}, \tag{3.16}$$

with the same parameters J_0, n, and T from the Shockley equation and the parasitic resistances R_s and R_{sh}. Note that in the case of current densities the unit of these resistances is $\Omega\,cm^2$ and not Ω, as in the case of absolute currents. Furthermore, owing to the occurrence of the term $J(V)R_s$ in the argument of the exponential function, the generalized Shockley equation is a transcendental equation. Simulation of J–V curves or fitting to experimental data therefore requires numerical methods.

In order to illustrate the effect of these two resistances, we consider the solar cell being operated at far-reverse and far-forward current. In reverse direction the current through the diode in the generalized Shockley equivalent circuit is virtually 0, and the total current is given by J_{ph} plus the current through $R_s + R_{sh}$. For good solar cells, $R_{sh} \gg R_s$ so that $R_s + R_{sh} \approx R_{sh}$. Therefore, R_{sh} can be approximated from the inverse slope of the $J-V$ curve at reverse bias. In the case of strong forward bias the internal resistance of the diode in the equivalent circuit is approaching 0. This current injected into the device due to the external bias together with J_{ph} then only goes through the series resistance, while the shunt resistance is effectively by passed by the diode in forward direction. Therefore, the inverse slope of the $J-V$ curve in forward bias is approximately R_s. For moderate biases between $V = 0$ and $V = V_{OC}$, both parasitic resistances play a role, potentially leading to a decrease in FF.

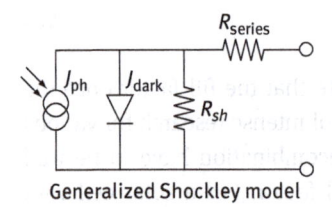

Fig. 3.13: Equivalent circuits for a solar cell as described by the generalized Shockley model. While the generalized Shockley model, in contrast to the Shockley model, contains equivalent resistances R_s and R_{sh}, the photocurrent J_{ph} is assumed to be voltage-independent in both models.

To further understand the impact of the magnitude of the two parasitic resistances in the generalized Shockley model, we will now look at diode curves for varying R_s and R_{sh}. Figure 3.14 illustrates the impact of decreasing R_{sh} on the diode curve. As mentioned before, the presence of a noninfinite shunt resistance leads to linearly decreasing currents at increasing reverse bias, i.e. the effect is mainly visible in the third quadrant. For moderate values of R_{sh}, this mainly results in a decrease in FF, while also the V_{OC} drops if R_{sh} is further decreased. Very small shunt resistances correspond to short cuts in the active layer of the solar cell, which still allow the generation of a photocurrent but hinder the built-up of a photoinduced potential. In real devices, low shunt resistances are commonly related to small defects, like pinholes, in the active film, that introduce tiny short cuts into the device.

Figure 3.15 shows how the $J-V$ curve changes with increasing series resistance. In contrast to the shunt resistance, R_s plays a major role in the first quadrant. Moderate values of R_s also lead to a decrease in FF, while initially V_{OC} and J_{SC} remain unchanged. For further increasing R_s, the J_{SC} is lowered along with the FF, leading to a further decrease in device efficiency.

In optimized solar cells and especially in modules, series resistances play a major role since shunts can be avoided by optimizing the processing of the photoactive layer. In contrast, the series resistance is mostly influenced by external connections and the transparent conductive front electrode, which typically has many orders of magnitude higher resistance than a metal film. In modules, where individual solar cells have to be electrically connected to large active areas, these connections are responsible for a significant series resistance, resulting in a lower FF of the entire module compared to the FF of optimized individual cells.

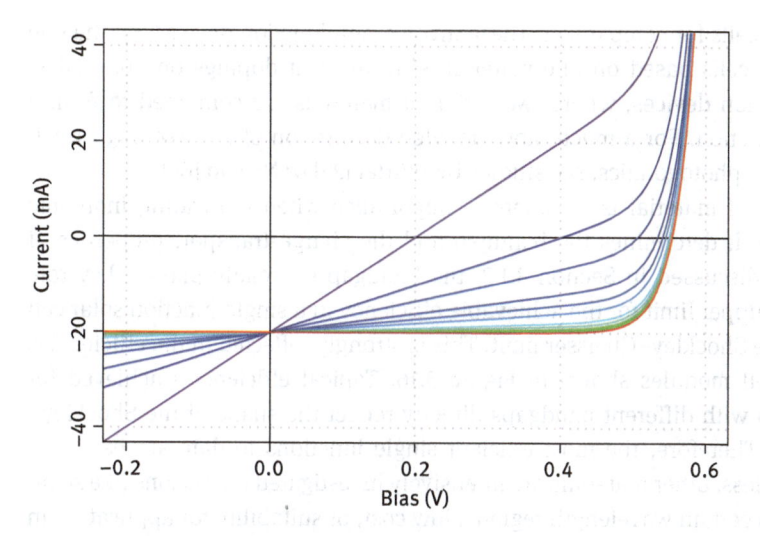

Fig. 3.14: Impact of decreasing shunt resistance on the shape of the current density–voltage curve. R_s is decreasing from red toward blue curves.

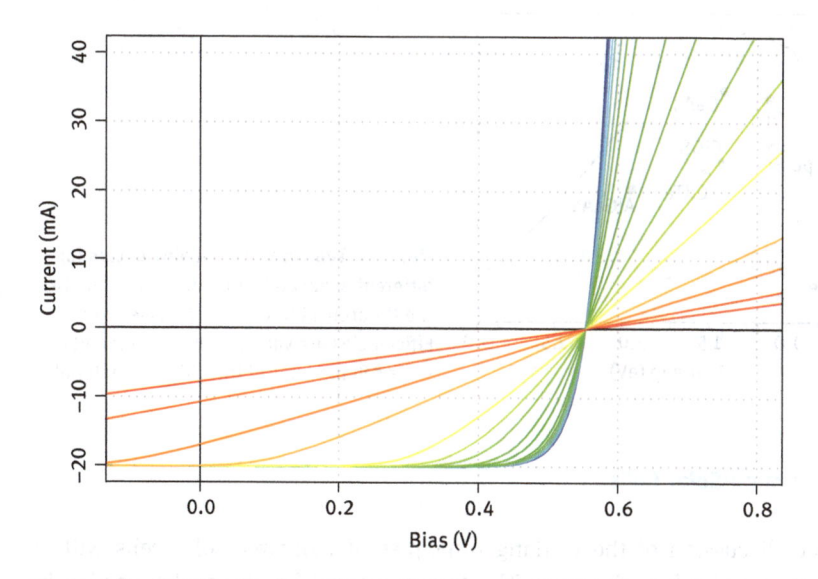

Fig. 3.15: Impact of increasing series resistance on the shape of the current density–voltage curve. R_s is increasing from blue to red curves.

3.2 Inorganic Junction Solar Cells

Even though we are mainly interested in solar cells based on organic and hybrid materials, we want to give a short overview of the working mechanisms of fully inorganic solar cells for comparison. These involve p–n junction devices, on the one hand, i.e. solar cells based on one material with different dopings on both sides, and heterojunction devices, where two different materials are combined to form a photovoltaic junction. For a much more detailed description of the working principles of inorganic photovoltaics, see studies by Würfel [29] or Nelson [64].

The choice of material is of immense importance when fabricating inorganic solar cells since it determines the bandgap and the charge transport properties of the device. As discussed in Section 3.1.2, the bandgap especially plays a key role since it sets an upper limit for the achievable efficiency in a single-junction solar cell according to the Shockley–Queisser limit. This is strongly reflected in the efficiencies of real solar cell modules shown in Figure 3.16. Typical efficiencies achieved for semiconductors with different bandgaps directly reflect the shape of the Shockley–Queisser limit. Therefore, the most efficient single junctions to date are based on GaAs. Nevertheless, other materials are intensively investigated for reasons like semitransparency in certain wavelength regions, low cost, or suitability for application in multijunction devices.

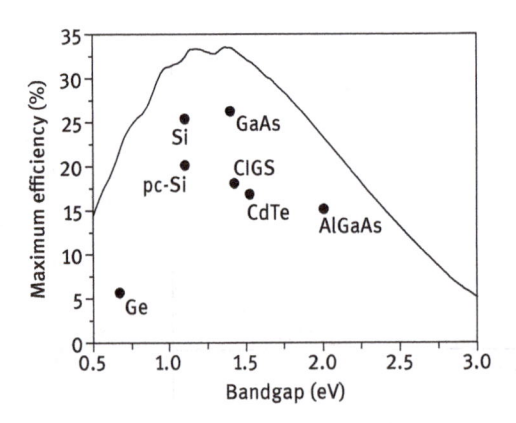

Fig. 3.16: Typical solar cell efficiencies for different semiconductors in comparison to the theoretical Shockley–Queisser limit. Efficiencies are values as typically found for solar cell modules of the different materials.

3.2.1 p–n Junction Solar Cells

We start our discussion of the working principles of inorganic solar cells with a device based on a p–n homojunction. The area of interest for photovoltaic action in such a device is the depletion region, where conduction band and valence band are bent. Figure 3.17 depicts the elementary process of charge generation in such a device (especially the external contacts to the metal electrodes are just schematically shown). For simplicity we first consider the situation at short-circuit current, i.e. without an applied external field. A photon that is absorbed in the semiconductor leads to the formation of an exciton, which typically directly splits into an electron and a hole polaron due to the low binding energy of Wannier excitons. Even though this is not necessarily the case for all inorganic semiconductors, materials with low binding energies are more commonly employed in inorganic p–n junction solar cells since exciton recombination losses can be avoided. In the case the exciton was generated within the depletion region, the resulting electron and hole polaron find themselves within the internal electric field of the depletion region, which drives them in opposite directions: the hole travels into the p-type region, while the electron goes into the n-type region. For charge generation within the field-free regions of the device, charges move diffusively. Charge carriers are then collected at the external metal electrodes, resulting in the generation of a photocurrent. Since it can be beneficial to have wide areas of the active layer within the internal electric field, the width of the depletion region is often controlled to assure that light absorption is efficiently translated into current generation. In many devices this is realized by combining a highly doped and thin n-type layer (the so-called *emitter*) with a thick and only lightly doped p-type region (the so-called *base*). As can be seen from Equation (2.118) and Equation (2.119), this results in a wide depletion region, which mainly extends into the p-type region.

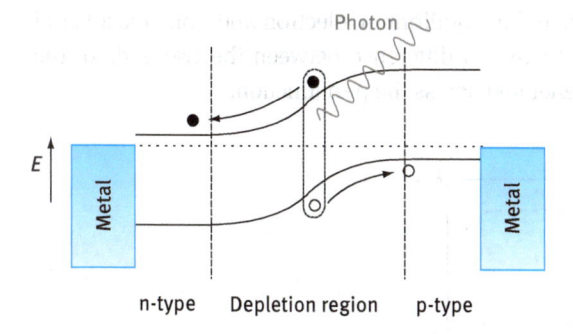

Fig. 3.17: Schematic of the elementary charge-generation process in an inorganic p–n homojunction. An exciton forming within the depletion region upon photon absorption rapidly dissociates into an electron and a hole, which are then driven in opposite directions due to the internal electric field. This results in the generation of a photocurrent.

Once the p–n junction is operated at an external bias, the band bending changes. We are mainly interested in the case of moderate forward bias (when the J–V curve is in the fourth quadrant) since this is the operating condition for a solar cell when it does work on an electric load. An applied forward bias reduces the band offset and with this the strength of the electric field that drives separated charges out of the depletion region. In good devices the forward bias can be close to the built-in voltage without showing a significant change in net current output of the device, reflected in a good fill factor (FF). For further increasing voltages, however, current is injected into the device, which exactly cancels the photocurrent at open-circuit voltage. Therefore, the maximum achievable V_{OC} is the built-in voltage (see Figure 2.46).

We will now further underline the necessity of using a p–n junction for photovoltaic applications rather than a single semiconductor. In principle, charge generation is also possible in a single chunk of a semiconductor without a built-in gradient in dopants as long as the exciton binding energy can be thermally overcome (which is the case in many semiconductors at room temperatures, as outlined earlier). However, without a p–n junction, no net current or voltage is generated. Figure 3.18 compares the situation at open-circuit conditions for an illuminated n-type semiconductor and an illuminated semiconductor with n- and p-type ends. Due to the incident photons there is a constant charge generation in the n-type semiconductor, resulting in a splitting of electron and hole quasi-Fermi levels in the center of the semiconductor. Since there is no driving force in either directions, electrons and holes diffuse through the semiconductor toward its edges in the same fashion, where they eventually recombine. Thus, no net voltage is generated between the two sides of the semiconductor. In contrast, due to the electric field inside the depletion region, electrons and holes experience different driving forces toward the two sides of the p–n semiconductor, so that electrons accumulate at the n-type side and holes at the p-type side. Due to surface recombination at the ends, there is a slight bending of the quasi-Fermi level of the majority carriers, and in equilibrium, electron and hole quasi-Fermi levels coincide. Overall, however, there is a difference between the two ends of the semiconductor so that voltage is generated across the p–n junction.

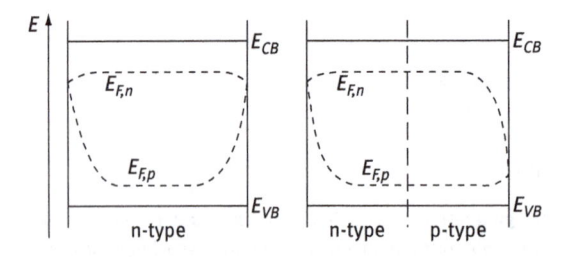

Fig. 3.18: Splitting of the quasi-Fermi levels in a single chunk of an n-type semiconductor and in a p–n homojunction at open-circuit conditions under illumination. Due to the constant charge generation inside the semiconductor, there is a splitting between electron and hole Fermi level also in a single semiconductor. However, charges are similarly driven to both ends of the semiconductor so that no net voltage is generated.

3.2.2 Inorganic Heterojunction Solar Cells

In Section 2.3.2, we have seen how the energy-level alignment proceeds in an inorganic heterojunction device in comparison to a p–n homojunction. Of course, such an inorganic heterojunction can also be employed as a photovoltaic cell if the materials used provide sufficient photon absorption and low exciton binding energies. Similar to the p–n homojunction, exciton separation leads to the generation of an electron and a hole, which are driven toward opposing sides of the device in the case they are created within the depletion region. Besides, width and location of the depletion region are determined by the concentration of dopants in either of the two materials.

Inorganic heterojunctions are often realized by combining a low-charge-density material with a bandgap favorable for strong light absorption with a wide-bandgap material that serves as an emitter. Such devices are often realized using low-cost materials like metal oxides, which are often intrinsically n-doped due to the presence of oxygen vacancies. Another commonly used type of heterojunction is a p–i–n structure, where an intrinsic layer responsible for light absorption and exciton separation is sandwiched between two strongly doped n-type and p-type semiconductors.

3.3 Organic Junction Solar Cells

After a short overview of charge separation in inorganic solar cells, we will now turn to the main subject of this book, which is photovoltaics that include organic semiconductors. We start with a detailed discussion of organic–organic junctions (Section 3.3), before we look into hybrid inorganic–organic junctions in Section 3.4

We first look at how charge separation works at organic–organic interfaces, before we discuss the impact of the interface morphology. This discussion includes flat heterojunctions as well as the so-called bulk heterojunctions, where donor and acceptor materials form a mixed phase. Since for bulk heterojunctions the molecular properties of donor and acceptor materials massively influence the internal morphology, Section 3.3.4 also features a discussion of crystallization of organic semiconductors.

3.3.1 Charge Separation in Organic Solar Cells

3.3.1.1 Donor–Acceptor Junctions for Exciton Splitting

As discussed in Chapter 2, excited states in organic materials are strongly bound with binding energies of at least a few 100 meV. This energy is high compared to the thermal energy at room temperature ($k_B T$), i.e. the product of Boltzmann's constant k_B and the temperature T, which is around 25 meV at room temperature. Thermal splitting of excitons in organic materials is therefore unlikely. However, for the function of a solar cell it is imperative to split excited states into free charge carriers or, more precisely, into polarons. Therefore, in photovoltaics based on organic semiconductors, mostly a donor–acceptor structure with two different organic materials is implemented. The energy offsets between the HOMOs (highest occupied molecular orbitals) and LUMOs (lowest unoccupied molecular orbitals) of the materials can thus provide the energy required for charge separation. As will be discussed later, organic semiconductors can also be combined with inorganic materials to realize a donor–acceptor junction. However, still the role of this junction is to provide the energy necessary to overcome the binding energy of excitons in the organic compound.

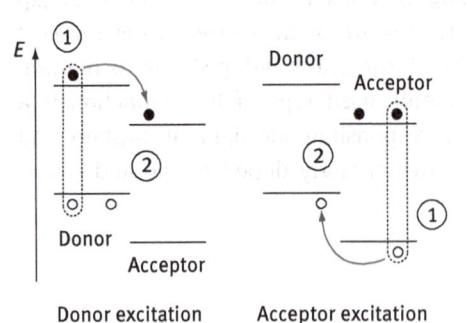

Fig. 3.19: Possible mechanisms of charge separation at an organic–organic heterojunction. Upon absorption of an incident photon in the donor (acceptor), an exciton is generated (1). The energy necessary to split the exciton into a hole polaron in the donor and an electron polaron in the acceptor (2) is provided by the offset between the LUMOs (HOMOs) of the materials.

The two fundamental routes toward charge generation in an organic solar cell are shown in Figure 3.19. A photon impinging on the solar cell can be absorbed by one of the two active materials, the donor or the acceptor. This absorption process results in the generation of an exciton. If the exciton can diffuse toward a donor–acceptor junction during its lifetime, it can split up, resulting in a hole polaron in the LUMO of the donor and an electron polaron in the HOMO of the acceptor. Provided suitable energy offsets, this process can occur in two fashions. For an exciton in the donor, the exciton splitting results in the injection of an electron into the LUMO of the acceptor, while the hole remains in the HOMO of the donor. On the other hand, if the acceptor is excited, a hole is injected into the HOMO of the donor, while the electron remains in the LUMO of the acceptor. Analogous, if either donor or acceptor is replaced by an inorganic semiconductor, the hole is injected into the valence band or the electron into the conduction band of the material, respectively.

3.3.1.2 Charge Transfer States

Since the introduction of the organic–organic heterojunction for efficient charge separation by Tang, the main steps in the process of charge generation in organic solar cells are known: exciton generation by a photon → exciton diffusion toward the organic–organic junction → splitting of the exciton into polarons [6]. However, the fundamentals of how the exciton is split are still not completely understood.

In the current picture there is an intermediate state populated at the interface, a so-called charge transfer state [76, 77]. Charge transfer states exist only at the interface between two different organic semiconductors and are occupied by an electron–hole pair with the electron residing in the acceptor and the hole in the donor.

The two charges form a coulombically bound polaron pair. Typically, these polarons are located on neighboring molecules and therefore separated by 0.5–2 nm depending on the structure of the organic semiconductors, their orientation with respect to each other, and the spatial location of the HOMO and LUMO levels. Using a point-charge approximation and assuming a relative permittivity of 3, this results in binding energies between 0.25 and 1 eV, which is in good accordance with experimental results. If this binding energy is overcome, this charge transfer state is separated, i.e. the polaron pair is split into an electron polaron in the acceptor's LUMO and a hole polaron in the donor's HOMO.

An exciton, a charge transfer state, and a charge separated state are illustrated in Figure 3.20. As introduced in Chapter 2, an exciton is a bound state of electron and hole that can be described as an electrically neutral quasiparticle with an energy that is lower than the optical bandgap of the organic semiconductor. Due to their high binding energy these excitons do not split within one organic semiconductor but instead diffuse as quasiparticle. Typical diffusion lengths are on the order of a few nanometers due to the short lifetime of excitons. However, if the exciton reaches an interface within its lifetime, a charge transfer state can be populated. This charge transfer state can then split into an electron and a hole, provided that this final charge separated state is energetically accessible.

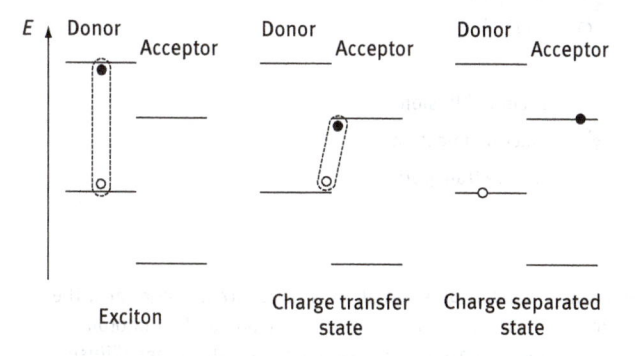

Fig. 3.20: Energy-level diagram showing difference between an exciton, a charge transfer state, and a charge separated state at an organic–organic heterojunction. The exciton is often illustrated as bound pair of electron and hole on the same molecule. The energy of the exciton is lower than the optical bandgap of the donor. At organic–organic interfaces a charge transfer state can be populated by a polaron pair, which is still subjected to Coulomb attraction. The charge transfer state can be split if the occupation of the final charge separated state (free electron and hole polaron in the acceptor and donor, respectively) is energetically accessible.

A schematic picture of the processes occurring upon photon absorption in proximity of an organic–organic heterojunction is depicted in Figure 3.21. The situation is illustrated exemplarily for a photoexcited polymeric donor and a fullerene acceptor (PCBM), the most common combination in organic solar cells. The initially formed exciton diffuses through the donor (blue trace) toward the interface. There, the excitation energy can transfer so that a charge transfer state between donor and acceptor is occupied. This state can be illustrated as a strongly bound electron–hole pair, where the hole is located on the donor side of the interface and the electron on the acceptor side. This initial separation of electron and hole increases the Coulomb radius between the two carriers, resulting in a lowered binding energy of the charge transfer exciton compared to the initial exciton formed in the donor. Since the charge transfer state is obtained by mixing the HOMO of the donor and the LUMO of the acceptor, its energy is also significantly reduced compared to the equilibrium exciton energy in the donor, which is in the order of the optical HOMO–LUMO gap in the donor. Due to its reduced binding energy the charge transfer state can be thermally split into an electron polaron residing in an acceptor molecule close to the interface and a hole polaron in the donor polymer. Depending on the operation conditions (external bias) of the device, these polarons can then migrate away from the interface toward the external contacts, i.e. result in the generation of a photocurrent.

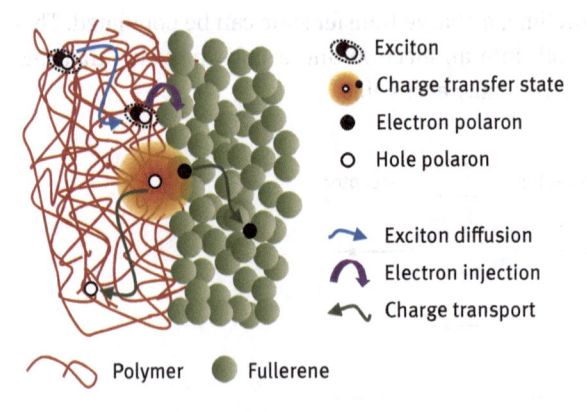

Fig. 3.21: Schematic snapshot of a charge separation event at an organic–organic interface. The figure illustrates the mechanism exemplarily for a polymer–fullerene junction and for photon absorption in the polymer. After photon absorption an exciton is generated, which then diffuses through the donor. Once it reaches the interface a charge transfer state is populated. This state is formed by a coulombically bound electron–hole pair with the hole residing in the donor and the electron already in the acceptor. The charge transfer state has an initial thermalization radius by which electron and hole are separated. This state can then split into free polarons once they escaped the Coulomb capture radius described by Onsager theory.

Onsager Model and Onsager–Braun Model. One of the most fundamental models to describe the separation of a charge transfer state into an electron and a hole was proposed by Onsager in 1938 [78]. This model treats a charge transfer state as a pair of electron and hole that are separated by a certain distance and paired due to the Coulomb force between them. The charge pair can then be split in the case the potential energy of the charge pair is smaller than the thermal energy $k_B T$.

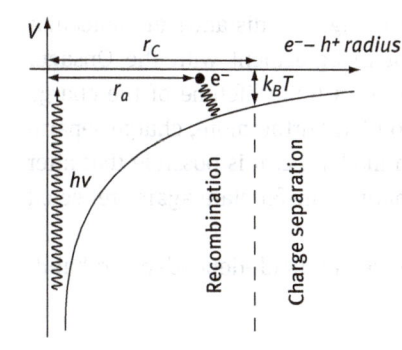

Fig. 3.22: Schematic of the potential landscape for an electron–hole pair in the Onsager model. Electron and hole are coulombically bound, i.e. there is an increasing attraction for a decreasing electron–hole radius. The thermal energy $k_B T$ determines a threshold radius r_C below which electron and hole eventually recombine instead of getting separated.

The fundamental idea of the Onsager model is depicted in Figure 3.22. An electron–hole pair is generated upon absorption of a photon and reaches a certain position on the potential landscape at an electron–hole radius r_a after initial thermalization. The Coulomb attraction between an electron and a hole in a medium with relative permittivity ε_r and thermal energy $k_B T$ defines a critical radius, the so-called *Onsager radius*

$$r_C = \frac{q^2}{4\pi\varepsilon_r\varepsilon_0 k_B T} \tag{3.17}$$

for which charges are considered free and not paired anymore. However, for $r_a < r_C$ there is the possibility of geminate recombination, i.e. electron and hole annihilate before they escape the attractive potential. Charge separation is then subject to a probability distribution P_{CS} that depends on the initial thermalization length r_a and the critical radius r_C. In the absence of an external electric field this probability distribution is of simple exponential form and is given as

$$P_{CS} = \exp\left(-\frac{r_C}{r_a}\right). \tag{3.18}$$

However, the presence of an electric field modifies this probability since it lowers the Coulomb potential in the direction of the field, making charge separation more likely. This effect was originally investigated by Tachiya in 1988 using Monte Carlo simulations [79]. For an electric field strength the probability for charge separation is given as

$$P(F) = \exp\left(-\frac{r_C}{r_a}\right)\left(1 + \frac{qr_C}{2k_B T}F\right), \tag{3.19}$$

which implies that there is a linear increase in charge separation probability with the external field.

The Onsager model was extended by Braun in 1984 in order to account for differences between typical thermalization lengths r_a predicted by the Onsager model (~3 nm) and typically observed electron–hole separation at interfaces, i.e. in charge transfer excitons. Since the latter is the nearest-neighbor distance at molecular interfaces, it is typically below 1 nm and is not in agreement with the Onsager model. This issue was solved by the introduction of a finite lifetime of the charge transfer state via a rate for geminate recombination k_{gr}. Furthermore, charge separation is a reversible process in the Onsager–Braun model; i.e. it is possible that after dissociation, the electron and hole populate a charge transfer state again, reflected in a charge transfer state regeneration rate k_r.

According to the Onsager–Braun model, the electric field–dependent probability for charge separation is given as

$$P(F) = \frac{k_d(F)}{k_{gr} + k_d(F)} = k_d(F)\tau(F), \tag{3.20}$$

where $\tau(F)$ is the lifetime of the charge transfer state. The dissociation rate $k_d(F)$ has a relatively complex form, including a Taylor expansion of a first-order Bessel function and reads

$$k_d(F) = \frac{3\langle\mu\rangle q}{4\pi\langle\varepsilon_r\rangle\varepsilon_0 r_a^3} \exp\left(-\frac{\Delta E}{k_B T}\right)\left(1 + b + \frac{b^2}{3} + \frac{b^3}{18} + \ldots\right), \tag{3.21}$$

where $\langle\mu\rangle$ is the spatially averaged sum of electron and hole mobilities, $\langle\varepsilon_r\rangle$ is the spatially averaged relative permittivity, ΔE is the Coulomb attraction between the initially formed electron–hole pair after thermalization, and $b = (q^3 E)/(8\pi\langle\varepsilon_r\rangle\varepsilon_0 k_B^2 T^2)$.

In contrast to the original Onsager model the thermalization radii found in the Onsager–Braun model are consistent with nearest-neighbor interactions as present, e.g. for charge transfer states between donor and acceptor molecules. It is also widely used in order to understand the impact of external electric fields on the efficiency of charge separation at functional interfaces. However, quantitative predictions based on the model are often not in very good agreement with experimentally found values. One of the most important reasons is that organic and hybrid systems especially often exhibit massive energetic disorder with energy state differences nearly exceeding $k_B T$ at room temperature, so that the effective Coulomb capture radius is reduced.

Electroluminescence of Charge Transfer States. For many material combinations, charge transfer states can be experimentally observed in steady-state spectroscopic experiments like photoluminescence or electroluminescence. This observation is possible if there is a non-negligible quantum efficiency for photon emission upon recombination of charge transfer states, i.e. in the case the electron–hole pair annihilates rather than separates into polarons. Electroluminescence spectra are obtained by applying a bias voltage to a thin film of organic material on a conductive substrate, which leads to charge injection into the material. In a pristine material this can result in photon emission if an electron in the LUMO of the material recombines with a hole in its HOMO since both carrier types travel through the same material.

However, in the case of mixed films with selective contacts, electrons are injected into the LUMO of the acceptor on one side of the device while holes are injected into the HOMO of the donor on the other side. These polarons then travel toward the interface and possibly populate a charge transfer state, which then can recombine radiatively.

Figure 3.23 illustrates the electroluminescence of a charge transfer state between a poly(p-phenylene vinylene) (PPV) donor and PCBM in comparison to the spectra of the pristine materials. Pristine samples of donor and acceptor show distinct emission spectra that overlap only at the spectral edges and exhibit peaks at around 600 nm (donor polymer) and 780 nm (acceptor fullerene). Interestingly, neither of these spectral features is observed in the mixed sample, which in this case is fabricated at a 1:4 weight ratio. However, there is prominent emission around 980 nm, which is further in the red, i.e. at lower energies than the emission features of both pristine materials. This emission is directly attributed to charge transfer states forming between the HOMO of the polymer and the LUMO of the fullerene, consistent with the observed lower energy compared to the optical bandgaps of the two pristine materials.

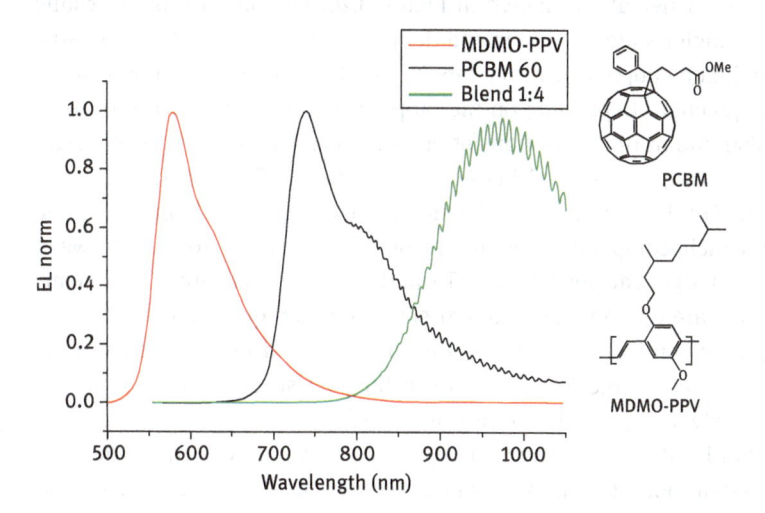

Fig. 3.23: Normalized electroluminescence spectra of a donor polymer, an acceptor fullerene, and the resulting charge transfer state when mixing the two materials. Since the charge transfer state is formed by the HOMO of the donor and the LUMO of the acceptor, the electroluminescence of the charge transfer state occurs at lower energy than the emission spectra of both donor and acceptor. Reprinted with permission from [80]. Copyright (2009) American Chemical Society.

The energetically lower position of the charge transfer state compared to excited states in both donor and acceptor also has important implications for the maximum achievable efficiency of organic solar cells. Typically, direct photon absorption by charge transfer states is relatively unlikely so that the effective optical bandgap of a donor–acceptor organic solar cell is given by the lower energy gap material rather than the energy of the charge transfer state. However, since charge separation is typically preceded by population of a charge transfer state of lower energy than the original excitation energy, there is an intrinsic energy loss that has to be taken into account. This energy loss also has to be considered for the Shockley–Queisser limit of organic solar cells [81].

3.3.1.3 Detailed Energetic Picture of Charge Transfer States

Having discussed the Onsager model and the fundamental principle of charge transfer states in organic donor–acceptor system, we will now have a closer look at the energetic situation. The exact nature of the positions of donor, acceptor, and charge transfer state energy levels, of course, is a function of the involved materials and their nanoscale molecular arrangement. Ongoing research in various labs is concerned with identification of charge transfer species. Quantification of involved energy and charge transfer rates has lead to a widely accepted generalized picture of the energy levels present at an organic–organic heterojunction.

A corresponding schematic is shown in Figure 3.24. Photon absorption results in excitation of a singlet state S_1, e.g. in the donor. As discussed earlier, a charge transfer state can then be populated with a rate k_{CTS} in the case the excitation reaches a donor–acceptor junction. Depending on the properties of the donor–acceptor junction, there is a distribution of charge transfer energy states with a lowest state CTS_0, often referred to as *cold*, and several higher excited states CTS_n (n = 1, 2, 3, . . .), often referred to as *hot*. Due to the delocalization of electron and hole in the charge transfer state and their comparably low interaction, intersystem crossing between singlet and triplet charge transfer states, ^1CTS and ^3CTS, respectively, are comparably probable compared to intersystem crossing in other quantum mechanical systems [83, 84]. A potential loss mechanism for organic solar cells is therefore the transition from a triplet charge transfer state to the lowest triplet state T_1 in the donor (or, analogously, acceptor) that occurs at a rate k_T.

More important for the operation of an organic solar cell, however, are the processes of charge separation at a rate k_{CS}, geminate recombination at a rate k_{GR}, and non-geminate charge carrier recombination k_{CR}. k_{CS} is strongly material-dependent. While for certain material systems charge separation is found to solely occur after the system goes through CTS_0, for many other systems it appears that there are charge separation events from higher energetic charge transfer states. Once the system is in one of many possible charge separated states (CS_0 to CS_n), it can readily fall back into the original charge transfer state and basically undergo oscillations. This is illustrated in the schematic with the two-directional arrow for charge separation. In the case charge separation is not completed within the lifetime of the charge transfer state, geminate recombination occurs and the system falls back into the ground state S_0, possibly accompanied by emission of a photon, even though this process in many instances is non-radiative. Furthermore, a state of completely separated charges can undergo non-geminate charge carrier recombination as indicated by the transition from a CS state directly to S_0. This recombination typically also goes through a charge transfer state but is commonly treated differently due to the different mechanism leading to charge recombination (k_{CR} is associated with recombination of charges that are not necessarily from the same generation event, whereas k_{GR} describes geminate recombination immediately after excitation).

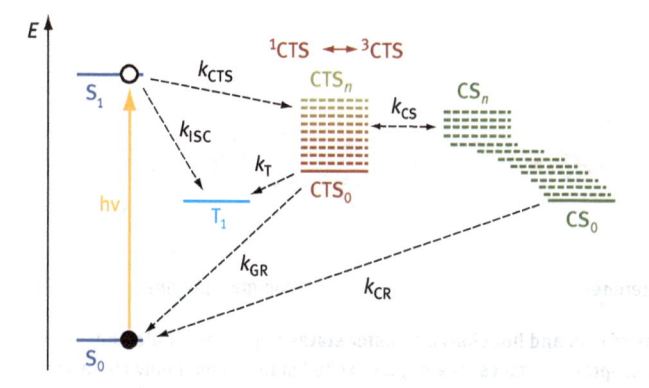

Fig. 3.24: **Energy-level diagram showing the fundamental processes occurring during charge separation in organic solar cells.** After excitation a charge transfer state can be populated, followed by separation of electron and hole into a charge separated state. Via intersystem crossing it is also possible that a singlet–triplet transition occurs. Depending on the donor–acceptor combination there is a certain energy distribution of charge transfer states and an activation energy is necessary in order to reach a charge separated state from the lowest charge transfer state. Adapted with permission from [82]. Copyright (2009) American Chemical Society.

In addition to the population of a charge transfer state upon photon absorption, there is a certain (low) rate k_{ISC} for intersystem crossing, upon which an excited triplet state (e.g. T_1) in the donor is formed. Typically, this state is energetically below all charge transfer states, so that charge separation is impossible once T_1 is reached. Therefore, the triplet exciton eventually undergoes another intersystem crossing and recombines back to the ground state S_0.

Pump–Push Experiments and Hot Charge Transfer States. After this schematic picture of different charge transfer states, we will now illustrate the difference between cold and hot charge transfer states in more detail and discuss experimental evidence for transitions between those. Figure 3.25 shows simulated charge densities of cold and hot charge transfer excitons in two different donor–acceptor systems, a polymer–fullerene combination and a polymer–polymer system. In both cases, the charge distribution found from electrostatic simulations indicate a localization of the hole portion of the charge transfer state on the donor while the electron sits on the acceptor. Furthermore, a higher degree of charge delocalization is found in the higher CTS_n compared to the lowest CTS_0, resulting in a weaker Coulomb interaction between electron and hole in the higher charge transfer state.

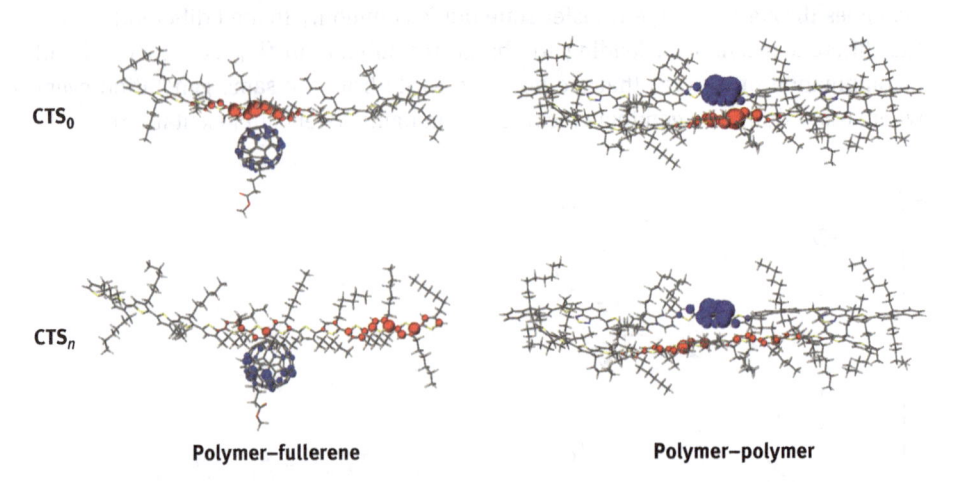

Fig. 3.25: Atomistic simulations of cold and hot charge transfer states at polymer–fullerene and polymer–polymer donor–acceptor interfaces. The higher excited state is commonly characterized by an enhanced delocalization of the charge transfer state wavefunction. Figure after [85].

As mentioned before, it has been found experimentally that a certain activation energy is necessary to overcome the binding energy of the lowest charge transfer state CTS_0, which has been interpreted as a higher probability for population of a charge separated state if the system comes from a hot charge transfer state. For different polymer–fullerene and polymer–polymer systems, this has been shown by Bakulin and coworkers in 2012 using a pump–push technique, where the system under investigation is pumped with visible light so that charge transfer states are generated and then hit with a second IR photon (the so-called *push photon*) of lower energy (lower than the energy gaps of both donor and acceptor). This second photon can push the CTS_0 into an energetically higher hot charge transfer state, from which charge separation is more likely, reflected in an increase in differential (additional) photocurrent upon IR pushing, which is measured during the pump–push experiment.

A schematic of the processes during a pump–push experiment is shown in Figure 3.26. Typically, samples are pumped (1) either via excitation across the optical energy gap of donor and acceptor or (2) directly into a charge transfer state when using photons of an energy smaller than the energy gaps of both donor and acceptor but equal to or higher than the energy gap between S_0 and CTS_0. For the push photon smaller energies are used, typically around 0.5 eV, which is below the energy required to lift the system from S_0 into a charge transfer state. However, the push energy can be absorbed by either lower charge transfer states that are then excited to hot states or directly by polarons. In order to distinguish the effects due to transitions between charge transfer states and absorption events by polarons, pump–push experiments are typically performed at varying time delays between pump and push. Due to the short lifetime of charge transfer states on the order of only a few 100 ps compared to µs-lifetimes of polarons, a dependency of the differential photocurrent on the time delay indicates an effect due to charge transfer states since polaron absorption would be virtually independent of the pump pulse.

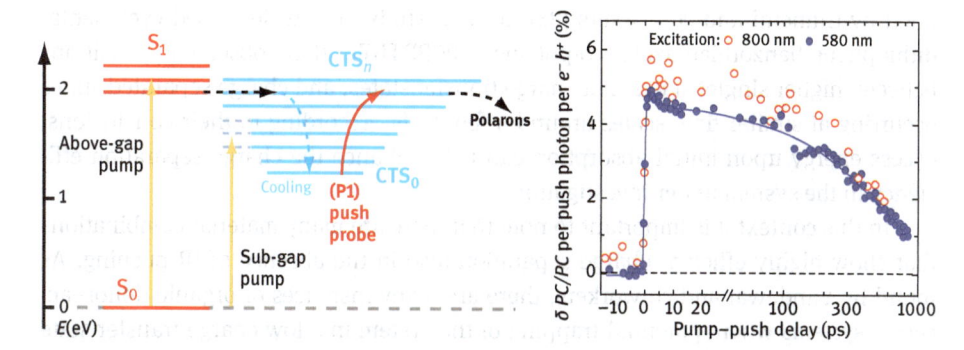

Fig. 3.26: Additional photocurrent in pump–push experiments as an indication for the presence of hot charge transfer states. The investigated system is excited with an fs visible light pulse and pushed with an IR pulse after a defined waiting times in the ps range (see left-hand-side schematic). The push photon lifts the excited charge transfer state from the lowest level to a hot state, leading to additional photocurrent for moderate waiting times (right-hand side). For longer waiting times, less and less charge transfer states are populated due to geminate recombination and the push photocurrent is reduced. Figure after [85].

Typical differential photocurrents found during a pump–push experiment are shown in Figure 3.26 for a combination of a polyphenylene vinylene (MDMO-PPV) as donor and a fullerene derivative ($PC_{70}BM$) as acceptor. The two curves shown correspond to pump wavelengths of 580 and 800 nm. Both curves show a strong dependency of the differential photocurrent on the pump–push delay, indicating that the push photon absorption by charge transfer states is responsible for the additional photocurrent. The decrease in differential photocurrent for delays exceeding 100 ps is directly associated with the lifetime of charge transfer states. For longer delays, geminate recombination comes into play and the push photon has a reduced probability of being absorbed by a charge transfer state. Direct comparison between the high- and low-energy pump shows that there is a slightly higher differential photocurrent for direct charge transfer excitation, indicating that the excess energy available when exciting the donor slightly enhances the efficiency of charge separation. More importantly, however, is the fact that the dynamics are the same in both cases. This indicates that the dynamics of transitions between charge transfer states (more precisely from higher to lower states) is independent of the initial energy and the system always tends to move toward the CTS_0 in the case charge separation does not occur. Furthermore, it is important to note that for many polymer systems the additional photocurrent per push photon is at most on the order of a few percent as in the present example and often even one or two orders of magnitude below. This means that the efficiency of charge separation is generally high in polymer–fullerene systems, and only a small number of states get trapped in low charge transfer states from where they can be freed with a push photon.

In 2013, the role of excess energy in the charge generation process was also outlined by Grancini and co-workers [86]. In their study on a widely used cyclopentadithiophene-benzothiadiazole co-polymer (PCPDTBT), they observed transitions between higher singlet states, hot charge transfer states, and charge separated states occurring at similar time scales around a few 10 fs. According to their conclusions, excess energy upon initial absorption can help enhance the charge separation efficiency in the system under investigation.

In this context it is important to note that there are many material combinations that show highly efficient charge separation also in the absence of IR pushing. As found by Vandewal and co-workers, there are many instances of organic donor–acceptor systems where potential trapping of the system in a low charge transfer state plays only a minimal role (if at all) for the photocurrent generation. This is apparent from bias-dependent charge extraction experiments at different photon energies above and below the optical energy gaps of donor and acceptor. Examples for two different polymers, a polyphenylene vinylene (MEH-PPV) and a thienopyrrolebenzodithiophene (PBDTTPD) co-polymer, in combination with a fullerene acceptor are shown in Figure 3.27. Monochromatic charge extraction data are shown in combination with current density–voltage curves measured under simulated solar illumination in order to illustrate that the bias dependency is virtually independent of the energy of the exciting photons. For both polymers the authors find the same

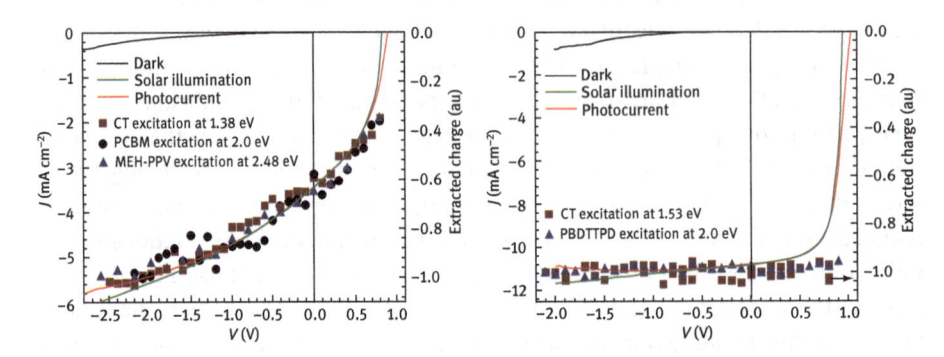

Fig. 3.27: Relative bias-dependent charge generation for two polymer–fullerene systems at different excitation energies. In both cases the energy of the exciting photon does not have a significant impact on the charge separation efficiency, indicating only a marginal additional contribution of hot charge transfer states to the photocurrent. Reprinted by permission from Macmillan Publishers Ltd: Nature Materials [87]. Copyright (2013).

behavior for (1) excitation of donor or acceptor and (2) direct excitation into the lowest charge transfer state. As is apparent from the results shown, charge separation is strongly field dependent in the case of the polyphenylene vinylene, but virtually field-independent for the co polymer. However, in the case of a given polymer the excitation energy does not seem to play a role in the field dependency.

Using transient absorption spectroscopy with time resolutions on the order of 10 fs, it was further possible to look at the charge generation process at times very early after excitation. In 2014, Falke and co-workers discussed oscillating features in transient absorption spectra of a donor–acceptor system with a polythiophene donor (P3HT) and a fullerene acceptor [88]. These oscillations occur at a periodicity of around 25 fs and are attributed to coherent vibrational motion of donor and acceptor. In agreement with first principle simulations, they describe a delocalization of charge between donor and acceptor due to the coherent vibronic coupling between electronic and nuclear degrees of freedom. As shown in Figure 3.28, there is rapid

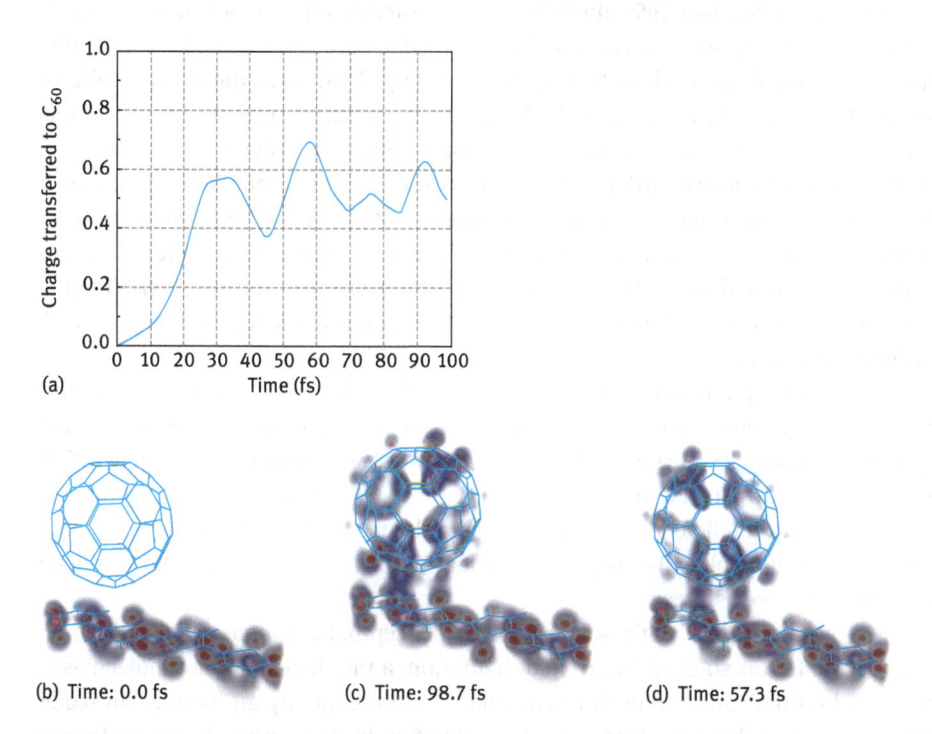

Fig. 3.28: Simulations of charge transfer dynamics between a short conjugated donor oligomer and a fullerene acceptor. (a) Due to coherent vibrations of donor and acceptor, there is an oscillating charge transfer probability to the acceptor. **(b)–(d)** Snapshots of the simulated time evolution of the charge density show how the initial state, where all charge is localized on the donor, moves toward a situation where the charge is shared between donor and acceptor. After [88].

transfer of a large portion of the electron from the donor to the acceptor during the first 25 fs. After that, the amount of transferred charges oscillates coherently with the vibrational motions of the involved molecules, in this case the carbon–carbon double-bond stretching mode in the polymer at $1,450\,\text{cm}^{-1}$ and the pentagonal pinch mode of the fullerene at $1,470\,\text{cm}^{-1}$.

3.3.2 Flat Heterojunctions

Even though the exact mechanism of charge separation at organic–organic junctions is not completely understood yet, all efficient organic photovoltaic systems to date rely on the combination of two materials, a donor and an acceptor. This is the case since only the junction between those two provides sufficient driving force for cracking of typically strongly bound excitons that are generated upon photon absorption.

Already in the late 1950 there had been research on a photovoltaic effect in devices with magnesium phthalocyanine dyes, which are metal-organic–conjugated molecules that form semiconducting crystals [89]. Soon after the conductivity of polyacetylene had been described by Heeger et al., research further started to focus on the photovoltaic implementation of semiconducting polymers. Interestingly, there is already a review article on organic solar cells by Chamberlain [4], in which study the efficiencies hardly reached 1%. However, it took until 1986 before the community fully acknowledged that the initially fabricated metal–semiconductor–metal Schottky-junction devices were performing so poorly due to their very limited quantum efficiency for charge separation, which is the main conceptual drawback of such a geometry.

In 1986, Tang introduced the first organic solar cell with a flat donor–acceptor heterojunction, which was showing impressive current densities and high external quantum efficiencies owing to the now much more efficient charge generation process. As shown in Figure 3.29, Tang was able to achieve high fill factors and power conversion efficiencies exceeding 1% under simulated AM 2 solar light using the very common blue dye copper phthalocyanine as donor and a red perylene diimide derivative as acceptor.

The geometry of Tang's solar cell was conceptually simple as illustrated in Figure 3.30. In this so-called *bilayer* configuration, a thin layer of donor molecules is deposited onto a transparent front electrode, most commonly an indium–tin oxide film on a glass substrate. Most commonly, this first layer is either deposited from a solution via spin casting or blade coating, or realized using thermal evaporation of the donor molecules in a vacuum system. On top of this donor layer a thin film of acceptor molecules is evaporated, resulting in a layered structure with a well-defined planar interface. The solar cell is finalized by thermal evaporation of a metallic top electrode. There, a metal with a low work function like aluminum or calcium is necessary in order to allow ohmic extraction of electrons from the LUMO of the acceptor.

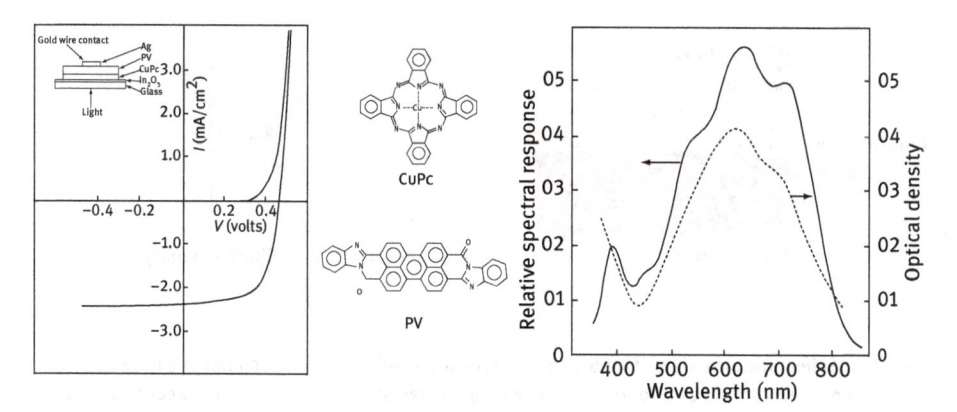

Fig. 3.29: Current density–voltage curve and external quantum efficiency spectra of the first organic donor–acceptor solar cells. The concept was introduced by Tang in 1986. Data are shown along with the chemical structures of the used donor and acceptor molecules. Reprinted with permission from [6]. Copyright (1986), AIP Publishing LLC.

Upon illumination of such a bilayer solar cell, excitons are generated throughout the whole active film (donor + acceptor). However, due to the typically low exciton diffusion lengths of organic semiconductors, which are on the order of only a few nanometers, only a small region in close proximity to the donor–acceptor interface can contribute to the photocurrent generation of the device. For exciton diffusion lengths $L_{D,D}$ and $L_{D,A}$ in the donor and the acceptor, respectively, this results in the contributing region, indicated in Figure 3.30. Excitons that are generated outside this region cannot diffuse far enough to reach the heterojunction and recombine, i.e. they are lost for charge generation. The figure further illustrates the charge generation profile of a bilayer device and the vertical charge density in the donor and acceptor film for the cell being operated at V_{OC} and J_{SC}. Charge separation occurs only in a very small region around the interface, depending on the delocalization of charge transfer states. In theoretical models, organic bilayers are therefore often described as a non-photoactive heterojunction with a very thin center layer that is solely responsible for charge generation [90]. The charge density is also dominated by the abrupt interface between donor and acceptor, since electrons are always injected into the upper part of the device and holes into the lower part.

The main limitation of simple bilayer devices is insufficient light harvesting; i.e. only a fraction of the incident photons can be converted to charge carriers. Even though the absorption depth of organic semiconductors can be on the order of only a few 100 nm due to their outstandingly high extinction coefficients, the contributing region is too small to absorb a large portion of the incident light, which conceptually limits the photocurrent in bilayer devices. We will see in the following sections how other concepts for donor–acceptor structures allow for the fabrication of more efficient organic photovoltaics.

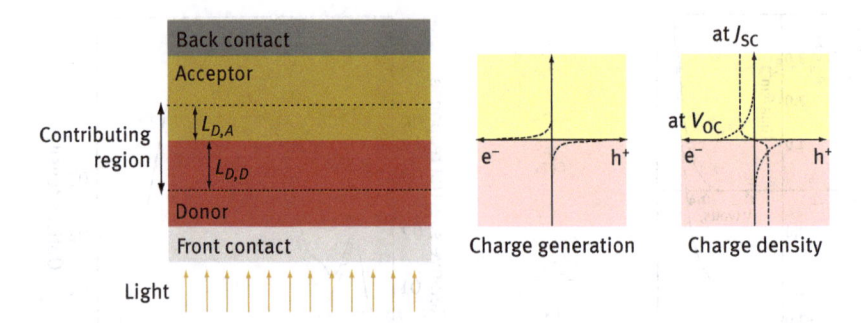

Fig. 3.30: Schematic side view of an organic bilayer solar cell. In the most common configuration a transparent front contact electrode is coated with a layered structure of donor and acceptor. Charge generation is possible only upon photon absorption in a region that has the size of the respective exciton diffusion lengths in the two materials. Charge generation occurs solely at the donor–acceptor junction, resulting in a charge profile in the device with a spontaneous change in sign at the interface.

In contrast to the limited photocurrent generation, bilayer devices exhibit excellent electronic properties, reflected in typically high FF and V_{OC}. Due to the well-defined geometry of the active layer, there are direct transport pathways for both electrons and holes from the charge-separating interface toward the external electrodes. Furthermore, the layered structure completely forbids electronic contact of both the donor with the electron-collecting electrode and the acceptor with the hole-collecting electrode. These two factors lead to reduced charge carrier recombination losses (non-geminate) compared to other active layer morphologies.

3.3.3 Bulk Heterojunctions

Different approaches are conceivable in order to address the issue of insufficient photocurrent generation in organic bilayer devices. One pathway is the design of materials with significantly increased exciton diffusion lengths, which would result in a thicker contributing region. Single materials with impressively high exciton diffusion lengths have been reported, as for example rubrene with $L_D = 1$ μm, as discussed in Section 2.2.2.4, or large single crystals of polythiophene [91]. However, when processed on a scale suitable for the fabrication of organic solar cells the crystallinity of organic semiconductors typically cannot be completely controlled, resulting in the presence of defect states that reduce the exciton diffusion length to a few nanometers. Such defects are present in particular when using high-throughput solution-based deposition methods like printing or blade coating, which, however, are the most interesting in terms of commercialization of organic photovoltaics.

The more commonly pursued strategy was originally introduced in 1995, separately by Yu et al. and Halls et al. [13, 14]. For the first time donor and acceptor were successfully processed together to form a mixed film rather than a layered structure. Such mixed films are commonly referred to as *bulk heterojunctions* since donor–acceptor heterojunctions are present throughout the whole film rather than only at one defined interface. Bulk heterojunctions are commonly realized either by co-evaporation of donor and acceptor materials or by deposition of a thin mixed film from a blended solution that contains donor and acceptor molecules at a defined ratio. The phase separation between donor and acceptor in the bulk heterojunction can be partly controlled via processing parameters, such as the evaporation rate or the solution concentration, the substrate temperature, potential post-treatments like thermal or solvent annealings, solvent additives, and most importantly material's properties like their tendency to crystallize or their hydrophobicity. However, direct control down to the nanometer scale is not possible so that the donor–acceptor morphology can be only indirectly adjusted and experimentally optimized.

We will now have a closer look at the fundamental processes that play a role in the operation of a bulk heterojunction solar cell. A schematic side view of the active film of such a device is shown in Figure 3.31. Typical bulk heterojunction films have a thickness on the order of 100 nm, i.e. a factor of 10 or more higher than the typical exciton diffusion lengths L_D. Due to the intimate mixing of donor and acceptor that ideally results in domains of diameters smaller than L_D, exciton dissociation is nevertheless highly efficient since a large fraction of excitons encounters a donor–acceptor heterojunction during their lifetime. The function of a bulk heterojunction is crucially dependent on the domain size and shape. While small diameters guarantee efficient charge generation, domains have to be anisotropically stretched in vertical direction in order to allow for charge transport throughout the whole active film toward the charge collecting front and back electrodes. Ideally, donor and acceptor form interpenetrating networks with a phase separation on the nanometer-length scale so that excitons always reach a heterojunction before they recombine while charges can be efficiently transported.

A successful charge generation and collection event in a bulk heterojunction is schematically depicted in the case (1) of Figure 3.31. After electron and hole are generated at a donor–acceptor heterojunction, they find consistent pathways to the external electrodes and thus contribute to the photocurrent of the device. In the case of inconsistent pathways, case (2), islands of either donor or acceptor function as charge traps, which eventually leads to non-geminate recombination of a free carrier with one of the trapped carriers. Additional charge recombination can be caused if either the donor, which transports holes, is in direct contact with the electron-collecting electrode, case (3), or the acceptor, which transports electrons, touches the hole-collecting electrode, case (4). This issue is commonly addressed by implementation of electron- or hole-selective interfacial layers at the external contacts.

In many instances the phase separation between donor and acceptor is not as clear as indicated in the schematic, especially in the very common case of polymeric donors in combination with fullerene acceptors. Many materials employed in organic solar cells show a more or less pronounced tendency to crystallize, which leads to the presence of amorphous and crystalline regions and more importantly to mixed phases of donor and acceptor where the materials are intimately mixed on a molecular level, as indicated in the zoom in Figure 3.31. The different phases and their structures (amorphous, aggregated) also play an important role for the energetic disorder in the film, which has a strong influence on the exciton diffusion, separation, and recombination of such mixed donor–acceptor films (see Section 2.2.2.4 and Section 3.3.4.2).

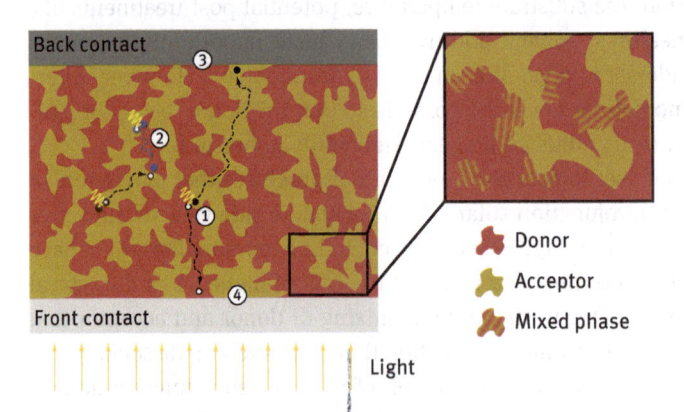

Fig. 3.31: Schematic side view of an organic bulk heterojunction active film. Donor and acceptor molecules are processed in a mixed film containing domains of pure donor, pure acceptor, and possible mixed phases at the interfaces. The internal morphology of bulk heterojunctions can be only indirectly controlled via processing parameters. (1) Ideally, electron and hole separate at a location from where direct charge percolation toward the respective electrodes is possible. (2) However, charge trapping can occur at islands, resulting in charge carrier recombination. (3), (4) The probability of charge carrier recombination is further enhanced at spots where donor and acceptor are in contact with the electron- and hole-collecting electrodes, respectively.

As we have seen now, the efficient operation of a bulk heterojunction depends on the different processes of exciton diffusion, charge separation, and charge transport that all depend on the material's properties, on the one hand, and the internal morphology, on the other hand. This is often summarized in writing the internal quantum efficiency η_{IQE} of a bulk heterojunction solar cell as

$$\eta_{IQE} = \eta_{ED} \cdot \eta_{CS} \cdot \eta_{CC}, \tag{3.22}$$

where η_{ED} is the efficiency at which excitons diffuse through the material and reach a donor–acceptor heterojunction within their lifetime, η_{CS} is the efficiency of charge separation once an exciton reaches the heterojunction, and η_{CC} is the efficiency for charge collection at the external electrodes, which is reduced in the case there is non-geminate recombination in the active layer. As we have seen in Section 3.3.1, η_{CS} is a function of the used materials and depends on the exact energetic landscape for charge transfer states and the transition rates between quantum states at the heterojunction. η_{CC} is sometimes written as a product of two separate efficiencies, η_{CT} and η_{coll}, that denote charge transport toward the external electrodes and charge extraction out of the active layer, respectively.

With this it is further possible to write down the overall quantum efficiency of the bulk heterojunction device, which is the product of the light absorption efficiency η_{abs} and η_{IQE}

$$\eta_{EQE} = \eta_{abs} \cdot \eta_{IQE} = \eta_{abs} \cdot \eta_{ED} \cdot \eta_{CS} \cdot \eta_{CC}. \tag{3.23}$$

$\eta_{IQE} = \eta_{IQE}(V)$ is a function of the external bias V of the solar cell since charge generation and, more importantly, charge collection vary in the presence of an external electric field. Furthermore, as discussed in Section 3.3.1 there are indications that charge separation is more efficient in the presence of excess energy in certain material systems, rendering η_{CS} also dependent on the wavelength of the incident photons. An additional and more obvious wavelength dependency is present in the case of η_{abs}, which is given by the absorption spectrum of the active film. Overall, $\eta_{EQE} = \eta_{EQE}(V, \lambda)$ is therefore a complicated function of both external field and wavelength of incident light, and understanding its individual components is the focus of ongoing research in the community.

3.3.4 Mixing of Polymers and Fullerenes

3.3.4.1 Pure Materials

As mentioned Section 3.3.3 the morphology of a bulk heterojunction and the electronic properties and function of a solar cell depend strongly on the choice of donor and acceptor materials. Since the introduction of the bulk heterojunction concept, a virtually unmanageable number of materials has been investigated in order to simultaneously meet the requirements of beneficial donor– acceptor phase separation, efficient exciton splitting, and loss-free charge transport. There is also a number of excellent review articles solely concerned with the impact of the bulk heterojunction morphology and organic solar cells in general, to which the reader is referred for details on specific materials.

Here we want to focus on the aspects of donor and acceptor crystallinity and the implications for mixed films that are of more general nature. We will therefore restrict our discussion to very widely used materials that are considered as model systems in organic solar cell research. In the following, first we will discuss the crystallization of P3HT, poly(3-hexylthiophene-2,5-diyl), for pure polymer films, before we discuss the crystallization of fullerenes. Chemical structures of P3HT and $PC_{60}BM$, the fullerene derivative [6,6]-phenyl-C_{61}-butyric acid methyl ester, which is the most commonly used organic solar cell acceptor, are shown in Figure 3.32. These discussions will serve as the basis for our analysis of donor–acceptor mixed films in the following section, where we will also have a look at bulk heterojunctions of fullerenes with the polymer PTB7, poly({4,8-bis[(2-ethylhexyl)oxy]benzo[1,2-b:4,5-b']dithiophene-2,6-diyl}{3-fluoro-2-[(2-ethylhexyl)carbonyl]thieno[3,4-b]thiophenediyl}), a high-efficiency donor polymer that has been intensively investigated since its introduction in 2010.

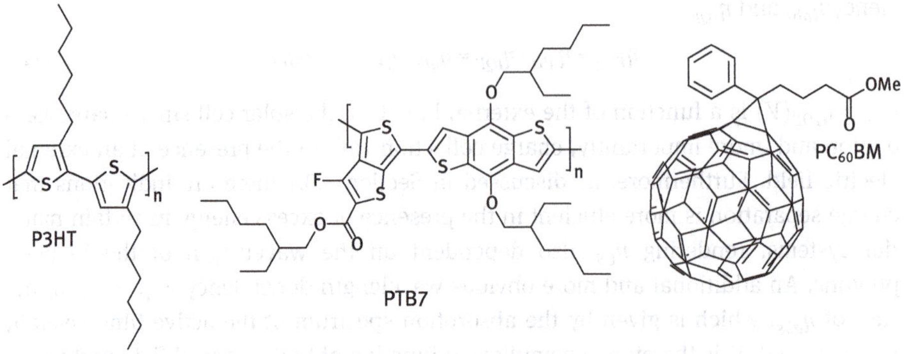

Fig. 3.32: Chemical structures of P3HT, PTB7 and PC$_{60}$BM. P3HT, poly(3-hexylthiophene-2,5-diyl), is one of the most commonly used donor polymers, PTB7 (poly({4,8-bis[(2-ethylhexyl) oxy]benzo[1,2-b: 4,5-b'] dithiophene-2,6-diyl}{3-fluoro-2-[(2-ethylhexyl)carbonyl] thieno[3,4-b]thiophenediyl})) a high-efficiency donor polymer, while PC$_{60}$BM, [6,6]-phenyl-C$_{61}$-butyric acid methyl ester, is a fullerene derivative that has proven to be highly efficient as an electron acceptor and is easy to process from various solvents, which makes it the standard acceptor material in organic photovoltaic research.

Crystallization of Poly(3-hexylthiophene-2,5-diyl) (P3HT). After its introduction into organic solar cell research in 2002, P3HT quickly became the benchmark polymer for polymer-PCBM bulk heterojunction devices due to the high efficiencies approaching 5% under simulated solar light. Furthermore, the strong tendency of P3HT to crystallize let it to play a key role in the morphology formation of a P3HT-PCBM bulk heterojunction. Hence, the bulk heterojunction morphology can be controlled via the P3HT processing parameters.

An indication for the dominant role P3HT plays in the formation of bulk heterojunction morphologies is illustrated in Figure 3.33. If processed from an appropriate solvent, P3HT tends to form thin films of three-dimensional network structures on substrates with domain sizes on the order of and below 100 nm. As discussed in Section 3.3.3 such structures are well-suited for bulk heterojunction since they allow sufficient donor–acceptor interface for exciton separation, but consistent pathways for charge transport. Even though the final bulk heterojunction morphology is also strongly influenced by the properties of the acceptor material, it is believed that similar three-dimensional P3HT networks form from blend solutions and are one of the reasons for the good performance of P3HT-based organic photovoltaics.

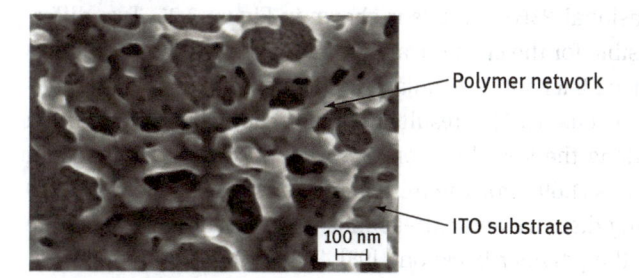

Fig. 3.33: Scanning electron microscopy top view of a P3HT film forming a network structure on a conducting glass substrate. The film was processed from a dilute solution in a low-boiling-point solvent, leading to rapid film formation during spin coating. Figure adapted from [92] with friendly permission by Wiley and Sons (2008).

By far, the most important parameter that determines the degree of crystallization of P3HT is the so-called *regioregularity*, a measure that tells how many side chains are attached to the thiophene backbone of the polymer in the same way. The case where the regioregularity is very low is often referred to as *regiorandom*. Chemical structures of short parts of regioregular (rr-P3HT) and regiorandom (ra-P3HT) polymers are shown in Figure 3.34. In the case of rr-P3HT, all side chains are attached to the same C atom in the respective thiophene, so that a well-ordered arrangement of the side chains is possible along the extended backbone. In contrast, for ra-P3HT the distance between the two side chains is not constant for all monomer repeat units. Since the crystallization of P3HT is to a large extent driven by the interaction between alkyl side chains, ra-P3HT forms amorphous films and rr-P3HT tends to form three-dimensional crystals if proper film formation parameters are used. For most applications in excitonic solar cells, only rr-P3HT is of interest, so we will exclude ra-P3HT from our discussion and from here on refer rr-P3HT to simply as *P3HT*.

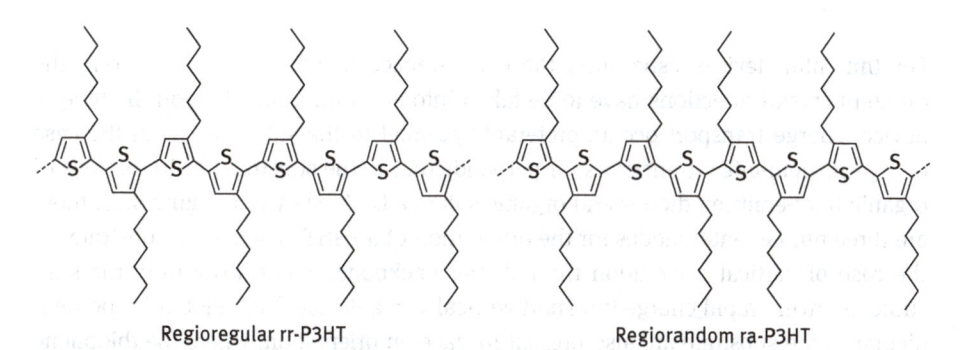

Fig. 3.34: Regioregular and regiorandom P3HT. In regioregular P3HT every thiophene carries the hexyl side group at the same C atom, while in regiorandom P3HT the side chains are randomly positioned.

A schematic of three-dimensional P3HT crystals is shown in Figure 3.35. Two different interactions are responsible for the crystallization of P3HT, namely (1) π–π interactions between the thiophene backbones of adjacent polymer strands and (2) interactions between the alkyl side chains. This results in the formation of crystals with a shorter stacking distance along the π–π direction (3.8 Å) and a longer chain–chain distance across the side chains (1.69 nm). The third characteristic distance (0.38 nm) is similar to the π–π stacking distance and corresponds to the distance between the two thiophene rings along the polymer backbone [93]. This composition of a P3HT crystal has important implications for the transport of charges through the material. While hole mobilities are high (on the order of 0.1 $cm^2V^{-1}s^{-1}$) along the polymer backbone and in the direction of the π–π stacking due to the close proximity of neighboring thiophenes, the mobility is significantly lower across the alkyl side chains (around $1 \cdot 10^{-4}$ $cm^2V^{-1}s^{-1}$).

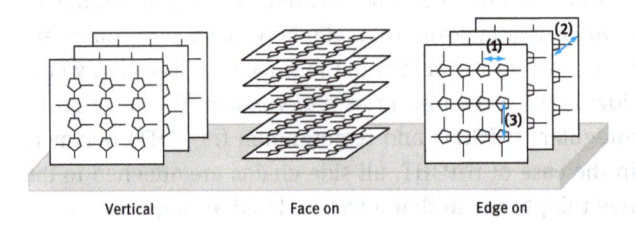

Fig. 3.35: Schematic drawing of crystallized P3HT and the possible crystal orientations on a substrate. P3HT aggregates into three-dimensional crystals, where sheets of parallel polymer chains that interact via their alkyl side chains form stacks due to interactions between the thiophene rings. Fundamentally, there are three possible orientations for such a crystallite on a substrate, namely, **vertical**, where the polymer chains are oriented perpendicular to the substrate; **face on**, with the π–π stacking direction perpendicular to the substrate; and **edge on**, with the alkyl chains pointing toward the substrate. Charge mobility in such crystals is high **(1)** along the polymer backbone and **(2)** along the π–π stacking direction, but low **(3)** across the alkyl side chains.

For thin film devices especially, these differences in charge mobility along the different crystal directions have to be taken into account. Depending on the type of device, charge transport occurs preferably parallel to the substrate, as in the case of organic field-effect transistors, or perpendicular to the substrate, as in the case of organic light-emitting diodes and organic solar cells. As shown in Figure 3.35, there are three fundamental modes for the orientation of a P3HT crystal on a substrate. In the case of vertical orientation the polymer backbones point away from the substrate, allowing rapid charge transport vertically in a device. High mobilities perpendicular to the substrate are also present for face-on orientation, where the thiophene rings align themselves flat on the substrate. The least favorable orientation for organic solar cells, where charges have to be collected at the front and back electrodes, is the edge-on orientation, since mobilities are only high parallel to the substrate, while vertical charge transport relies on charge hopping across the alkyl side chains of P3HT.

In 2005, Coakley et al. were able to show that controlled reorientation of P3HT molecules from the edge on to the vertical configuration leads to significantly enhanced macroscopic charge mobilities perpendicular to a substrate (Figure 3.36). By infiltration of P3HT into a nanoporous alumina membrane, a net vertical orientation of P3HT was achieved, compared to the edge-on case on a flat substrate. Charge mobility was then extracted from dark diode curves obtained after evaporation of metallic top contacts. The striking result is that the device current is higher for the sample with alumina membrane, even though the membrane itself is insulating and thus the cross section for charge transport of the P3HT is reduced compared to the flat substrate. This higher current is directly attributed to the higher hole mobility along the polymer backbone.

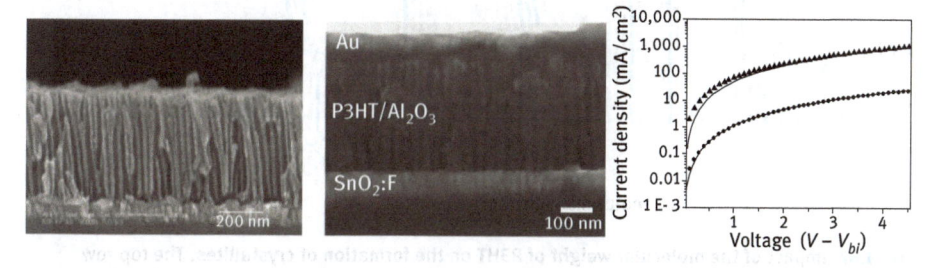

Fig. 3.36: **Impact of P3HT crystallinity on charge mobility found in an experiment where the polymer is confined in a nanoporous structure.** The left image shows a scanning electron microscopy cross section of a porous Al_2O_3 membrane. This membrane is infiltrated with P3HT (middle image), resulting in crystallization of the P3HT vertically with respect to the substrate. The plot on the right hand side shows how the polymer in the nanostructure exhibits a higher conductivity compared to a compact film, even though the active area is reduce by the Al_2O_3 membrane. The higher conductivity is directly related to the higher mobility in vertical direction. Figure adapted from [94] with friendly permission by Wiley and Sons (2005).

Besides the orientation of P3HT crystallites with respect to a substrate, there are also differences in the degree and type of crystallization, depending on the molecular weight, i.e. the length of polymer strands [95, 96]. Figure 3.37 illustrates for P3HT how increasing molecular weight influence the crystallization behaviour after annealing above the melting temperature. For lower molecular weight the polymer strands tend to form crystals where they are fully extended such that very high average crystallinities can be obtained. As the molecular weight increases, the polymer chains start to fold back on themselves or extend into adjacent crystallites. This is commonly accompanied by the formation of amorphous P3HT regions, separating crystalline domains, especially in the case of high-molecular-weight P3HT, where entanglement of chains is present in the melt. Thus, the molecular weight, in combination with the processing conditions, determines the size and distribution of crystallites in P3HT films.

Direct evidence for the anisotropic charge mobility of P3HT on a macroscopic level in films with crystalline and amorphous regions has been provided by Crossland

and co-workers [97]. P3HT can be grown in large spherulites that are separated by amorphous domains, as illustrated in Figure 3.38. Charge transport is thus fast along the stacking direction but slow across amorphous polymer regions between lamellar crystallites within one spherulite and at the spherulite boundaries.

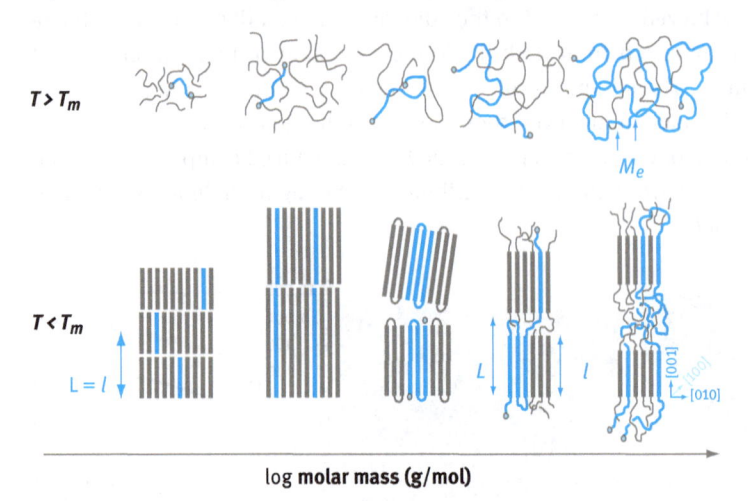

Fig. 3.37: Impact of the molecular weight of P3HT on the formation of crystallites. The top row shows the system above its melting temperature, with a single polymer strand indicated in blue. For high-molecular-weight P3HT entanglements (M_e) are present. Depending on the molecular weight, different crystallites form with either fully extended chains, where the spacing between crystallites L equals the crystallite size l, or with chains folding back on themselves or even extending into adjacent crystallites, that are separated by amorphous regions. Figure after [95].

By contacting such spherulites with a point grid of metallic contacts, an angle-dependence of the charge mobility is found, as shown in Figure 3.39. Macroscopically, the charge mobility is higher in the direction of the π–π stacking than perpendicular to it due to the longer extension of P3HT crystallites in this direction.

It is interesting to note that the optoelectronic properties of a polymer aggregate further depend on the type of interaction of the chromophores. While interaction of monomers along the polymer chain, i.e. across chemical bonds, is that of a J-aggregate, chain–chain interactions result in H-aggregate behavior (compare Section 2.2.2.3). Accordingly, the crystallinity of a conjugated polymer film can be readily analyzed via the absorption and emission spectra. Frank Spano developed a combined *HJ-aggregate model* based on a Franck–Condon progression, which allows us to predict the shape of these spectra based on the interaction between chromophores (intrachain and interchain) [99]. As shown in Figure 3.40 for simulations of P3HT, increasing interchain interaction enhances the H-aggregate behavior of the material, reflected in a more and more forbidden 0–0 transition. Experimentally, it is therefore common to normalize absorption spectra to the 0–1 peak intensity such that by comparing the 0–0 and the 0–1 intensities, we can directly assess how prominent H-aggregates are in the material

3.3 Organic Junction Solar Cells — 141

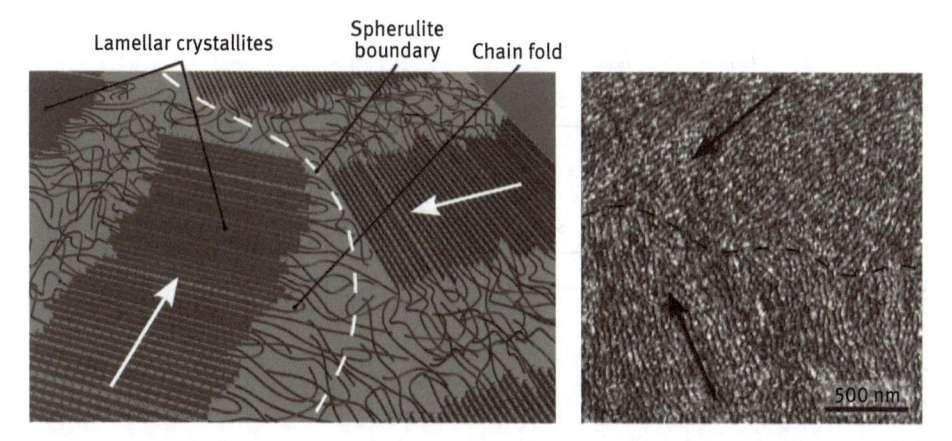

Fig. 3.38: Visualization of crystalline and amorphous domains in a thin P3HT film along with an experimental electron micrograph. Due to the strong π–π interactions between the thiophene backbones, P3HT form lamellae, which are separated by amorphous regions formed by dangling polymer strands. Figure adapted from [97] with friendly permission by Wiley and Sons (2012).

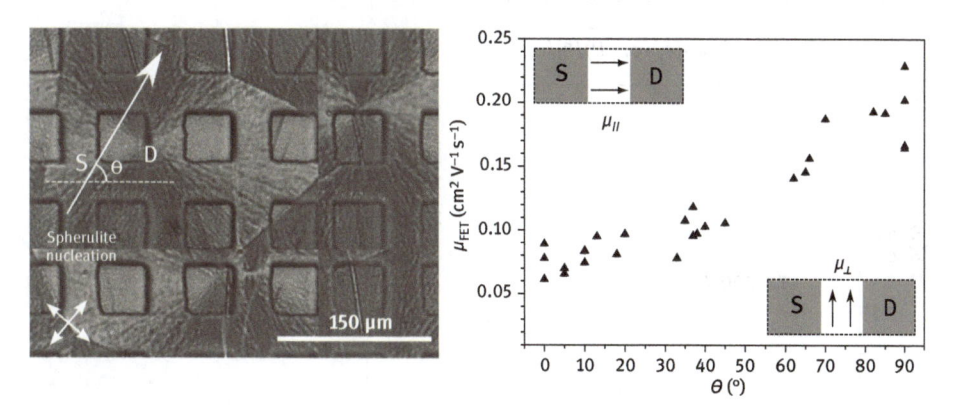

Fig. 3.39: Hole mobility measurements on spherulitic P3HT crystallites. Large spherulitic crystallites are contacted via evaporation of a grid of metal squares in order to measure the hole mobility in different directions. The observed mobilities depend on the angle with respect to the crystal growth direction, where transport is faster along the π–π stacking direction than along the polymer backbones. Figure adapted from [97] with friendly permission by Wiley and Sons (2012).

Crystallization of Fullerenes. By far the most common electron acceptors in organic solar cells are derivatives of the fullerenes C_{60} and C_{70}, with the latter being used especially in combination with small bandgap donors due to its higher light extinction in the visible range. As mentioned before, fullerenes have a strong tendency to crystallize, which in many polymer–fullerene combinations is the key driving force for the formation of the bulk heterojunction morphology [100].

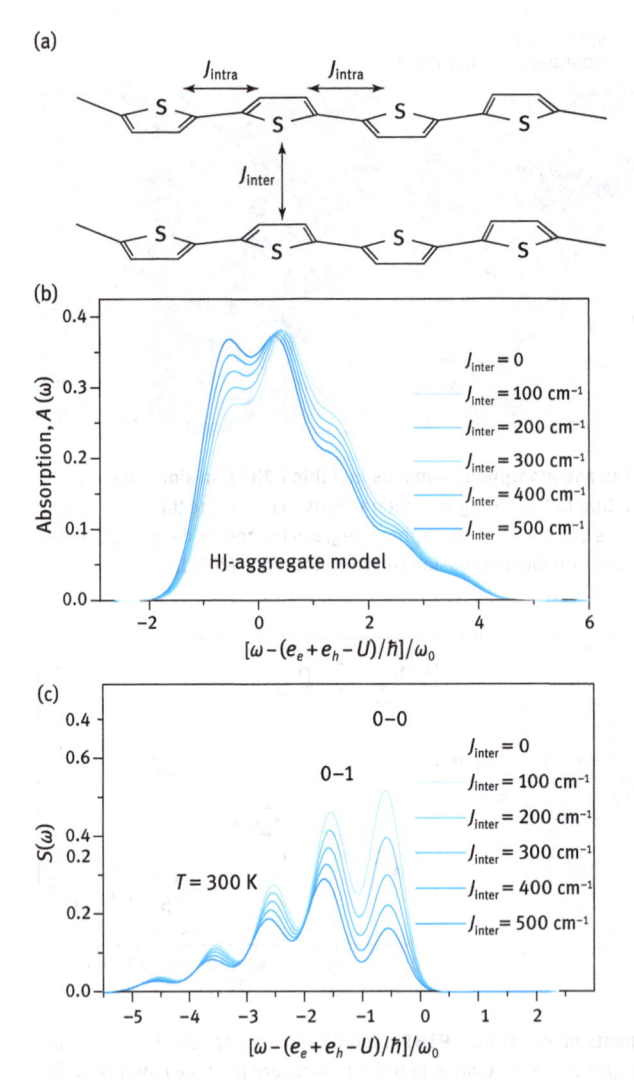

Fig. 3.40: Impact of inter- and intramolecular interactions between P3HT chains. (a) Schematic of two neighboring polymer chains. **(b)** Simulated absorption spectra of P3HT for different intermolecular interaction energies. **(c)** Corresponding simulated emission spectra. Figure after [98].

Due to van der Waals interactions, C_{60} forms molecular crystals with a face-centered cubic (fcc) crystal structure, as shown in Figure 3.41. Below a critical temperature of around 260 K, the C_{60} molecules cannot rotate freely. Instead, the symmetry axes of the molecule have certain fixed relative orientations with respect to each other. These axes are the fivefold and threefold symmetry axes through two opposing pentagonal and hexagonal faces, respectively. Above this critical temperature these specific orientations are lost since the fullerenes rotate freely, leading to a situation where the C_{60} molecules are effectively equivalent [101].

An fcc crystal of fullerenes is further characterized by large holes between the packed fullerenes, with radii of 3.06 Å for the four octahedral interstitials and 1.12 Å for the eight tetrahedral interstitials per unit cell. These holes allow the introduction of intercalated dopant atoms that alter the electronic properties but leave the crystal structure virtually unchanged.

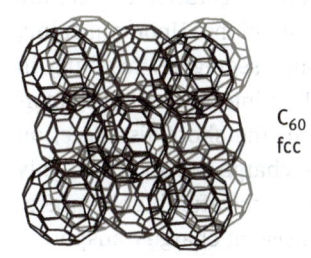

Fig. 3.41: Crystal structure of C_{60}. Fullerenes have a strong tendency to crystallize in an fcc motif with a lattice constant of 1.415 nm for temperature above 260 K, where the fullerenes rotate almost freely so that their symmetry does not play a role.

As shown by Giudice et al. [102], C_{60} crystallization can be observed not only indirectly, as for instance, in X-ray diffraction experiments, but also directly in a scanning tunneling microscope. Crystal domains are typically rotated by 90° with respect to each other and extend over large areas, while being built up of layers of closely packed fullerenes.

In principle, crystallization of fullerene derivatives is driven in the same manner as for pure C_{60}. Thus, a strong tendency to crystallize is also present for the two most commonly used fullerene derivatives $PC_{60}BM$ and the larger $PC_{70}BM$, which has the same side groups but is a C_{70} fullerene. However, crystallization of fullerene derivatives is also a function of the moiety attached to the fullerene, and bulky side groups can significantly hinder crystallization.

3.3.4.2 Mixed Films

In Section 3.3.3, we saw the important role the donor–acceptor morphology played in the efficient operation of an organic solar cell. The morphology of the active film determines how efficiently excitons can diffuse to charge-separating interfaces and how separated charges can migrate toward the external electrodes without being lost to charge carrier recombination. However, in the most common approach to synthesize polymer–fullerene bulk heterojunctions, which is processing from a blended solution, the donor–acceptor morphology can be controlled only indirectly and partly, for instance via the choice of solvent, the deposition process, the wettability of the substrate, and the concentration of the materials in the solution. In addition, different post-processing treatments are possible, most prominently, thermal annealing at temperatures that allow partial reorganization in the active layer.

We will now have a closer look at two examples, where the mixing of donor and acceptor is influenced by such a thermal annealing. Figure 3.42 shows an example published by Peumans and co-workers in 2003, where demixing of donor and acceptor was observed during thermal annealing of films with top electrodes, so that the material is confined and the surface roughness of the film is not significantly changed during annealing [103]. In this study a mixed film of copper phthalocyanine as donor and a perylene derivative as acceptor was used, which was deposited via thermal co-evaporation. The as-prepared film is characterized by a fine mixing of donor and acceptor molecules, as indicated by the virtually featureless scanning electron micrograph, and shown in three-dimensional simulations. Upon annealing at increasing temperatures, donor and acceptor molecules segregate, resulting in increasing donor and acceptor domain size. While the fine intermixed initial morphology allows highly efficient exciton separation due to the short distances, an exciton has to travel to reach a donor–acceptor interface, charge transport is highly limited by charge carrier recombination. The opposite is the case after high-temperature annealing, where coarse phase separation allows consistent charge transport to the external electrodes, but exciton recombination is a severe loss mechanism. Consequently, and consistent with the considerations in the context of the working mechanism of a bulk heterojunction in Section 3.3.3, the optimum device efficiency for the intermediate case is found in the study by Peumans et al., where the donor–acceptor phase separation is fine enough to allow for a high exciton separation yield, but yet not too fine to hamper charge collection.

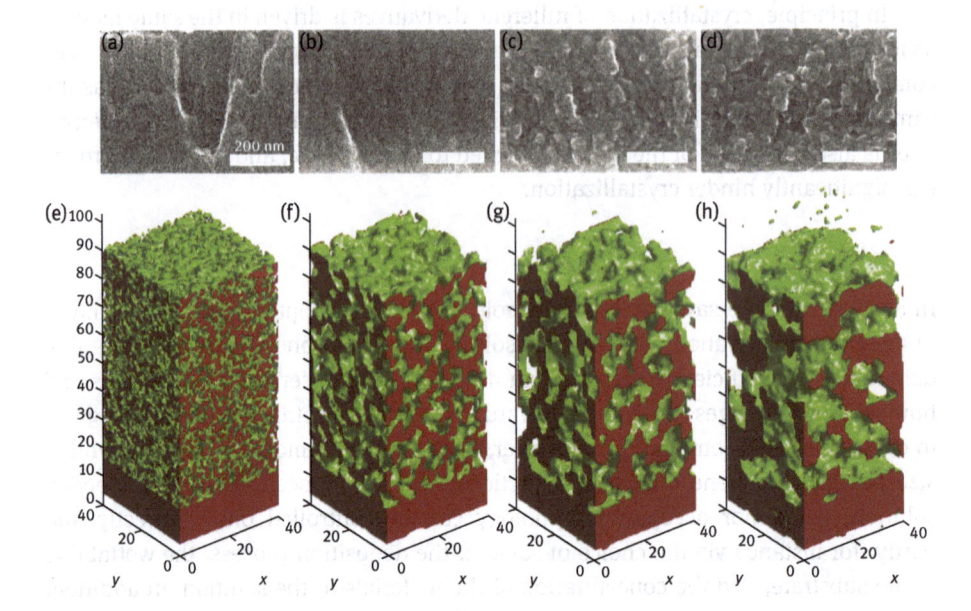

Fig. 3.42: Thermal annealing of a small molecular organic solar cell. The top row shows scanning electron microscopy cross sections of mixed films of copper phthalocyanine and a perylene derivative before (left) and after (three images on the right side) thermal treatments for 15 min at increasing temperatures between 450 and 550 K. The lower row shows the respective three-dimensional simulations, indicating how there is a demixing of donor and acceptor from an initial state of very fine intermixing. Reprinted by permission from Macmillan Publishers Ltd: Nature [103]. Copyright (2003).

The second example we want to highlight is a study by Treat et al. [104], which illustrates the strong tendency of typical organic solar cell materials to alter their spatial distribution upon thermal annealing. As summarized in Figure 3.43, the study investigates interdiffusion of $PC_{60}BM$ into P3HT at different temperatures, starting from a strict bilayer structure where a P3HT film is deposited onto a $PC_{60}BM$ film on a silicon substrate via a lift-off process. In order to probe the vertical distribution of polymer and fullerene, the authors employ a deuterated $PC_{60}BM$ and acquire ^2H depth profiles. Such profiles along with exemplary scanning electron micrographs are shown in Figure 3.43. While the as-cast sample cleary indicates a strict bilayer structure, there is an increasing amount of fullerene present in the polymer phase after thermal annealing at increasing temperatures. After 5 min at 50°C, there is a significant increase in $PC_{60}BM$ concentration in the top polymer layer and a completely homogeneous vertical P3HT–$PC_{60}BM$ distribution when annealed at 150°C. At these elevated temperatures a similar result is found when the annealing is carried out for only 30 s.

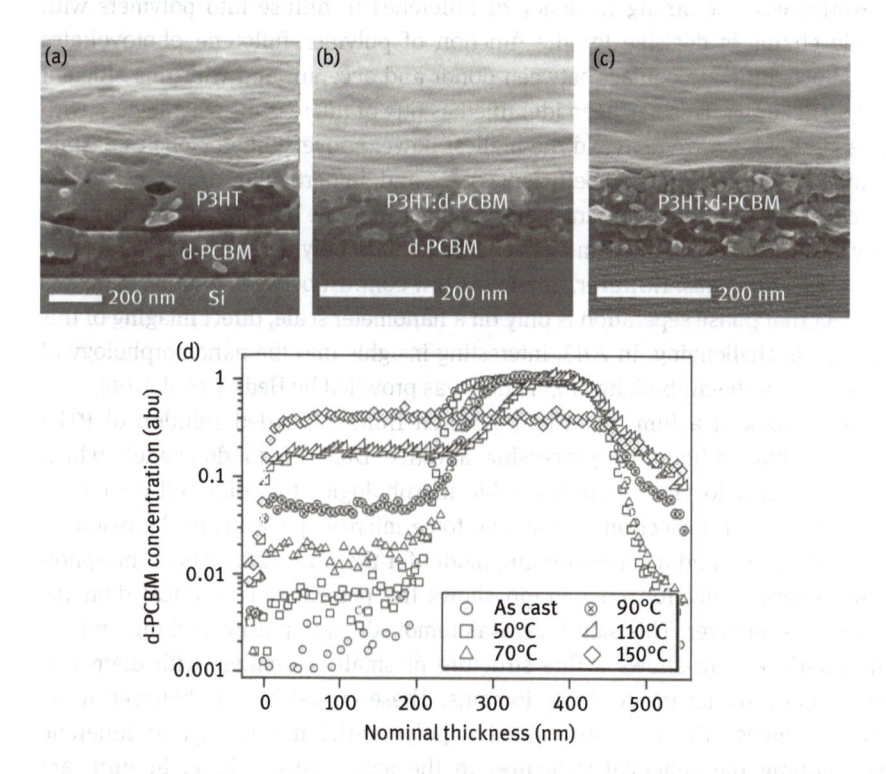

Fig. 3.43: Interdiffusion of $PC_{60}BM$ into a P3HT film. A strict bilayer structure of P3HT and PCBM shown in **(a)** is annealed at **(b)** 110°C and **(c)** 150°C for 5 min. Deuterized PCBM is used so that secondary ion mass spectroscopy depth profiles can be acquired for various annealing temperatures. These profiles, as shown in **(d)**, indicate the strong tendency of $PC_{60}BM$ to diffuse into P3HT films. Figure adapted from [104] with friendly permission by Wiley and Sons (2011).

This observed reorganization of the P3HT–PC$_{60}$BM morphology is consistent with calorimetric results of Zhao and co-workers [105]. Their analysis of P3HT–PC$_{60}$BM mixed films shows that there is a common glass transition temperature in such mixed films, i.e. the individual glass transitions for P3HT and PC$_{60}$BM are not observed in such films. The glass transition temperature is a function of the mixing ratio and increases for decreasing polymer content. At P3HT:PC$_{60}$BM ratios around 1:1, which are most relevant since being most efficient in organic solar cells, the glass transition temperature is only around 80°C, which explains the rapid interdiffusion of fullerene into the polymer.

Even though P3HT and PC$_{60}$BM intermix strongly, the nanomorphology of a P3HT–PC$_{60}$BM bulk heterojunction is not characterized at a molecular level of intermixing, but rather exhibits domains of pure polymer and pure fullerene phases. The morphology further depends on the exact processing parameters, as for instance the cooling rate after thermal annealing or the use of processing additives like high-boiling–point solvents that specifically dissolve only one of the two involved materials. Nevertheless, the strong tendency of fullerenes to diffuse into polymers with alkyl side chains is decisive for the function of polymer–fullerene photovoltaics since it allows intimate contact between donor and acceptor and with this efficient splitting of excitons. This has been identified as one of the reasons why devices with fullerene acceptors currently yield the highest power conversion efficiencies despite the comparably low extinction coefficients of C$_{70}$ and, in particular, C$_{60}$.

Due to its crucial role in organic solar cell operation the bulk heterojunction morphology has been investigated intensively for various polymer–fullerene and other donor–acceptor systems. However, due to the low contrast between organic materials and the fact that phase separation is only on a nanometer scale, direct imaging of this morphology is challenging. In 2013, interesting insights into the nanomorphology of a current state-of-the-art bulk heterojunction was provided by Hedley et al. [106].

First we look at a film that was processed from a blended solution of PTB7 and PC$_{70}$BM but without the processing additive DIO (1,8-diiodooctane), which is commonly used to yield more favorable morphologies for solar cell operation. Figure 3.44 shows the electron and atomic force microscope scans of the resulting bulk heterojunction and a corresponding model for the three-dimensional morphology. The scanning electron micrograph shows the presence of spherical domains covered by a skin layer. This skin layer was removed using plasma etching and the resulting surface scan shows a fine structure of small crystallites with diameters between 20 and 60 nm inside these domains. These crystallites are believed to be PC$_{70}$BM aggregates, which are surrounded by polymer-rich mixed polymer–fullerene phases, forming the spherical structures in the active layer. These, in turn, are separated and covered on top with a thin skin layer, which is a finely mixed PTB7:PC$_{70}$BM phase. According to the considerations in Section 3.3.3, it is clear that such a structure does not allow high-power conversion efficiencies. First, some of the PC$_{70}$BM crystallites are larger than typical exciton diffusion lengths, limiting the exciton separation yield for absorption in the absorber. On the other hand, good exciton separation is expected inside the present mixed phases. However, charge transport away from the charge-separating interfaces is subject to pronounced charge recombination losses, as directly apparent from relatively low efficiencies of PTB7:$PC_{70}BM$ solar cells with active layers processed without DIO. This is the case since there are consistent pathways neither for electrons nor for holes toward the external electrodes. Furthermore, the $PC_{70}BM$ crystallites inside the spherical domains act as electron traps that are detrimental for charge extraction and thus for efficient device operation.

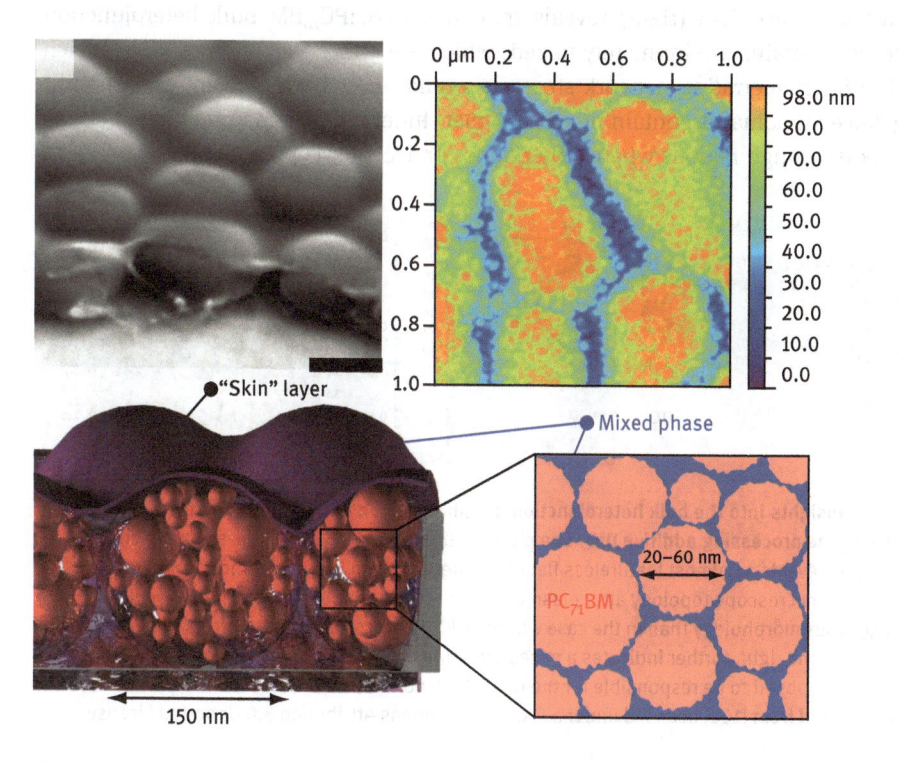

Fig. 3.44: Insights into the bulk heterojunction morphology of a PTB7–$PC_{70}BM$ active layer, deposited without the use of the processing additive DIO. The top left image shows a scanning electron microscopy cross section indicating the presence of large spherical domains with a skin layer on top. The upper right image shows an atomic force micrograph of the surface of such a film after plasma removal of the skin layer. The schematic shows the current picture of the PTB7–$PC_{70}BM$ morphology, where fullerene crystallites are surrounded by PTB7-rich mixed polymer–fullerene phases. These mixed phases are also present in the skin layer. Figure adapted from [106] licensed under a Creative Commons Attribution 3.0 Unported License.

A different situation is observed for PTB7:$PC_{70}BM$ bulk heterojunctions processed with DIO. Since the additive has a high boiling point and relatively selectively dissolves $PC_{70}BM$, fullerene crystallization is slowed down and the resulting morphology is well-suited for implementation into high-efficiency organic solar cells (approaching 10% under AM 1.5G-simulated solar light). As shown in Figure 3.45, the first indication of an altered film formation process is that the polymer–fullerene layer is much smoother and does not exhibit the spherical domains observed in films processed without DIO. The atomic force microscope topology further reveals less and smaller crystallites in the film and points toward a more homogeneous mixing of PTB7 and $PC_{70}BM$. Hedley and co-workers also employed local current–voltage measurements that allowed us to conclude on the composition of the contacted nanosized spot. The resulting photoconductive atomic force micrograph shown in Figure 3.45 (right) reveals that in a PTB7:PC_{70}BM bulk heterojunction, fiber-like domains are intimately mixed, which are either PTB7-rich or PC_{70}BM-rich. Such a interpenetrating network structure would allow both efficient exciton splitting since all domains contain mixed polymer–fullerene phases and efficient charge transport through regions where either donor or acceptor is predominant.

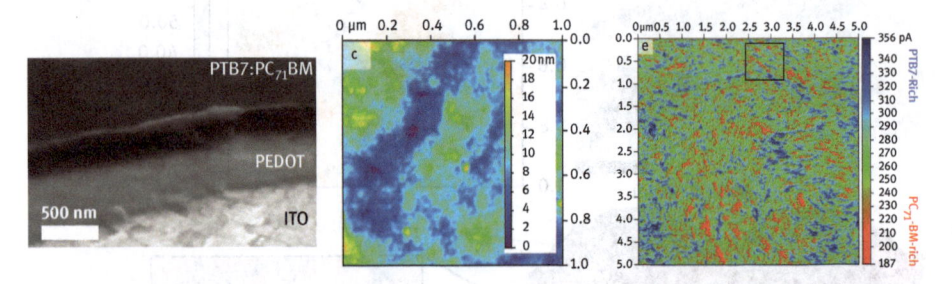

Fig. 3.45: Insights into the bulk heterojunction morphology of a PTB7–PC_{70}BM active layer, deposited using the processing additive DIO. Compared to the case without DIO, the scanning electron micrograph reveals an almost featureless flat bulk heterojunction film. The middle image shows the atomic force microscope topology after plasma removal of the top 20 nm of the film, revealing a more homogeneous morphology than in the case without DIO. Photoconductive atomic force microscopy, as shown on the right, further indicates a mixed structure of fiber-like donor and acceptor domains, which are supposed to be responsible for the high efficiencies observed for PTB7:PC_{70}BM devices. Figure adapted from [106] licensed under a Creative Commons Attribution 3.0 Unported License.

Having these examples, we will now examine a more generalized picture of polymer– fullerene bulk heterojunction morphologies. After an introduction on the bulk heterojunction concept in 1995, the common view on the internal morphology was relatively simple and the active layer was considered only as a two-phase system with donor and acceptor domains. While this is true for some small molecular systems, polymer– fullerene bulk heterojunctions are commonly much more complex since the crystallinity of the different materials and the potential presence of mixed phases have to be considered. As illustrated in Figure 3.46, there are at least five different phases present in most polymer–fullerene bulk heterojunctions, where all pairs of two can be in direct contact. These phases are (1) crystalline regions of pure donor, (2) less-ordered and amorphous regions of pure donor, (3) crystalline regions of pure acceptor, (4) less-ordered and amorphous regions of pure acceptor, and (5) mixed phases of donor and acceptor. While the exact morphology of a bulk heterojunction depends on the materials used and the processing parameters, a common property of all bulk heterojunctions is that the presence of such different phases results in an energy landscape that exhibits a driving force toward more crystalline domains for both electrons and holes. Mixed phases are commonly characterized by a fine intermixing of donor and acceptor, resulting in a rather amorphous arrangement. This results in a typical charge generation event as follows: Excitons preferentially separate in mixed phases due to the large donor– acceptor interfacial area; subsequently, electrons migrate toward pure acceptor phases and are driven into crystalline domains, while holes migrate toward pure donor phases and also end up inside the crystalline phases. From there onward, both types of charge carriers can be efficiently transported toward the external electrodes due to very high charge mobilities in crystalline organic semiconductors. Apparently, a suitable phase separation and contact of the different domains with the respective electrodes is a prerequisite for such an efficient charge generation process and requires well-adjusted processing parameters and fine-tuned material properties.

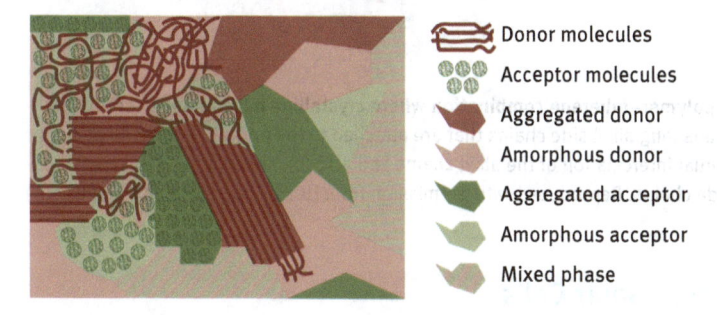

Fig. 3.46: Schematic view of a mixed polymer–fullerene film. Typically five types of domains are present in bulk heterojunctions with polymeric donors and fullerene acceptors, namely crystalline domains of donor, crystalline domains of acceptor, amorphous domains of donor, amorphous domains of acceptor, and mixed phases, where donor and acceptor are intermixed on a molecular level.

In this context it has to be noted that not all polymer–fullerene mixed phases are disordered in nature. In the case the spacing between polymer side chains allows incorporation of fullerenes, co-crystals can form, with properties that are only implemented into a functional device due to the morphological interplay of donor and acceptor. An example of a photoactive polymer–fullerene co-crystal is shown in Figure 3.47. This co-crystal is based on the polymer PBTTT (poly(2,5-bis(3-hexadecylthiophen-2-yl)thieno[3,2-b]thiophene)), which allows intercalation of $PC_{60}BM$ between its long alkyl side chains. The resulting mixed crystal exhibits fascinating properties like the formation of molecular fullerene nanowires inside the crystal (in out-of-plane direction), and a common melt temperature that exceeds that of pure PBTTT and PC$_{60}$BM.

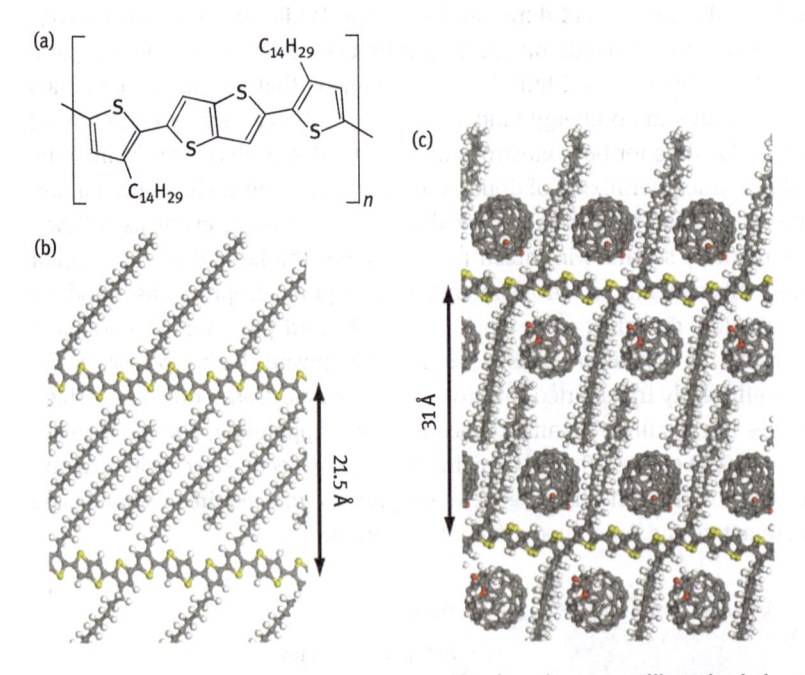

Fig. 3.47: Example of a polymer–fullerene combination where crystalline mixed phases occur. The model polymer **(a)** contains long alkyl side chains that are attached to the polymer backbone at a distance that allows **(b)** mutual intercalation of the alkyl chains and also **(c)** intercalation of fullerenes between the polymer side chains. Reproduced with permission from [107]. Copyright (2009), SPIE.

3.4 Hybrid Junction Solar Cells

Having discussed on solar cells based on fully organic active films, i.e. devices in which donor and acceptor materials are conjugated molecules, we will now highlight a second class of second-generation photovoltaics, namely hybrid solar cells, that combine inorganic with organic materials. Metal oxides are very attractive materials for applications in optoelectronic devices since they are abundant, are chemically stable in many cases, easy to purify (especially in terms of the energy required for purification), and often exhibit n-type semiconducting properties due to an intrinsic doping with oxygen vacancies.

We will first discuss dye-sensitized solar cells that rely on a monolayer of high-extinction dye molecules covering a high surface area metal oxide nanostructure (Section 3.4.1). In Section 3.4.2 we will then deal with metal oxide–polymer hybrid devices, where either conjugated polymers take over the role of the light absorber or thin crystalline inorganic absorbers are employed on metal oxide nanostructures.

3.4.1 Dye-Sensitized Solar Cells

Dye-sensitized solar cells are a well-established technology developed in the early 1990s [9]. Initially, these devices were always using a liquid electrolyte as hole transporter, until a solid-state version of this type of solar cell was introduced in 1998 [15]. Here liquid electrolyte dye-sensitized solar cells will be discussed first before we discuss on the solid-state dye-sensitized solar cells. In the following sections we will use the abbreviations *DSC* for *dye-sensitized solar cell* and *SDSC* for *solid-state dye-sensitized solar cell*.

3.4.1.1 Liquid Electrolyte Dye-Sensitized Solar Cells

The fundamental mechanism of charge separation in a DSC is similar to the mechanisms in an organic solar cell, where charge separation occurs upon exciton splitting via injection of an electron from the excited donor into the acceptor. However, while exciton diffusion plays a crucial role in the function of organic solar cells, DSCs feature monolayers of absorbers that are chemically attached to the electron-accepting metal oxide electrode. Thus, the exciton is confined to the excited molecule and located close to the acceptor throughout its whole lifetime. This conceptual advantage results in very high internal quantum efficiencies and virtually no excitons recombine before they get split at the dye–metal oxide interface, i.e. η_{ED} from Equation (3.23) is unity. However, before we discuss this in more detail, we will first have to take a look at the geometry of a typical DSC.

Most commonly DSCs employ TiO_2 as n-type metal oxide since it has a wide bandgap around 3.2 eV, which renders it transparent for visible light. Furthermore, it is an abundant material and chemically highly stable (TiO_2 resists very high and very low pH values and dissolves only in HF or highly concentrated sulfuric acid and some fluorine-containing solutions). Due to the presence of oxygen vacancies, TiO_2 is intrinsically n-type. An oxygen vacancy results in two Ti^{3+} instead of Ti^{4+} ions being present in the neighborhood of the vacancy, which one weakly bound electron. These electrons occupy states within the bandgap of TiO_2 that are located close to the conduction band, hence are n-dopants. Furthermore, TiO_2 has a comparably low conduction band that allows efficient electron transfer from the LUMO of most donor materials.

The three fundamental assembly steps for a DSC and the resulting device geometry are shown in Figure 3.48. DSCs are built on conductive glass substrates so that light can be coupled into the device through this front contact. The conductive film is covered with a compact TiO_2 layer, which serves as a hole blocker so that the front contact is selective for electrons. On top of this blocking layer lies a highly porous TiO_2 network of approximately 10–20 μm thickness. This network is assembled from nanoparticles with approximately 20 nm diameter and surface areas on the order of $50 \, m^2 g^{-1}$ that are sintered together to form a polycrystalline, porous, n-type semiconducting matrix.

In the next step this nanoparticle film is sensitized with a self-assembled monolayer of dye molecules by immersion in a dye solution and subsequent washing in an appropriate solvent. Most dye molecules carry a functional COOH-group attached to their chromophore, which enables the self-assembly of a monolayer since this moiety, forms a covalent bond to the metal oxide. Such a sensitization reaction is schematically depicted in Figure 3.49. Like all metal oxides TiO_2 is commonly terminated with OH groups. By substituting one of these terminating groups with the COO^- moiety, the dye can chemisorb and thus bind to the TiO_2, either as depicted in the figure or via both oxygen atoms with the electrons from the double bond being delocalized. Besides, physisorption via hydrogen bonding is also possible [108]. In all cases the binding energy between the dye and the metal oxide is high enough that the dye sticks to the surface during washing, so that ideally a monolayer can be obtained (in real devices there are often dye aggregates present). The color of the DSC can thus be tuned by the choice of dye molecules with an appropriate absorption spectrum since the dye is the only relevant absorber in the device. This is the case since TiO_2 absorbs only UV light, which plays only a minor role in the energy conversion of a solar cell since the vast portion of sunlight on the surface of the earth is in the visible and near IR regions (compare Section 3.1.1). Due to the tremendously high surface area of the mesoporous TiO_2 film (As rule of thumb: a 10-µm-thick mesoporous TiO_2 film has around a 1000× higher surface area compared to a flat layer.), the overall light extinction in the active film of a DSC is close to unity at and around the absorption peak(s) of the dye molecules even though the dye is employed only as a monolayer.

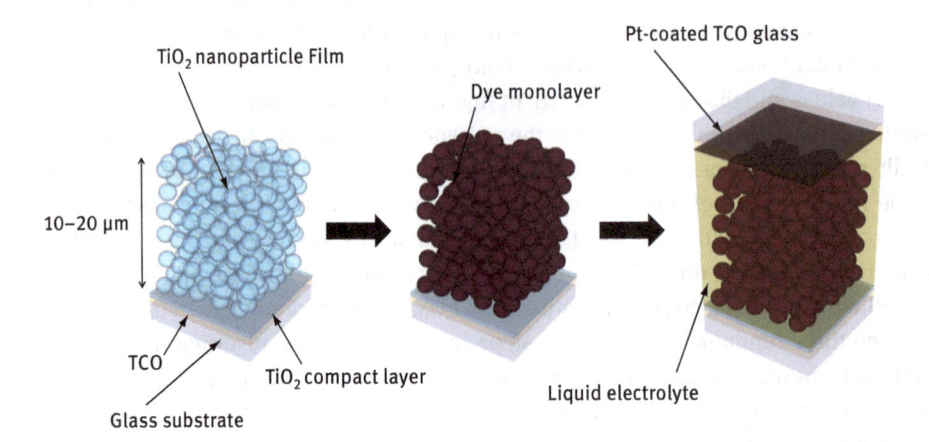

Fig. 3.48: Assembly of a dye-sensitized solar cell. A 10–20 µm thick film of TiO_2 nanoparticles with approximately 20 nm diameter is deposited on a thin compact TiO_2 film on a transparent conductive substrate. This nanostructured film is then sensitized with a self-assembled monolayer of dye molecules in order to render the wide-bandgap TiO_2 visible light absorbing. The device is finalized by

infiltration of a liquid electrolyte solution into the nanoparticle matrix and subsequent encapsulation with a platinized top electrode.

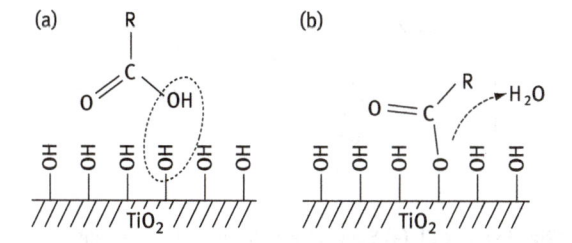

Fig. 3.49: Sensitization of a TiO_2 surface with dye molecules. Most commonly the chromophore (R) of the dye has a COOH-group attached, which then binds to the metal oxide by substitution of one of the terminating OH groups at the surface of the metal oxide.

The dye molecule is essentially the most decisive compound of a DSC and the choice of the dye determines the time scales and efficiencies of all processes occurring during charge generation and recombination.

As an alternative to self-assembled dye monolayers, there have been attempts to use nanometer-thin coatings of strong inorganic absorbers on the TiO_2 electrode or to sensitize it with quantum dots. Especially the latter enables full control over the absorption spectrum similar to the case of dye molecules since the quantum dot material and size determine their bandgap.

After dye sensitization the nanoparticle film is filled with a liquid electrolyte that regenerates the oxidized dye molecules after electron injection into the TiO_2. This regeneration (reduction) of the dye ion is an oxidation of the ionic species in the electrolyte, which in turn is reduced at the platinized top electrode of the device. This electrode commonly consists of a second transparent conducting glass coated with a nanometer-thin platinum film that catalyzes the electrolyte regeneration. A liquid-proof sealing is employed between the front and back substrates that defines the active volume of the device.

Figure 3.50 depicts the working mechanism of a DSC with a geometry as discussed so far. Light is coupled into the active film through the bottom conductive glass substrate and impinging upon the dye-sensitized TiO_2 electrode. Once a photon is absorbed in a dye molecule, an exciton forms that typically exhibits a binding energy comparable to the exciton binding energies in organic semiconductors. This exciton can then be rapidly split via injection of an electron into the conduction band of the TiO_2, through which it then migrates to the front electrode. The now oxidized dye molecule is then reduced by the electron shuttle in the liquid electrolyte. Due to diffusion of the charge shuttles in the electrolytes, holes are transported toward the platinized top electrode and a photocurrent or photovoltage is generated.

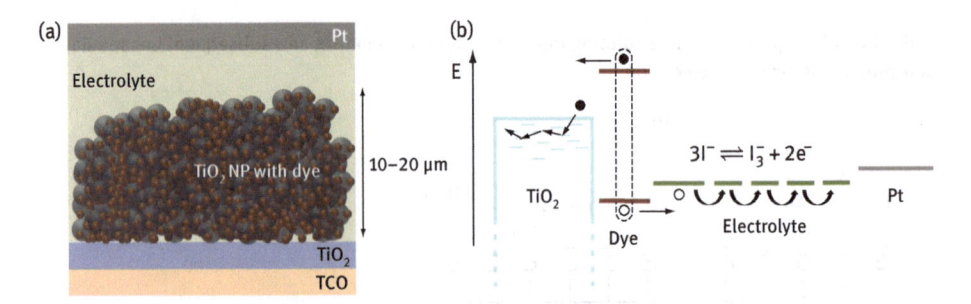

Fig. 3.50: Working principle of a dye-sensitized solar cell. (a) Schematic side view of a dye-sensitized solar cell and **(b)** respective energy-level diagram.

Typical time scales for the different processes during charge generation and collection in a DSC are visualized in Figure 3.51. Due to the intimate contact between the dye and the TiO_2, exciton splitting via electron injection into the TiO_2 occurs on a 100-fs time scale and typically gets completes within 1 ps. This process is fast enough to compete with relaxation of the dye to the lowest vibrational state. Commonly, however, exciton splitting is also fast for this state, so the efficiency for exciton splitting is virtually 1 since the actually competing mechanism, exciton recombination, occurs approximately 10,000 times slower. Owing to this rapid charge separation, the main loss mechanism in DSCs is not exciton recombination, but non-geminate charge recombination of an electron in the TiO_2 either with the hole in a dye cation or directly in the electrolyte that occur on 0.1 and 10 ms time scales, respectively. The former typically plays only a minor role since dye regeneration is completed within μs. Recombination in turn competes with the transport of charges away from the TiO_2–dye interface toward the two external electrodes. While charge transport is diffusive for both electrons and holes, characteristic time constants are 1 ms for electrons and only 0.1 s for holes, both of which are relatively slow.

In the case of holes, the transport is slow since it is dominated by diffusion of ionic charge shuttles through the electrolyte compartment of the DSC. In contrast, electrons migrate directly through the solid TiO_2 phase. Nevertheless, the slow electron transport is a result of the pronounced polycrystallinity of the TiO_2 electrode. The fabrication of this nanoporous network relies on the sintering of nanoparticles, which results in the presence of crystal grain boundaries between these nanoparticles. Grain boundaries in turn introduce sub-bandgap states into the TiO_2, which trap mobile electrons in the conduction band. Charge transport through a TiO_2 nanoparticle network therefore is characterized by a sequence of electron trapping and detrapping events, somewhat comparable to charge hopping in organic semiconductors. This transport process and the respective energy diagram is shown in Figure 3.52. This results in comparably low electron mobilities in DSCs, which are on the order of 10^{-4} $cm^2 V^{-1} s^{-1}$.

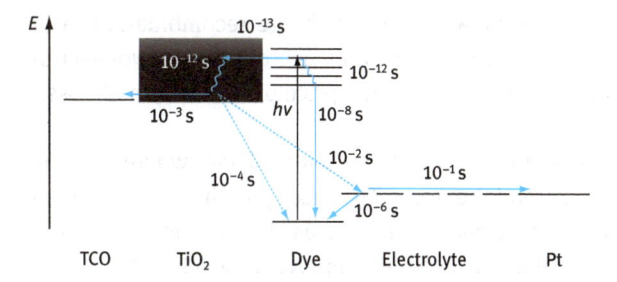

Fig. 3.51: Illustration of the time constants for various processes in a dye-sensitized solar cell. Photon absorption, electron injection, and electron relaxation are very fast processes and typically completes within ps, while electron and hole transport, and, in particular, charge carrier recombination are several orders of magnitude slower. Time scales according to the review by Hagfeldt et al. [109].

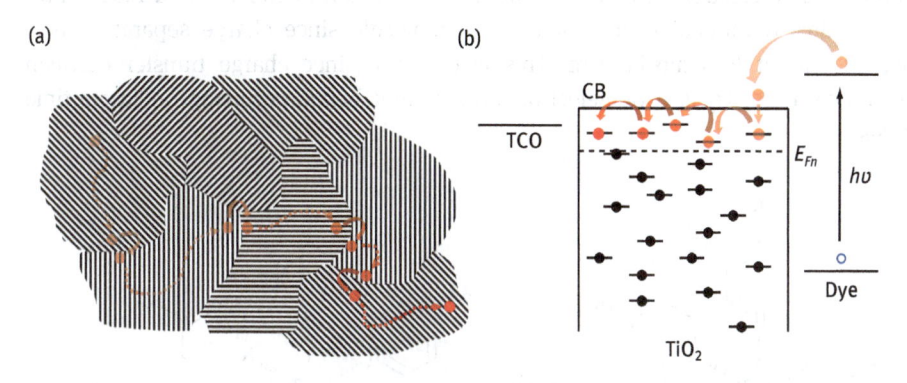

Fig. 3.52: Electron transport through polycrystalline TiO$_2$. (a) Illustration of an electron migrating through several TiO$_2$ crystallites and across several grain boundaries. (b) Energetic view of electron hopping transport through TiO$_2$ in a dye-sensitized solar cell. Figure adapted from [110] with friendly permission by Wiley and Sons (2014).

Competition between charge transport and charge recombination is a key issue in DSCs and is addressed using a variety of approaches. These involve modification of the TiO$_2$–dye interface, use of chemically modified dyes with suitable side chains, dyes with a built-in donor–acceptor structure, and synthesis of ordered TiO$_2$ nanostructures with improved crystallinity.

For now, however, we will conclude this introduction into the working principles of DSCs by taking a look at the charge shuttles employed in the liquid electrolyte. For a long time after the introduction of the first DSC in 1991, an iodine-based electrolyte was used due to the very good efficiencies that could be accomplished with it. The oxidation/reduction cycle in this electrolyte can be summarized as

$$3I^- \rightleftharpoons I_3^- + 2e^-,$$

where iodide reacts to form iodate during dye regeneration and iodate to iodide at the platinized top electrode. One of the major advantages of the I$^-$/I$_3^-$ system is that

DSCs with this electrolyte commonly show very slow charge recombination rates. This is the case since charge transfer between TiO_2 and electrolyte is a two-electron process. At the same time, however, this comes at the price of relatively high losses in open-circuit potential.

This issue has been addressed by using different electrolytes, where the hole transfer is less lossy in terms of potential energy. The most prominent example are electrolytes based on a cobalt-complex charge shuttle, as depicted in Figure 3.53. Molecules with different side chains have been employed, which offer certain advantages and disadvantages depending on the application. However, the fundamental charge transfer mechanism is a oxidation and reduction reaction between the Co^{2+} and Co^{3+} center ion. Due to the more favorable energy levels of holes in this electrolyte, open-circuit voltages are typically 200–300 mV higher than in DSCs with an iodide electrolyte. However, the implementation of cobalt-based redox shuttles requires a rational design of the dye molecule since charge separation is a more dominant loss mechanism. This is the case since charge transfer between electrolyte and TiO_2 is a one-electron process and hence occurring on faster time scales.

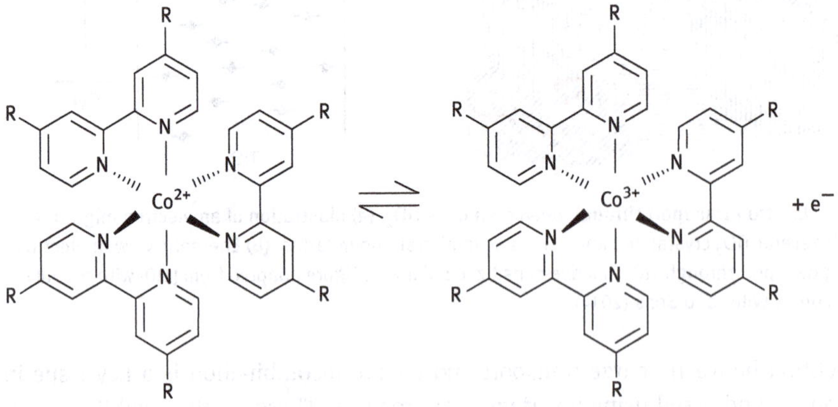

Fig. 3.53: Chemical structure of the most commonly used cobalt–complex electrolyte. The central cobalt ion can undergo oxidation and reduction reactions between Co^{3+} and Co^{2+} and thus function as charge shuttle.

3.4.1.2 Solid-State Dye-Sensitized Solar Cells

The major drawback of conventional DSCs is the presence of the liquid electrolyte in the cell architecture, which demands for long-term stable solvent-proof encapsulation. In contrast to organic solar cells, it is therefore unpractical to realize DSCs on flexible substrates since encapsulation between two glass substrates has proven to be the most promising strategy to realize long-term stable DSCs on large areas, both of which are requirements for commercialization of DSCs. This issue has been addressed by substituting the liquid electrolyte with a solid-state hole transporter in the so-called SDSC. In the original paper by Bach et al. [15] used the organic small-molecule 2,2′,7,7′-Tetrakis-[N,N-di(4-methoxyphenyl)amino]-9,9′-spirobifluorene, commonly referred to as *Spiro-OMeTAD* or sometimes *Spiro-MeOTAD*. Due to its small size and solubility in organic solvents, Spiro-OMeTAD can be infiltrated into TiO_2 nanoparticle films so that a bicontinuous network of the electron-transporting metal oxide and the hole-transporting material is formed. Spiro-OMeTAD has a HOMO–LUMO gap of approximately 3 eV; i.e. it is transparent for visible light so that the dye monolayer is the only absorber in an SDSC, analogous to the liquid electrolyte counterpart.

The working principle of an SDSC is shown in Figure 3.54, along with the structural formula of Spiro-OMeTAD. After photon absorption and exciton formation in the dye, charge separation occur upon electron injection into the TiO_2. However, dye regeneration takes place via a direct hole transfer between the HOMOs of the dye and an adjacent Spiro-OMeTAD molecule. The holes are then transported via a series of charge hoppings through the TiO_2–Spiro-OMeTAD network and the Spiro-OMeTAD capping layer on top toward the metallic top electrode, while electrons are collected at the transparent front electrode.

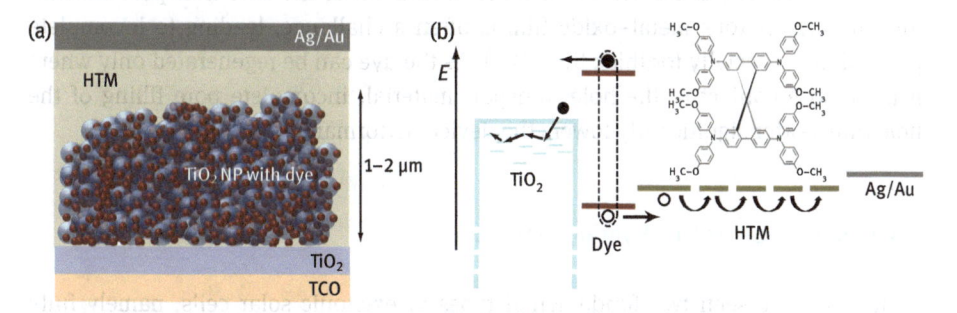

Fig. 3.54: Working principle of a solid-state dye-sensitized solar cell. The schematic sideview in **(a)** shows that commonly much thinner TiO_2 nanoparticle networks are used owing to the limited capability of the hole transporter molecules to infiltrate into the nanoporous film and **(b)** shows the energy landscape in a solid-state dye-sensitized solar cell along with the chemical structure of Spiro-OMeTAD, the most commonly employed organic hole transporter.

The presence of a solid-state hole transporter that substitutes the liquid electrolyte with its ionic charge shuttles has important implications for the fundamental processes of charge generation and recombination since the respective time scales are different. Hole transfer between dye and Spiro-OMeTAD occurs at rates similar to electron injection from the dye into the TiO_2, while in conventional DSCs the electron transfer is several orders of magnitude faster than dye regeneration. Therefore, in addition to the charge separation mechanism described before, it is also possible that the hole is transferred to the Spiro-OMeTAD first and the dye remains in reduced state. In this case the dye is regenerated by the TiO$_2$ through the subsequent electron transfer. This competition of charge separation mechanisms demands dye molecules that are optimized for this particular application. While dyes for conventional DSCs have to be designed in a way that guarantees the spatial location of the hole in the oxidized dye (i.e. the HOMO of the oxidized dye) on the electrolyte side, the reduced dye does not have to be considered since it typically does not play any role. In contrast, both oxidized and reduced dyes can be present in SDSCs, which has to be considered in the development of high-efficiency materials for SDSCs.

The solid-state nature of Spiro-OMeTAD further changes the charge recombination dynamics. Charge recombination in SDSCs does not involve a slow diffusion process of a shuttle ion and is always a single-electron process (in contrast to DSCs with iodide electrolytes, where charge recombination involves a two-electron process). Charge recombination occurs on time scales on the order of 10^{-4} s, which leads to smaller electron and hole diffusion lengths in SDSCs compared to DSCs. This is reflected in optimized SDSC active-layer thicknesses of only 1–2 µm, since thicker active films result in pronounced recombination losses. The main loss mechanism in SDSCs is therefore incomplete photon absorption compared to conventional DSCs, where the slow recombination allows 10 times thicker and thus stronger absorbing active layers. Also the infiltration of the hole transport material into the mesoporous metal–oxide film is often a challenge, leading to incomplete pore filling, especially for thick films [111]. As the dye can be regenerated only where it is in close contact to the hole transport material, incomplete pore filling of the hole transporter significantly lowers the device performance.

3.4.2 Inorganic–Organic Hybrid Solar Cells

So far we have seen two fundamental types of excitonic solar cells, namely fully organic photovoltaics, with conjugated donor and acceptor materials in intimate contact, and dye-sensitized solar cells, which combine monolayers of conjugated absorbers with an inorganic electron transporter and a solid organic or a liquid electrolyte hole transporter. Since the mid-2000s, there is an increasing focus on a third type of excitonic solar cells, where concepts of both organic and dye-sensitized solar cells are combined. This class of devices is often referred to as *hybrid solar cells* since they combine inorganic and organic materials. This definition also fits to SCDCs, which are also sometimes referred to as hybrid solar cells. However, historically they are a further development coming from the DSCs, and usually in SDSCs the function of the organic material is limited to the hole transport and does not contribute to charge generation (as it generally does not absorb light in the visible region as, e.g. Spiro-OMeTAD), which is usually considered as an additional function of the organic semiconductor in hybrid solar cells.

Hybrid solar cells have been reported with a manifold of different geometries and material combinations, which, however, mostly share similar working mechanisms. We will therefore discuss the three most important concepts in the following, section since they serve as the basis for various other hybrid devices. They are (1) bulk heterojunctions of metal oxide nanoparticles and conjugated polymers, (2) devices with dye-sensitized nanostructured metal oxide electrodes and polymer absorbers as hole transporters, and (3) hybrid solar cells with inorganic absorbers and organic hole transporters.

3.4.2.1 Metal Oxide–Polymer Bulk Heterojunctions

The conceptually simplest hybrid solar cell is a polymer–metal oxide bulk heterojunction solar cell. The function of such a device is analogous to a fully organic bulk heterojunction, where donor and acceptor materials are in intimate contact due to parallel deposition. However, instead of fullerenes or other conjugated acceptor molecules, metal oxide nanoparticles are mixed with donor polymers, most commonly ZnO nanocrystals with diameters on the order of 10–20 nm and in some instances TiO_2, which typically leads only to lower efficiencies. Due to the presence of oxygen vacancies in the metal oxide lattice, these particles are intrinsic n-type semiconductors, which function as electron acceptors due to their low conduction band energies. Furthermore, these metal oxides exhibit wide-bandgap energies on the order of 3 eV, so that the polymer is the only visible light absorber in the device and as such in large part responsible for photocurrent generation.

A typical charge generation event thus is analogous to the case in an organic solar cell for photon absorption in the donor. An exciton in the polymer diffuses to the metal oxide–polymer junction, where it can be split via electron injection from the LUMO of the polymer into the conduction band of the polymer. Electron injection is preceded by population of a hybrid charge transfer exciton, which is pictured as a paired state of a comparably localized hole on the polymer and a rather delocalized electron in the metal oxide [112]. Subsequent to charge separation, electron and hole then migrate through metal oxide and polymer domains, respectively, and get collected at the external electrodes.

A schematic of a typical metal oxide–polymer bulk heterojunction device is shown in Figure 3.55. Due to the internal disordered donor–acceptor morphology, metal oxide–polymer bulk heterojunctions are subject to the same loss mechanisms as organic bulk heterojunctions, where consistent pathways from the charge generation site through both donor and acceptor are required in order to allow loss-free charge collection (case (1) in Figure 3.55). As a consequence, morphology control is as important in metal oxide–polymer hybrid solar cells as in fully organic devices, since formation of islands (2) and contact of donor (3) and acceptor (4) to the wrong external electrodes result in increased charge carrier recombination.

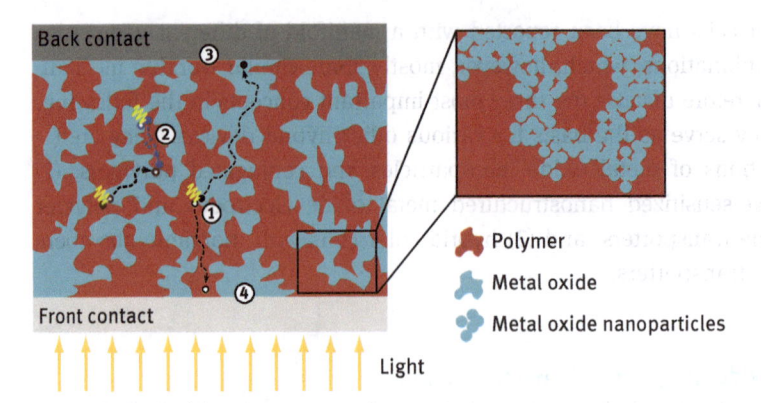

Fig. 3.55: Schematic cross-sectional view of a metal oxide–polymer bulk heterojunction hybrid solar cell. Analogous to all-organic bulk heterojunctions, a disordered donor–acceptor morphology is formed via co-processing of donor polymers and metal oxide nanoparticles, which function as electron acceptors.

Another crucial point in such devices is the electric contact between the metal oxide nanocrystals, which cannot be sintered as in dye-sensitized solar cells since high temperatures lead to oxidative burning of the conjugated donor polymers. In order to yield sufficient contact throughout the metal oxide nanoparticle network in the bulk heterojunction and thus pathways for electrons toward the external electrodes, commonly polymer–metal oxide mixing ratios with high metal oxide excess (on the order of 1:3 wt:wt) are used. Bulk heterojunction formation is then mostly dominated by the metal oxide, which forms a network structure that is infiltrated by the polymer donor.

While for ZnO–P3HT hybrid solar cells the bulk heterojunction has been shown to be close to an island-free bicontinuous donor–acceptor network in particular cases, internal quantum efficiencies remain comparably low at around 50% at maximum [113]. This is ascribed to a rather inefficient charge separation mechanism compared to polymer–fullerene devices and has been identified as the main limitation of metal oxide–polymer bulk heterojunction solar cells.

3.4.2.2 Dye-Sensitized Metal Oxide–Polymer Hybrid Solar Cells

In order to address the issue of inefficient exciton splitting at ZnO–polymer and in particular at TiO_2–polymer interfaces, molecular modifiers like dye monolayers have been implemented in hybrid solar cells. Since these molecules can affect the electrical contact between modified metal oxide particles, it is practical to first fabricate a sintered metal oxide nanonetwork, which can then be decorated with a self-assembled monolayer, analogous to the TiO_2 nanoparticle electrode of a dye-sensitized solar cell. In the most common case, where the modifier is a dye molecule, the function of such a dye-sensitized metal oxide–polymer hybrid solar cell is similar to that of a solid-state dye-sensitized solar cell. However, instead of Spiro-OMeTAD, which is transparent for visible light, the polymeric hole transporter is strongly absorbing and can thus contribute to the photocurrent generation. Efficient operation of such a hybrid solar cell is a function of the ternary interplay between the metal oxide electron acceptor, the dye monolayer, and the hole-transporting polymer.

The typical geometry of a dye-sensitized metal oxide–polymer hybrid solar cell is shown in Figure 3.56 (a). Since polymer infiltration into metal oxide nanostructures often is challenging and charge carrier lifetimes are comparable to solid-state dye-sensitized solar cells, comparably thin metal oxide nanoparticle films are used, typically with thicknesses between 0.5 and 2 µm. After dye sensitization, the nanoparticle film is filled with the donor polymer, commonly via solution processing, which further yields a polymer-capping layer on top of the dye-sensitized particle film. This capping layer is essential to avoid shortening of the device via direct contact between metal oxide and metal top electrode.

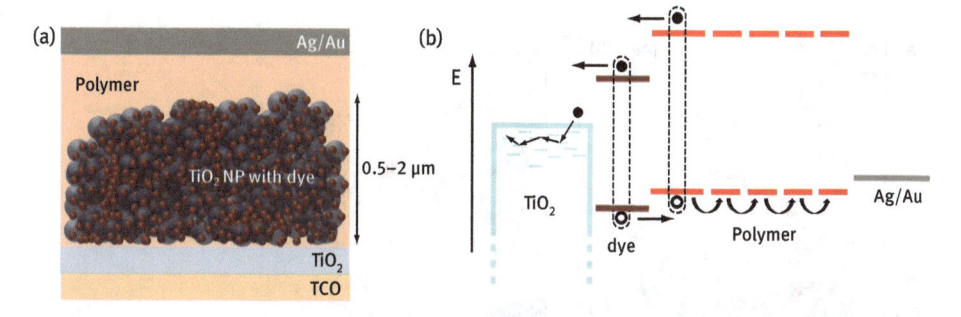

Fig. 3.56: **Working principle of a hybrid solar cell based on a dye-sensitized metal oxide–polymer junction.** Analogous to a solid-state dye-sensitized solar cell, a TiO_2 nanonetwork is decorated with a self-assembled monolayer of dye molecules and infiltrated with a hole transporter, in this case with a strongly absorbing conjugated polymer like P3HT. Charge generation is then fundamentally possible upon splitting of excitons in the dye monolayer and in the polymer.

Figure 3.56(b) illustrates the working principle of the device. Visible light photons are absorbed both in the dye monolayer and in the donor polymer, which ideally are energetically aligned so that a cascading structure is formed for both electrons and holes. Thus, electrons can migrate through the dye into the conduction band of the metal oxide, while holes move toward the HOMO of the polymer.

Since we are now dealing with excitons in both the dye and the polymer, we have to consider the possible mechanisms of charge generation for both cases. These are illustrated in Figure 3.57. Upon photon absorption in the dye, exciton separation occurs analogous to the case in a solid-state dye-sensitized solar cell, i.e. commonly via electron injection into the metal oxide and subsequent regeneration of the dye by the polymer. Additionally, it is possible that this mechanism competes with a reductive quenching of the dye, where the hole gets injected into the polymer and dye regeneration occurs through the metal oxide (not shown in the figure).

In the case of photon absorption in the polymer two fundamentally different processes are conceivable. First, an energy transfer can occur from the polymer to the dye in the case this is energetically allowed and quantum mechanically likely. While such a migration of the exciton from the polymer to the dye is possible via both Förster resonance energy transfer and Dexter energy transfer, a Förster-type transfer is more likely in most devices due to its longer characteristic radius. After the energy transfer the exciton in the dye molecule can then be split as described before. Beside energy transfer it is also possible that the exciton gets separated directly between polymer and dye via population of a dye–polymer charge transfer state and injection of an electron into the LUMO of the dye, from where it further moves into the conduction band of the metal oxides. Furthermore, in the case the dye monolayer is not complete and allows direct contact between polymer and metal oxide, excitons can be separated at this interface, analogous to the mechanism in metal oxide– polymer bulk heterojunctions discussed in Section 3.4.2.1.

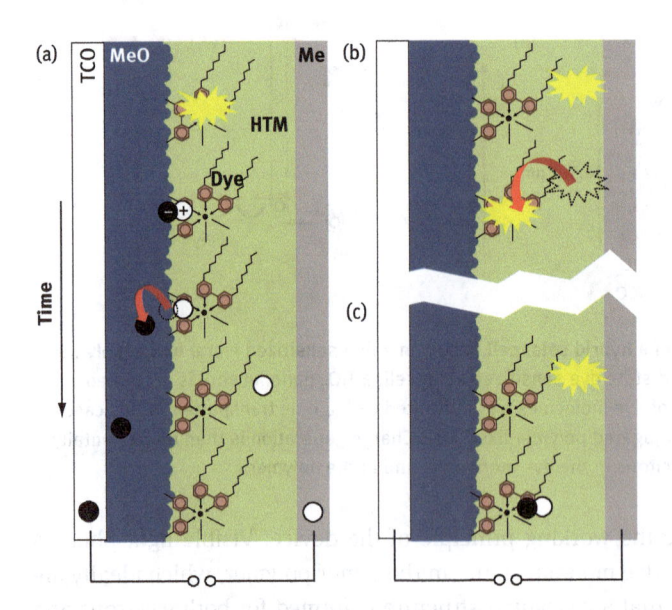

Fig. 3.57: **Possible mechanisms for charge generation at a dye-sensitized metal oxide–polymer interface. (a)** Upon photon absorption in a dye molecule, charges are separated after electron injection into the metal oxide or hole injection into the hole transporter, analogous to a solid-state dye-sensitized solar cell. Excitons in the polymer can contribute to charge generation via either **(b)** energy transfer to the dye and subsequent charge generation as in the previous case or **(c)** charge separation between dye and polymer. Figure adapted from [114] with friendly permission by Wiley and Sons (2013).

Depending on the energy-level alignment of the involved materials and the spacing between dye and polymer, all of the described charge separation mechanisms can occur concurrently. In many hybrid systems, however, especially splitting of excitons in the dye is efficient, while in particular charge separation between polymer and metal oxide is comparably inefficient. This has been illustrated by Vaynzof and co-workers using pump–push spectroscopy on ZnO–polymer hybrid interfaces, as summarized in Figure 3.58. After excitation of the polymer and subsequent population of a hybrid charge transfer state, an additional push photon yields significant additional photocurrent, indicating trapping of the excitation in an energetically low-charge transfer state, which tends to recombine rather than split into electron and hole. Interestingly, the additional photocurrent is reduced by roughly a factor of 2 for two different polymer systems in the presence of a fullerene sensitizer, which forms a monolayer on the ZnO. With this fullerene sensitizer, charge separation occurs between polymer and sensitizer, resulting in less energetically trapped charge transfer states. This fundamental example illustrates the necessity of intelligent design of dye molecules for hybrid junction solar cells, which currently are mostly limited by a nonoptimized contribution of the hole-transporting polymer to the photocurrent of the device.

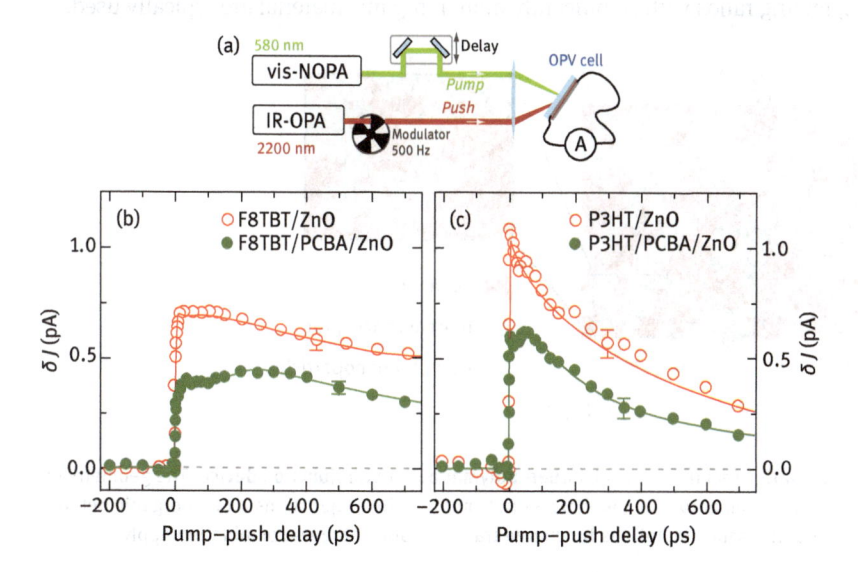

Fig. 3.58: Pump–push results for ZnO–polymer devices with and without a fullerene sensitizer as interface modification. The additionally released push photocurrent is reduced by roughly a factor of 2 in the samples where the fullerene derivative PCBA is present at the hybrid junction, indicating that more charge transfer states rapidly separate into free charge carriers upon absorption of the push photons. Reprinted with permission from [115]. Copyright (2012) American Physical Society.

3.4.2.3 Hybrid Solar Cells with Inorganic Absorbers

Due to the limited efficiency of charge separation at metal oxide–polymer interfaces, another type of hybrid solar cells has attracted tremendous interest over the past decade: hybrid solar cells that combine conjugated hole transporters with inorganic absorbers. The focus of this particular research direction is on low-cost processable materials, so that commonly either inorganic nanoparticles or quantum dots are used or thin absorber coatings are realized using inexpensive methods like electrodeposition or chemical bath deposition.

High efficiencies have been realized in absorber–polymer bulk heterojunction devices, where n-type materials like CdTe, PbS, or CdSe are processed to form nanoparticles, which can then be blended with hole-transporting polymers like P3HT. A schematic of such a bulk heterojunction is shown in Figure 3.59, exemplarily for a device based on low aspect ratio absorber nanowires. Since the same loss mechanisms due to inconsistent charge percolation pathways are present as in other bulk heterojunction devices, morphology control also plays a key role in inorganic absorber–polymer bulk heterojunctions. One commonly used approach to achieve better interconnectivity throughout the electron-transporting nanoparticle network is the use of nanowires rather than nanospheres, which allow more efficient charge transport along their long axis. Furthermore, similar to metal oxide–polymer bulk heterojunctions, mixing ratios with significantly more inorganic material are typically used.

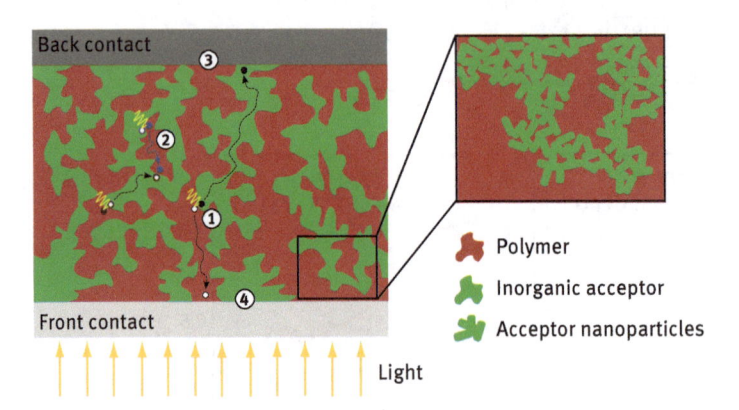

Fig. 3.59: Schematic of an inorganic absorber–polymer bulk heterojunction device. The geometry of such a solar cell is comparable to metal oxide–polymer bulk heterojunctions and all-organic heterojunctions so that the same limitations for charge transport and charge recombination apply.

Charge separation in such absorber–polymer bulk heterojunctions is observed mostly upon photon absorption in the inorganic absorber, either via direct splitting of excitons in materials with low exciton-binding energies and subsequent transport of holes into the polymer phase or hole injection into the polymer, resulting in exciton splitting at the absorber–polymer junction for materials where exciton migration in the inorganic absorber plays a role.

The second type of hybrid photovoltaics with inorganic absorbers we want to highlight here are the so-called *extremely thin absorber solar cells*. These devices combine nanostructured metal oxide electrodes as employed in solid-state dye-sensitized solar cells with coatings of inorganic absorbers like Sb_2S_3, $CuInS_2$, or CdSe that are only a few nanometers thin. Therefore they can be assigned as another class of device architectures in the family of dye-sensitized solar cells. The thin inorganic coatings play a role similar to the dye monolayer in solid-state dye-sensitized solar cells. Similar to dye sensitization, a small amount of absorber is sufficient to allow for high device extinction due to the high surface area that is covered. Since only thin absorber coatings are required, low-cost solution-based deposition techniques like chemical bath deposition or electrodeposition can be employed, which are characterized by comparably low crystal quality of the resulting films but are compatible with high-throughput fabrication techniques.

A schematic of a typical extremely thin absorber solar cell based on a coated TiO_2 nanoparticle network is shown systematically in Figure 3.60. For a well-working device the absorber shows Type II band alignment with the metal oxide so that electrons and holes are driven away from the interface in opposite directions. The absorber-coated nanostructure is infiltrated with an organic hole transporter, most commonly materials transparent to visible light like Spiro-OMeTAD or CuSCN, but also conjugated polymers like P3HT absorbing in the visible. In the latter case additional contribution to the photocurrent by the hole transporter is possible in the case the absorber–polymer interface promotes exciton splitting. This, however, is not the case in many instances, often due to the formation of thin oxide layers on top of the absorber. In such a case the device suffers from parasitic light absorption in the polymer, which reduces the power conversion efficiency compared to the case of a transparent hole transporter.

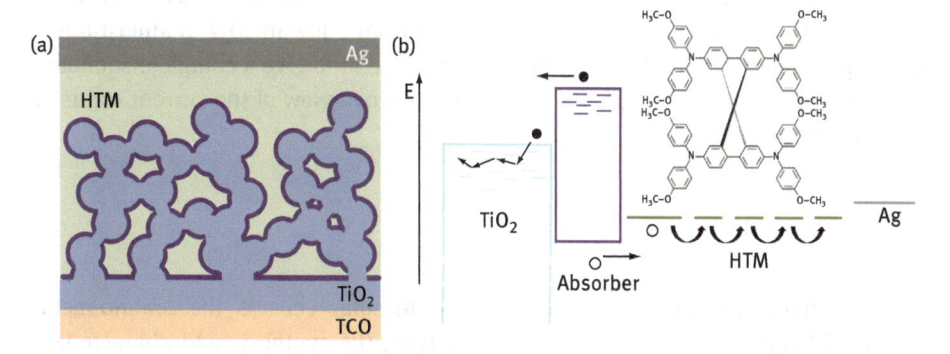

Fig. 3.60: Working principle of an extremely thin absorber solar cell. The geometry of extremely thin absorber devices is comparable to that of a solid-state dye-sensitized solar cell, with a nanometer-thin absorber coating as a substitute for the self-assembled dye monolayer.

3.5 Perovskite Solar Cells

A new class of solar cells based on an organo-metal halide perovskite emerged in the past years. It started off as semiconductor sensitizer in *extremely thin absorber cells*. In the beginning it was therefore seen as part of the dye-sensitized solar cell family. However, it turned out that it functions considerably different than dye-sensitized or organic solar cells. As described before, these cells are of excitonic nature, meaning that excitons and their separation plays a major role in the way the cells, functions. A donor–acceptor interface is needed to overcome the exciton binding energy of 0.1 to over 1.0 eV and split the exciton in free polarons. Therefore, excitons need to migrate to the interface or, as in the case of dye-sensitized solar cells, the exciton is directly located at such an interface as the dye is attached to a mesoporous TiO_2 layer. In both cases exciton separation is possible only due to the energy offset at this donor–acceptor junction as described in Section 3.3 and Section 3.4.1.

In perovskite solar cells the exciton-binding energy is much smaller and, as also found in inorganic semiconductor solar cells, in perovskite solar cells too we find Wannier-type excitons. However, still perovskite solar cells are considered as hybrid solar cells as the material itself, an organo-metal halide, is a hybrid material, and in most perovskite architectures an organic hole transport material, is applied as hole-selective contact.

In this chapter we will summarize the current status of perovskite solar cells. We start with a brief historic background and describe the organo-metal halide perovskite structure. Important for the preparation of perovskite solar cells is the fabrication technique. We will summarize the different techniques that have been developed to prepare the perovskite thin films leading to high-efficiency solar cells. Finally, we will address the current understanding of the device physics and look into the initial results on stability. Currently this field is one of the fastest growing fields and the number of publications has risen from a few in 2012 to almost a few new papers every day. In this chapter we do not intend to give a complete summary over the current literature, but give a very selective overview of the current status of perovskite solar cells.

3.5.1 History of Perovskite Solar Cells

Currently the dominant material for perovskite solar cells is the organo-metal halide $CH_3NH_3MX_3$ (with M=Pb and X=I). In 1978, the synthesis and characterization of this material were first described by Weber [116, 117]. More than a decade ago it has been investigated as a material in thin-film field-effect transistors [118]. Their successful implementation in solar cells began in 2009 with the first attempt to use methylammonium lead-halide perovskite ($CH_3NH_3PbX_3$ (X=Br and I)) as semiconductor sensitizer in liquid-electrolyte dye-sensitized solar cells [16]. As these first cells suffered from relatively low efficiencies (around 3–4%) and strong degradation issues, they had been overlooked for a while until 2011 when another report came out where an efficiency of 6.5% was presented [119]. Now the efficiency had reached values comparable to that of other extremely thin absorber cells.

However, still the cells suffered from low stability as the lead-halide perovskites are highly unstable in the liquid electrolyte used in the dye-sensitized solar cells. The breakthrough for these cells came in 2012 when the groups of Snaith and Grätzel [17, 18] reported the use of perovskite layers as thin absorber in solid-state dye-sensitized solar cell architectures. These initial cells showed a significantly improved stability and performance close to 10%. Interestingly, in the results of the Snaith group the mesoporous TiO_2 had been replaced by a mesoporous Al_2O_3 layer, which already gave an indication that the material allows ambipolar charge transport and can outperform the extremely thin absorbers layers used before in several aspects.

Since then, the field of perovskite solar cells developed extremely fast, and many different groups started to research on this exciting material. The efficiency increased to over 20% and we have already learned a lot about this exciting material. However, there still remain many open questions, including a full explanation of why this solution-processed material functions so extraordinary well in solar cell devices.

3.5.1.1 Perovskite Solar Cell Architectures

Over time, different perovskite solar cell architectures have been developed. Starting with an extremely thin absorber structures (see Figure 3.60), it turned out that thicker absorber layers allow higher charge generation without a large increase in recombination, leading to a structure as presented in Figure 3.61. This structure is still very common with a mesoporous TiO_2 layer, which has a thickness of ~200 nm (much thinner than in dye-sensitized solar cells), and a perovskite overlayer on top of this mesoporous layer. In 2012, Lee et al. reported that the mesoporous layer of TiO_2 can be replaced by a mesoporous Al_2O_3 layer [17]. In this early report the device based on the Al_2O_3 structure showed and even improved efficiency due to a higher open-circuit voltage as compared to the identical cell with a TiO_2 scaffold. This is explained by the energy loss when electrons drop from the perovskite layer into the conduction band of the TiO_2. Even though this structure is now rarely used, it is scientifically very interesting, as it shows that the perovskite layer is ambipolar and can conduct electrons and holes. A donor–acceptor interface (as for excitonic solar cells) is not needed for efficient charge separation.

One conclusion of this finding is that the cancelation of the mesoporous layer should be possible. And indeed, planar p–i–n and p–n type solar cells have been fabricated and show impressive efficiency not lacking behind mesoporous films. In Figure 3.62 schematics of the planar architectures are shown.

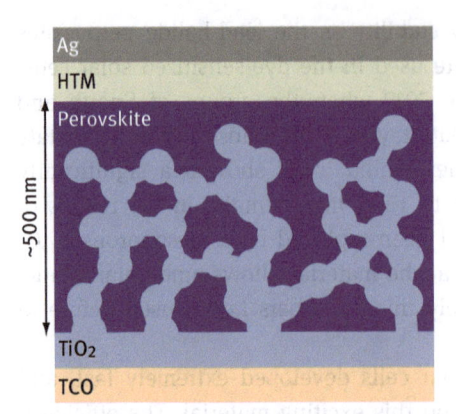

Fig. 3.61: Schematic of a perovskite solar cell with mesoporous TiO_2. The geometry of the first perovskite solar cell devices is comparable to that of a extremely thin absorber solar cell, with a much thicker absorber layer, which often exceeds the mesoporous layer considerably. In some cases the mesoporous TiO_2 is replaced by a mesoporous Al_2O_3 layer.

The cell architecture can be also inverted. Replacing the TiO_2 with a p-type material such as PEDOT:PSS and the hole-transport layer with an n-type material such as a PCBM also leads to efficient perovskite solar cells where the current flow is inverted [120].

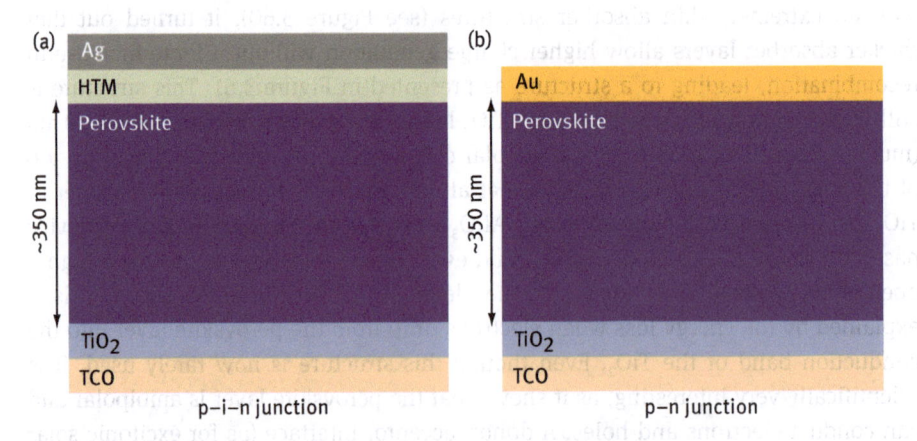

Fig. 3.62: Schematic of a perovskite solar cell with planar architecture. (a) p–i–n junction-type solar cell with TiO_2 as n-type, perovskite as i-type, and the hole-transport layer as p-type material. (b) For the p–n type architecture the n-type layer is represented by the TiO_2 and the p-type layer by the perovskite layer. No hole-transport layer is used in this solar cell architecture.

3.5.2 Organo-Metal Halide Perovskite

In 1839, the first perovskite was found by Gustav Rose, named after the mineralogist Lev Perovski. Its chemical formula is $CaTiO_3$. Later the full class of compounds with the stoichiometry AMX_3 has been called perovskites. They cover a wide range of ternary oxides, nitrides, halides, and other compositions. They form a three-dimensional structure based on corner-sharing anionic MX_6 octahedra, the X-atoms in the corners, the M-atom in the middle of the octahedra. The A-cations are located at the interstices, surrounded by eight octahedra [121]. Many of the interesting properties of different perovskites are based on slight deviations of this structure, such as - structural distortions of the octahedral sub-lattice, which can lead to magnetism or non-stoichiometries paired with electron correlations, which has lead to high T_C superconductors such as $YBa_2Cu_3O_{7-x}$. Other properties, such as ferro- and piezo-electricity or giant magneto-resistance have made perovskites materials for many device applications.

For solar cells the most prominent class of perovskites are the lead halide perovskite, incorporating an organic cation such as methylammonium ($CH_3NH_3^+$). In Section 3.5.2.1, we will focus on the organo-metal halide perovskites.

3.5.2.1 Structure and Phase Transitions

In the case of lead halide perovskites, the lead and halide atoms form the inorganic octahedron acting as the anion, while the organic cation resides in the interstices as shown in Figure 3.63.

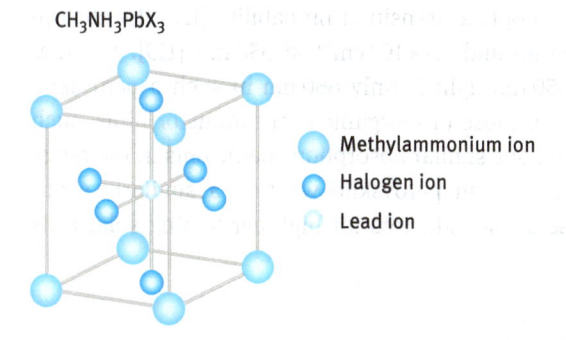

Fig. 3.63: Structure of an unit cell of the commonly employed perovskite absorber in solar cells: methylammonium lead trihalide ($CH_3NH_3PbX_3$, where X is a halogen ion, usually I⁻, sometimes Br⁻, Cl⁻ or a combination of them).

The structure of this inorganic/organic perovskite is strongly influenced by the size of the organic cation. While small cations (such as the commonly used methylammonium) allow to maintain the three-dimensional structure of the perovskite, larger cations will result in a layered structure with inorganic sheets alternating with the organic layer connected by van der Waals forces [122]. The three-dimensional lead halide compound $CH_3NH_3PbI_3$ has four solid phases (α, β, and γ, all perovskites, and a non-perovskite δ phase). At high temperatures with $T > 327$ K, we find the α phase with pseudo-cubical crystal structures [123]. For temperatures below 327 K, the perovskite undergoes a phase transition to the tetragonal β phase [124]. In the α and β phases the methylammonium cations are disordered. Ferroelectric response can be attributed to the reorientation of the methylammonium cations (and their molecular dipole moments) in an external field. Due to the tilting of the octahedra during the phase transition from α to β phase the unit cell doubles its length. At temperature below 162 K, the perovskite has its phase transition to a orthorhombic γ phase, where the methylammonium cations are ordered. The transition to the non-perovskite δ phase occurs only in the presence of solvents, whereas other transitions occur in the solid phase. The bromine equivalent $CH_3NH_3PbBr_3$ has a cubic structure at room temperature due to the different ion size of bromine and iodine in a sixfold coordination [125]. At different phases the degree of rotation of the $CH_3NH_3^+$ has been investigated and rapid rotation has been observed at high-temperature phases, while the rotation is restricted at lower temperatures [126]. This quenching of the molecular motion of the cation leads to a highly ordered arrangement along the C–N axis, which is seen as the reason for the phase transition from cubic to tetragonal to the orthorhombic phase as the temperature is lowered [127].

3.5.2.2 Band Positions and Engineering

The lead halide perovskite is a direct bandgap material with an exceptionally strong absorption due to the high lead p–p optical transition probability [128]. Its absorption coefficient is estimated to be around 1.5×10^4 cm^{-1} at 550 nm [119]. In other words, the penetration depth of 550 nm light is only 660 nm in such a perovskite film. These values are comparable to those of inorganic semiconductor films, such as GaAs, CdTe, and CIGS, where we have similar absorption coefficients in the range of 10^4–10^5 cm^{-1}. Also important is that in perovskite we find a similarly sharp absorption onset as found in these semiconductors for high-performing solar cells (see Figure 3.64) [129].

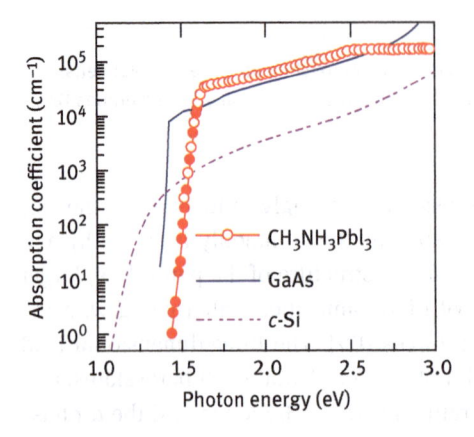

Fig. 3.64: Absorption onset for perovskite, c-Si, and GaAs solar cells. We find comparable strong absorption onset in perovskite solar cells and high-performing GaAs solar cells. Reprinted with permission from [129]. Copyright (2015) American Chemical Society.

The distribution of methylammonium ions in the lattice does not directly affect the valence and conduction band maxima and minima, respectively, but influences the Pb–I bond length, Pb–I–Pb bond angle, and bandgap energy. Therefore it plays an important role in the geometrical stability and electronic structure [127]. The bandgap of $CH_3NH_3PbI_3$ is around 1.55 eV, which is not ideal for panchromatic absorption with its absorption onset around 800 nm. Therefore, a tuning of this bandgap to shift the absorption onset to lower energy without sacrificing on the high absorption coefficient would be beneficial. Replacing the methylammonium cation with other cations allows such tuning of the bandgap. One common replacement for methylammonium is formamidinium ($HC(NH_2)_2^+$), which reduces the bandgap by about 0.07 eV, which allows to extend the absorption wavelength by about 40 nm [124]. The difference in the methylammonium compared to formamidinium are the different ionic radii (1.8 versus 1.9–2.2) [130]. While $CH_3NH_3PbI_3$ shows n-type behavior, $HC(NH_2)_2PbI_3$ has p-type behavior [124]. Solar cells with high efficiencies exceeding 20% have been demonstrated also with formamidinium [131].

3.5.3 Layer Preparation Methods

The tremendous progress of perovskite solar cells in the last years is strongly correlated to the optimization of the preparation method of the perovskite layer. It all started with solution-processed perovskite layers, where the layer is prepared in a single step by spin-coating a mixed solution of CH_3NH_3I and PbI_2 in appropriate solvent. Soon after these first reports of the one-step processing, a two-step processing method was introduced, where it was claimed that the results are more reproducible. In this method, first a PbI_2 layer has been deposited from solution followed by a second deposition of CH_3NH_3I from 2-propanol on the PbI_2 film. Nowadays different variation of these methods exists. In many cases the second step, the deposition of methylammonium-iodide on the PbI_2 film, has been done as vapor phase deposition. Besides, the evaporation of the layers has successfully been demonstrated [132, 133].

This simple processing at low temperature can yield crystalline films of high quality. Due to a low formation enthalpy the conversion happens at room temperature as soon as the reactants are in direct contact. Another common method is co-evaporation of the two precursor ions in vacuum, which has shown high-efficiency devices with good film coverage and uniformity [133].

It is difficult to judge which method is the best as the ideal preparation method might also strongly depend on the device geometry and architecture (such as planar or mesoporous and inverted, noninverted), including the type of contacting layers (see Section 3.5.1.1). In all cases the coating parameters need to be adjusted, which includes spinning rate and times, solution wettability and viscosity, solvent or solvent mixture, temperature of solution and substrate, and post-annealing processes. In the crystallization process of the perovskite film moisture should be excluded, but a small amount of moisture might be beneficial. In the next paragraph we will shortly introduce the most common solution deposition methods.

3.5.3.1 One-Step Preparation Method
A common approach to form the perovskite film is to mix the reactants, such as PbI_2 and CH_3NH_3I, in polar aprotic solvents like y-butyrolactone (GBL), N,N-dimethylformamide (DMF), or dimethyl sulfoxide (DMSO). In most recipes there is a strong excess of methylammonium and halide compared to the lead content used in the precursor solution (2:1 or even 3:1 molar ratio of methylammonium-iodide to PbCl), which mostly evaporates when the layer is thermally annealed above a certain threshold [134]. The film is prepared by spin-coating the solution onto a mesoporous or planar substrate. Important for a good solar cell performance is a pin-hole-free film. Unfortunately, often dewetting leads to islands and clusters of perovskites on the substrate, which is causing strong charge recombination in the device. For the planar architecture especially, full coverage of the layer is a challenge, but needs to be achieved for high-performing cells. This has been demonstrated by Eperon et al., which can be seen in Figure 3.65, where the coverage has been varied depending on the processing conditions. The most complete coverage by the perovskite layer has given the highest efficiency values for the solar cells [137].

The film formation process during deposition of the perovskite is mainly based on nucleation and crystal growth. As a result of the evaporation of the solvent of the perovskite precursor, a saturated or even oversaturated solution will be obtained, which triggers the precipitation of crystal nuclei of the perovskite and growth of these initial crystals. The density of the crystal nuclei decides to a large extent on the final film morphology and crystal size of the solid perovskite film. Therefore, control over the film formation and growth dynamics helps improve the perovskite film's quality.

Solvent engineering is an important tool to influence this dynamics and different approaches have been applied to improve the control over and quality of the perovskite layers. There are many different approaches and we will pick out only a few examples to demonstrate how solvent engineering can improve the perovskite film quality leading to improved device performance. Jeon et al. reported the use of a mixed solvent of y-butyrolactone (GBL) and dimethyl sulfoxide (DMSO) followed by toluene drop-casting, which leads to uniform and dense perovskite layers via a CH_3NH_3I–PbI_2–DMSO intermediate phase. They observed a significant improved device performance [135]. At the initial stage of spin-coating the film, it is composed of methylammonium-iodide (MAI) and PbI_2 dissolved in the GBL/DMSO solvent mixture. Then in an intermediate stage after evaporation of GBL, the composition of the film is concentrated. Subsequent drop-casting of toluene leads to an immediate freezing of the constituents on spinning via quick removal of the excess of DMSO solvent and rapid formation of an MAI–PBI_2–DMSO phase, which then forms

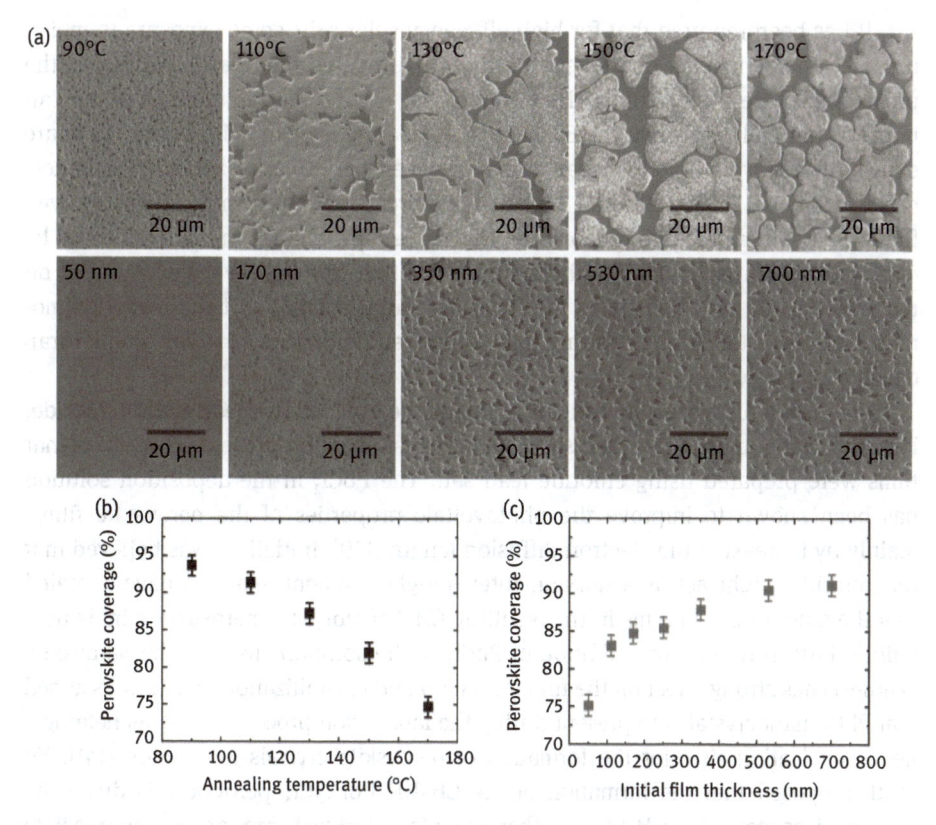

Fig. 3.65: Dependence of device performance on perovskite coverage. Top row: SEM images showing dependence of perovskite coverage on annealing temperature. Bottom row: effect of initial perovskite film thickness with annealing temperature fixed at 95°C. Perovskite surface coverage as function of **(b)** annealing temperature and **(c)** initial film thickness. Figure adapted from [134] with friendly permission by Wiley and Sons (2015).

a uniform transparent thin layer, which is converted to a pure crystalline perovskite layer after annealing to 100°C. The role of the DMSO is to slow down the rapid reaction between PbI_2 and MAI during the evaporation of the solvent in the spin-coating process. Another recipe reported by Xioa et al. is based on spin-coating of $CH_3NH_3PbI_3$ from DMF solution, followed immediately by exposure to chlorobenzene to induce crystallization [136]. Here the role of the second solvent is to rapidly reduce the solubility of $CH_3NH_3PbI_3$ in the mixed solvent and therefore promote fast nucleation and growth of the crystals in the film. Another approach is the incorporation of additives into the precursor solution. Liang et al. added 1,8-diiodooctane (DIO), which temporarily chelated with the Pb^{2+} during the crystal growth leading to an improved solubility of $PbCl_2$ when DIO is added. This allows a more homogeneous nucleation by changing the kinetics of the crystal growth [137].

It has been reported that for high-efficiency solar cells some exposure to moisture might be beneficial. Eperon et al. further investigated the role of moisture in the processing of perovskite films. They found that a controlled amount of water can improve the cell performance. Even by post-treatment of dry films with moisture exposure they were able to enhance the solar cell performance and photoluminescence behavior. Such cells had higher open-circuit voltages, indicating improved film quality after exposure to moisture. No microscopic change in the films could be observed, which is why the authors concluded that this improvement is based on reduction of the trap density in the films by partial solvation of the methylammonium and recrystallization, leading to an improved perovskite film with smaller trap density [138].

There has also been investigation about the **role of the lead salts** (chloride, iodide, nitrate, and acetate lead salts) on the crystallization process. The first planar films were prepared using chloride lead salt. The $PbCl_2$ in the deposition solution has been shown to improve the photovoltaic properties of the perovskite films, mainly by increasing the electron diffusion length [139]. Initially it was believed that the chloride might act as a dopant. Interestingly, element analyses have revealed that the amount of Cl atoms in the resulting $CH_3NH_3PbI_{3-x}Cl_x$ perovskite film is negligible [140]. However, the addition of $PbCl_2$ to the solutions used in the perovskite synthesis has strong effect on the film formation and crystallization. It is now assumed that $PbCl_2$ nanocrystals are present during the fabrication process, acting as heterogeneous nucleation sites for the formation of perovskite crystals in solution [141]. Yu et al. proposed that the formation of the $CH_3NH_3PbI_{3-x}Cl_x$ perovskite is driven by release of gaseous CH_3NH_3Cl (or other organic chlorides) through an intermediate organometal mixed halide phase. The initial introduction of a $CH_3NH_3^+$-rich environment retards the perovskite formation process, which leads to an improved growth of the crystal domains during annealing. So the function of the Cl^- is to facilitate the release of excess of $CH_3NH_3^+$ at relatively low temperatures [140]. An additional effect might be a preferential location of the chloride at the TiO_2 interface inducing a band bending, promoting charge collection at this interface [142].

Moore et al. performed in-situ X-ray scattering studies to investigate the isothermal transformation of perovskite films derived from chloride, iodide, nitrate, and acetate lead salts. The activation energy for crystallization differs depending on the used salt. The formation toward the perovskite formation is based on different steps indicated in Figure 3.66 [143]. Initially the solvent evaporates ("Evap. Solvent," dark blue), followed by the diffusion of the excess of spectator salt out of the precursor structure ("Diffuse (MA)SP," cyan), the removal of the spectator salt from the film ("Sublime(MA)SP," light cyan), and the removal of stoichiometric (MA)I from the perovskite lattice ("Diffuse (MA)I," light blue). The diffusion of (MA)I from the perovskite lattice is a disadvantageous process as it corresponds to a decomposition to PbI_2 [143]. The acetate system does allow a decoupling of crystal growth and salt removal and decomposition in time. It shows coarsening of the crystal domains after complete crystallization, which is advantageous for the perovskite film quality and leads to more reproducible and high-efficiency planar perovskite solar cells [144].

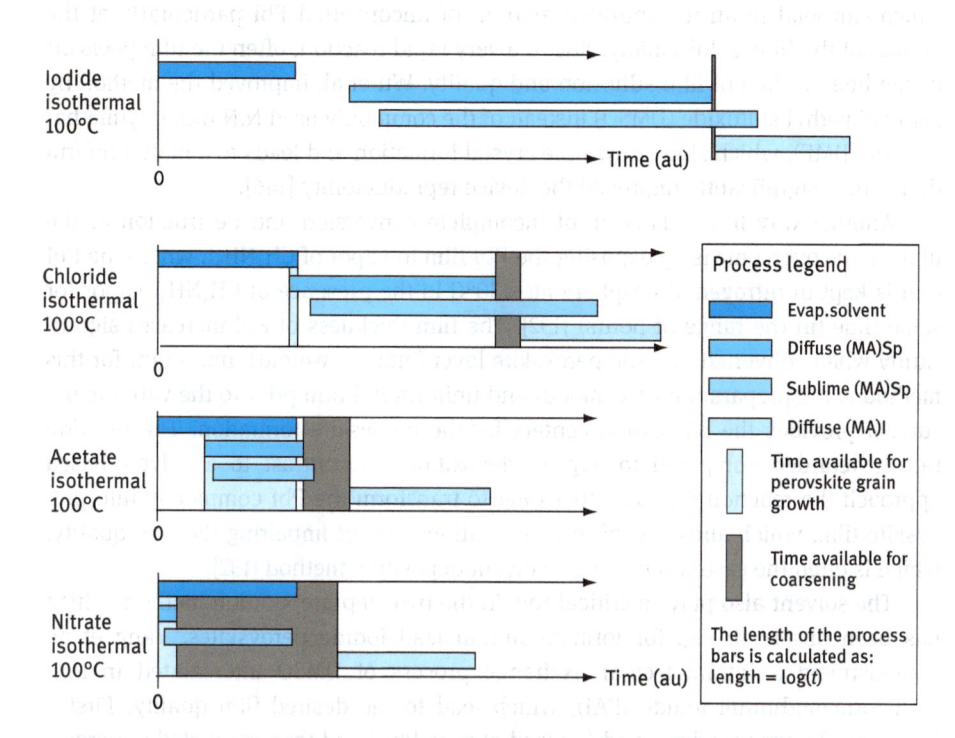

Fig. 3.66: Schematic representation of the perovskite crystallization and growth pathway as a function of lead anion. Schematic timeline for lead salt systems. The legend shows the title for each colored bar. Although the pathway timeline for each process is intended to be schematic, the start time and length for each process are calculated from experimental data. Reprinted with permission from [143]. Copyright (2015) American Chemical Society.

3.5.3.2 Two-Step Preparation Method

A common method to prepare the perovskite film is a two-step process. In a first step a PbI_2 film is deposited and then in a second step the PbI_2 layer is exposed to CH_3NH_3I. The idea behind this two-step preparation method is that there are no suitable solvents that can dissolve both components equally well and the high reaction rate often leading to nonuniform films with pinholes and incomplete surface coverage, which has detrimental effects on the device performance. A first report about this two-step preparation method for perovskite solar cells has been published by Burschka et al. [145]. This technique involves fabrication of solid-state mesoscopic solar cells to increase the reproducibility of their performance and to achieve high-power conversion efficiencies. The PbI_2 layer is transformed to perovskite as soon as it is in contact with the CH_3NH_3I, which has been supplied from solution in the first reports [145]. This method is, however, less suitable for planar perovskite cells, as a compact film of several hundred nanometers requires a long reaction time due to the limited access of CH_3NH_3I through the top of already converted $CH_3NH_3PbI_3$ film, which can lead to an uncontrolled amount of unconverted PbI particularly at the bottom of the film. Additionally, due to a very rapid reaction, often the film peels off depending on the PbI film adhesion and quality. Wu et al. improved the method by using dimethyl sulfoxide (DMSO) instead of the commonly used N,N-dimethylmethanamide (DMF), which slows down the crystal formation and leads to a more uniform film, which significantly improved the device reproducibility [146].

Another way how this issue of incomplete conversion and destruction of the film can be overcome is by exposing the PbI film to vapor of CH_3NH_3I, where the PbI film is kept in nitrogen atmosphere at ~150°C in the presence of CH_3NH_3I vapor for some time (in the range of hours) [132]. The film thickness of PbI increases significantly when converted into the perovskite layer (almost twofold). Important for this method is the preparation of a smooth and uniform PbI film prior to the vapor exposure. It provides the nucleation centers for the perovskite formation. The reaction rate is retarded compared to, e.g. co-deposition. In contrast to a solvent-based approach the reaction time is rather long to transform the PbI completely into perovskite film, which allows a soft transformation without impairing the film quality, which is often the case when using a solvent deposition method [132].

The solvent also plays a critical role in the two-step preparation method, which has been observed, e.g. for formamidinium lead iodide perovskites. Yang et al. demonstrated a intramolecular exchange process of DMSO intercalated in PbI_2 with formamidinium iodide (FAI), which lead to the desired film quality. First a PbI_2–DMSO layer was deposited (instead of pure PbI_2) and then converted to perovskite through exchange of DMSO with the FAI molecules. As the FAI molecules experience ionic interactions, they replace the DMSO molecules, which are held with lower affinity based only on van der Waals interactions. Therefore, highly uniform and dense PBI_2–DMSO layers could be directly converted to perovskite films without further volume expansion as the size of FAI and DMSO is similar [131]. Devices prepared by this method gave maximum efficiencies exceeding 20%.

3.5.3.3 Flexible Perovskite Solar Cells

One major technological aspect of perovskite solar cells is fabrication of flexible cells on foil, This will allow the use of roll-to-roll fabrication processes and easier handling, as such cells will be of lightweight, can be of delivered in rolls, and will be integrated into different shapes, which will open up additional applications.

Currently, the TiO_2 compact and mesoporous layer employed in most perovskite cells needs sintering at temperatures of ~500°C, which is not compatible with transparent flexible substrates such as PEN or PET (see Chapter 5 for more details). Therefore, for the fabrication of flexible perovskite solar cells there are mainly two general approaches: in one approach the TiO_2 is replaced by other metal oxides, which can be processed at low temperatures (or using pin-hole-free, low-temperature deposition methods for amorphous TiO_x layers as possible with atomic layer deposition [147]); another approach is using organic layers, such as PCBM as hole-blocking and PEDOT:PSS as electron-blocking contact. Both approaches have been successful and we give only examples here for both types of structures.

Liu et al. demonstrated a planar cell architecture with a 25-nm-thick ZnO layer, leading to an efficiency of over 14% on glass substrates and over 10% on flexible substrates [148]. An example of this cell is shown in Figure 3.67, which also demonstrates only a small decrease in efficiency with bending. Shin et al. demonstrated flexible solar cells by using highly dispersed Zn_2SnO_4 nanoparticles on flexible ITO substrates, which resulted in maximum efficiencies of over 14% [149].

Docampo et al. sandwiched the perovskite layer between organic contacts. They achieved with this configuration devices with power-conversion efficiency of up to 10% on glass substrates and over 6% on flexible polymer substrates using PEDOT:PSS and PCBM as the organic layers [150].

Efficiencies exceeding 10% are very impressive for flexible devices. However, currently long-term stability tests are missing and will certainly be a challenge for these cells, even more than for glass-encapsulated perovskite solar cells. For more details on stability of perovskite solar cells in general, the reader is pointed to Section 3.5.5.

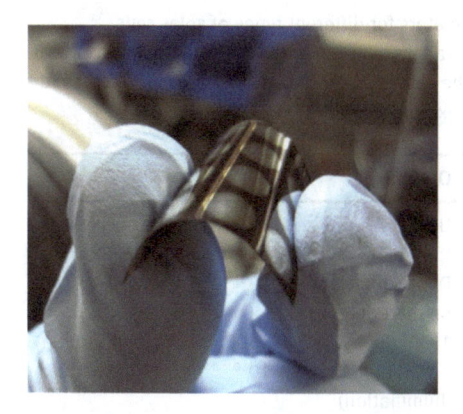

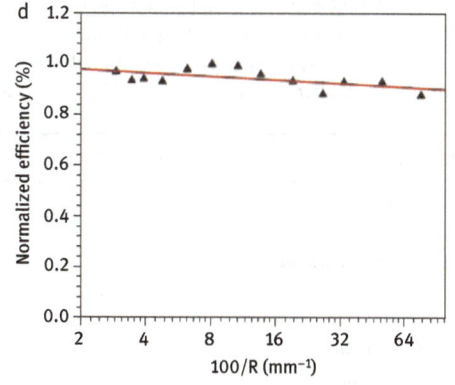

Fig. 3.67: **Photograph of an ITO/ZnO/$CH_3NH_3PbI_3$/spiro-OMeTAD/Ag device** prepared on a flexible PET substrate and normalized device power conversion efficiency (PCE) after first bending the substrate around a cylindrical object of the specified radius (R). All measurements were performed on a single device and were measured from the highest to the lowest radius of curvature. Reprinted with permission from Macmillan Publishers Ltd: Nature Photonics [148]. Copyright (2014).

3.5.4 Device Physics

The device physics of perovskite solar cells is quite different than the organic and hybrid solar cells. In Table 3.2 an overview of some important solar cell indicators of different types of cells is given, which clearly shows that perovskite solar cells come in many aspects very close to GaAs, the leading material for single-junction solar cells in terms of efficiency. It should be noted that we compare maximum lab values for the perovskite and organic photovoltaics, which have not necessarily reproduced or independently confirmed, and textbook values [170] for the inorganic solar cell materials.

Currently the understanding of perovskite solar cells is far from being complete, even though, next to the improved film formation, the understanding of the device mechanisms in perovskite cells has significantly improved. Nevertheless, still many questions on the detailed mechanisms are undiscovered and need to be revealed in the future for systematic improvements and a full picture. Here we will give a brief overview of the current understanding of the physical mechanisms of perovskite solar cells. Before we begin with the aspects of photoexcited states and their relaxation (Section 3.5.4.2), the role of excitons and their binding energy (Section 3.5.4.3), charge transport and carrier diffusion length (Section 3.5.4.4), interfaces (Section 3.5.4.5), and finally the potential role of the ferroelectricity (Section 3.5.4.6), we will introduce a measurement protocol of perovskite solar cells (Section 3.5.4.1), which is extremely important to allow a fair comparison between the results of different laboratories.

Tab. 3.2: Overview of some important solar cell parameters for different types of solar cells (The values for perovskites and organic photovoltaics are taken from different literature sources and are not necessarily independently confirmed, while the values for the inorganic solar cells are based mostly on established values published in the textbook by Sze and Ng [170].)

Material	Perovskite	Organic PV	Si	GaAs
Bandgap (eV)	1.5 (tunable)	UV-NIR (tunable)	Fixed 1.1	Fixed 1.43
Absorption coefficient (1/cm)	10^{4-5}	Depending on material	10^3	10^{4-5}
Carrier mobility (cm^2/Vs)	Up to 2000	1	1500	8500
Intrinsic carrier concentration (1/cm^3)	10^{16-17}	~ 10^{17} (under illumination)	10^{10}	10^6
Carrier lifetime	>100 ms	~μs	~ms	~100 ns
V_{oc} (V)	1.1–1.2	0.8	0.74	1.12
V_{oc} deficit (V)	0.3–0.45	≥0.6	0.3–0.4	0.3
J_{sc} (mA/cm^2)	~ 24	≤ 20	~ 42	~ 29
Fill factor	~ 80%	~ 70%	~ 80%	~ 85%
PCE	~ 20%	~ 10%	~ 25%	~ 30%
Film thickness	~350 nm	100–200 nm	100–200 μm	4 μm

3.5.4.1 Perovskite Solar Cell Efficiency Measurements

As the efficiency of perovskite solar cells has risen very quickly to over 20%, it is safe to say that further significant improvement will be more difficult. The efficiency based on the Shockley–Queisser limit of perovskite solar cells will not exceed values between 25% and 30% depending on the specific perovskite in use and its bandgap. Therefore, it is extremely important that the expected incremental improvements are based on correct measurements. As described in Section 4.2.2, care has to be taken to ensure correct values. This is even more critical for perovskite solar cells as these devices in many cases show a hysteretic effect when scanning the J–V characteristic in forward and backward directions. The hysteresis does not only depend on the scan direction and the scan speed, but also on many additional factors, such as preparation method of the perovskite film, the solar cell architecture, the treatment history of the device (e.g. light soaking or applying a bias to the cell prior to the measurement), and others. One example for strong hysteretic behavior is shown in Figure 3.68.

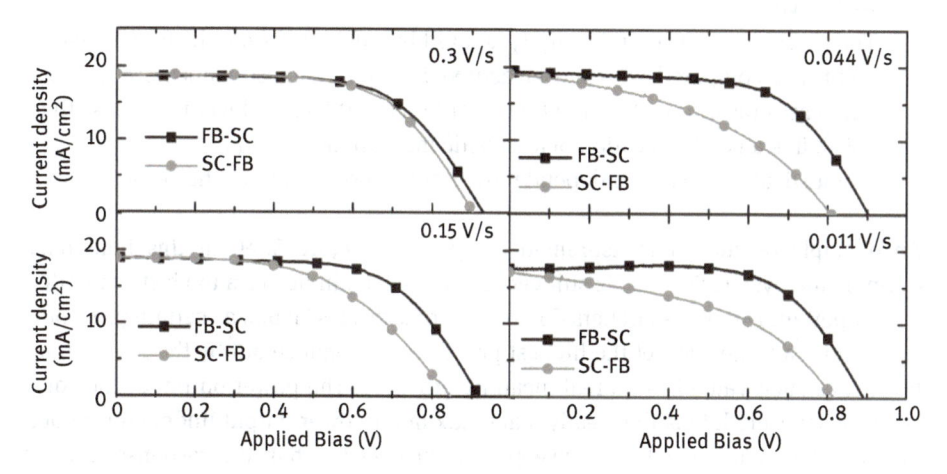

Fig. 3.68: Effect of sweep time and sweep direction on J–V curve measurements. Data are shown for sweeps from forward bias to J_{sc} (FB-SC, black) and in the opposite direction (SC-FB, gray) for sweep times between 0.3 V/s and 0.011 V/s for planar heterojunction perovskite solar cells. Reprinted with permission from [151]. Copyright 2014 American Chemical Society.

Therefore, the resulting efficiency value depends strongly on how the cell has been measured. However, only steady-state values are correct and meaningful values, and therefore it is very important to have a strict measurement protocol, which gives a real efficiency value and is reproducible independently in which laboratory it has been measured. This is of critical importance since perovskite cells show a strong variety in their behavior, often even cells where there has been no obvious difference in the cell fabrication. Perovskite solar cells can behave very differently. Some cells might not show any hysteretic behavior and the scan rate might have no influence on the measurement results; in others even unrealistically slow scan rates might still not give a hysteresis-free measurement. Therefore, it is difficult to give a strict measurement protocol, which is valid for all cells. However, there are some rules additionally to the standard measurement rules for solar cells given in Section 4.2.2 that should be considered when measuring perovskite solar cells (based on suggestions by Snaith et al. and Christians et al. [151, 152]).

- Maximum power-point P_{max} tracking should be performed until the value does not change anymore.
- The cell should be measured starting from the steady-state maximum power-point P_{max} to V_{OC} and then in reverse direction to J_{SC}. Both forward and reverse J–V curves should be plotted, especially if they do not superimpose exactly. Time-resolved J–V characteristics should be used to find a scan rate (delay time between two measurement points), which allows measuring stable equilibrium values.
- Scan rate and scan directions should always be accompanied by the recorded J–V curve.
- The short-circuit current density J_{SC} should be calculated from the EQE measurement and compared with the measured J_{SC}. Only if these values are in close agreement (as a rule of thumb they should have at least a deviation of less than 20%), it is a good indication of a realistic measurement of J_{SC}.
- A statistical analysis of independently produced solar cell batches is necessary.

An example of such a measurement is given in Figure 3.69. In this imaginary example the hysteresis effect is still very strong, which indicates a too fast scan rate.

Important is that researchers do not use some "tricks" (such as certain preconditions and scan rates) to get the highest possible performance out by the J–V characteristic, as such value has no real meaning anymore. The power output of the solar cell or a solar module is the steady-state maximum power output under continuous illumination. When a potential sweep is applied faster than the response time of the solar cell, this is no longer a steady-state measurement. Unfortunately, perovskites solar cells can have extremely long stabilization times of minutes rather than seconds. Therefore, the appropriate sweep rate must be selected based on the response time of each measured cell. It seems that high-efficiency, "well"-performing perovskite cells usually show a smaller or even no hysteresis effect, and it might be that if processing further improves, this issue will be resolved. However, currently it seems that efficiencies of perovskite solar cells are sometimes overestimated and wrongly reported as the same cell allows very different efficiency numbers depending on how it has been measured. We therefore suggest the above protocol, which makes the cell behavior very transparent and allows a better estimation of the validity of the given solar cell efficiency value. Also the reproducibility of perovskite solar cell results is often very difficult, so we find a strong deviation of efficiency values for "identically" processed cells. Therefore, more and more researchers also provide a statistical analysis of independently produced cell batches, which we see as important information that should always be provided for perovskite solar cell results, especially where the focus is on improved processing conditions.

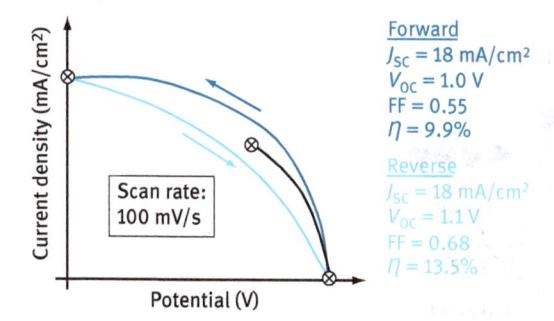

Fig. 3.69: Example for reporting a J–V characteristic of a perovskite solar cell. In the characteristic a value for a steady-state photocurrent density at short-circuit, open-circuit voltage and close to the maximum power-point is given (cross in circle) and scan direction and rate are reported. The scan started with the steady-state value close to the maximum power point in forward direction to V_{OC} and reverse to J_{SC}, before going in forward direction to V_{OC} again.

Forward
$J_{SC} = 18$ mA/cm^2
$V_{OC} = 1.0$ V
FF = 0.55
$\eta = 9.9\%$

Reverse
$J_{SC} = 18$ mA/cm^2
$V_{OC} = 1.1$ V
FF = 0.68
$\eta = 13.5\%$

Scan rate: 100 mV/s

3.5.4.2 Photoexcited States and Relaxation Dynamics

Knowing the photoexcited states in perovskites and their relaxation dynamics will help understand the photogeneration mechanisms better. We find in perovskite films radiative and nonradiative recombination processes. In the ideal case the nonradiative decay channels can be prevented either through the preparation of a defect-free perovskite film or maybe through filling of subgap states through doping or chemical passivation. If only radiative recombination exists, the solar cell efficiency will be closer to the Shockley–Queisser limit. Rau reported about this reciprocity relation between the photovoltaic quantum efficiency and the electroluminescent emission of solar cells [153]. Tvingsted et al. and Tress et al. recently applied this reciprocity relation to perovskite solar cells [154, 155]. The emission of the perovskite films is dominated by the band-to-band transitions with a factor 200 higher radiative efficiency than in organic solar cells, which leads to a much smaller loss in V_{OC} compared to the organic solar cells. At higher illumination intensities, bimolecular recombination has been observed, where the recombination rate is proportional to the square of the electron density: $\Gamma \propto n_e^2$. However, at low fluences, monomolecular recombination has been observed, where $\Gamma \propto n_e$ [156, 157]. Under high excitation a high photoluminescence quantum efficiency is observed. Stranks et al. explained this behavior with a distribution of subgap trap states that becomes filled. Holes are present as "background" or photo-doped charges as some of the electronic traps are filled. These have a very low depopulation rate. At low fluences the number of photoinduced electrons is now much lower than the total concentration of free holes, which leads to an almost mono-molecular recombination behavior. However, at higher fluences the concentration of the photoexcited electrons becomes similar to the hole concentration and bimolecular behavior becomes dominant [158]. This behavior is schematically shown in Figure 3.70.

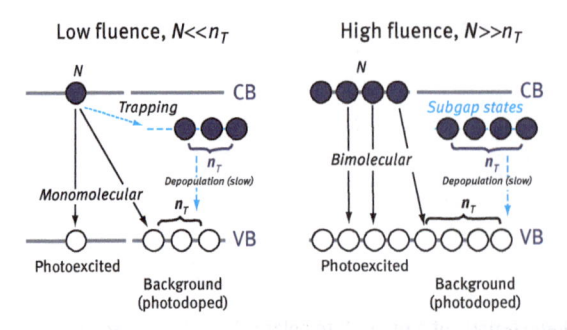

Fig. 3.70: Schematic illustrating the recombination mechanism for low and high fluences in perovskite films. For simplicity the subgap states are all drawn at the same energy even though a distribution of subgap states can be expected. Figure after [158].

In temperature-dependent photoluminescence measurements the transitions of the different phases of the perovskite crystal can be observed. Above ~300 K the structure forms the cubic phase, at room temperature the tetragonal phase, and at ~160 K the orthorhombic phase. A shift of the absorption band edge of 0.1 eV is observed at the tetragonal-to-orthorhombic phase transition. Interestingly, it has been observed by Wehrenfennig et al. that the tetragonal phase present at room temperature is quite independent of excitation intensity, which indicates that it appears to be either free of inclusions of other phases with lower bandgap as observed for the lower and higher temperature regimes or ineffective as charge carrier recombination centers [159].

3.5.4.3 Excitons and Exciton Binding Energy

Other than dye-sensitized organic or hybrid solar cells, perovskite solar cells are not excitonic solar cells. No donor–acceptor junction is necessary to split excitons. In 1994, theoretical studies by Hirasawa et al. have reported exciton binding energy close to kT at room temperature. In $CH_3NH_3PbI_3$, the binding energy was calculated to be of 37 meV, indicating a Mott–Wannier-type exciton with large Bohr radius of 28 Å [160]. Experimental values for the exciton binding energy in perovskite films have confirmed values below 50 meV [161, 162]. In contrast to this, the exciton binding energy in organic semiconductors is in the range of 0.2–1.0 eV, leading to their nature as Frenkel excitons. The low dielectric constant of organic semiconductors (usually ~3) does not allow efficient charge screening and is the origin for the large exciton binding energy. The dielectric constant of perovskite is much larger with a value of 32 in the dark [163] increasing after illumination to values of >500–1000 [163, 164]. This giant dielectric constant can lead to ultra-low binding energies of ~2 meV under illumination [165].

3.5.4.4 Charge Transport and Carrier Diffusion Length

One important aspect of the perovskite solar cells is the ambipolar charge transport in the perovskite layer. Heo et al. reported thin-film transistor measurements of $CH_3NH_3PbI_3$ with ambipolar character with a predominant p-type behavior [166]. Therefore, the material can act as both, depending on the junction formed with the neighboring semiconductor. In combination with TiO_2, it acts as p-type conductor; at the interface with the hole-transport layer, it acts as an n-type conductor. So it is possible to separate charges at both interfaces. This is the reason why an HTM-free p–n-type perovskite solar cell architecture is possible, where the perovskite is directly in contact with a gold electrode on one side (Figure 3.62). In the p–i–n junction-type the perovskite acts as an undoped intrinsic semiconductor sandwiched between a p- and an n-type semiconductor. In the p–n-type architecture the perovskite usually takes over the role of the p-type semiconductor, even though there are reports that it also can be used as an n-type layer. You et al. reported that mixed halides $PbI_{3-x}Cl_x$ perovskite films show week n-type behavior [120]. The role of the chloride and the precise amount of chlorine incorporated in the perovskite lattice is still under debate. In any case it is a very small or even negligible amount and it is assumed that the main effect is a different film crystallization in the presence of chloride ions [140, 141]. However, this has a large effect on the lifetime and charge mobility. PL-quenching experiments showed an electron and hole diffusion length of over 1 μm in the mixed halide compared to roughly 10 times smaller diffusion length in the pure $CH_3NH_3PbI_3$ film [139]. One reason for this improved lifetime is seen in the role of the chlorine when forming the perovskite crystal. The $CH_3NH_3PbI_{3-x}Cl_x$ formation is driven by the release of gaseous CH_3NH_3Cl or other organic chlorides. In the crystal-forming process a $CH_3NH_3^+$-rich environment is important to slow down the perovskite crystallization to lead to improved growth of crystal domains during an annealing step. The function of Cl^- is then to facilitate the release of this excess of $CH_3NH_3^+$ already at relatively low temperatures [140, 167].

Eames et al. [168] suggested that hybrid halide perovskites are mixed ionic and electronic conductors. Ionic transport seems also an important factor in these cells and might explain the unusual behavior concerning the current–voltage hysteresis and also the low-frequency giant dielectric response. How much ion diffusion contributes to the current depends strongly on the intrinsic iodide ion vacancies in the material, which differ from cell to cell as the vacancy density strongly depends on the fabrication conditions and sample history. They calculated the activation energies for ionic migration in the perovskites and found an activation energy of 0.6 eV for facile vacancy-assisted migration of iodide ions, which is in agreement with kinetic measurements [168].

3.5.4.5 Interfaces in PSCs

In all solar cells, interfaces play an important role as charges have to get across these interfaces to be collected at the electrodes. Often charge recombination at interfacial defects reduces the efficiency of the cells, significantly and needs to be minimized. By suppressing surface recombination using appropriate interfacial layers the open-circuit voltage of the cell can be increased. Additionally, at interfaces we find natural interfacial defects due to the breaking of the periodicity, which can act as charge carrier traps. In $CH_3NH_3PbI_3$ perovskite solar cells, these will consist of under-coordinated iodide or lead ions at the surface of the perovskite film. This may cause static charge accumulation and lead to increased recombination.

We know from inorganic solar cells that special care is taken to the metal–semiconductor interface, which is modified by doping, interfacial buffers, passivation layers, and other methods to overcome the limitations induced by the interface [169]. So far major improvements in the efficiency of these types of solar cells have been achieved by improving the perovskite film quality (see Section 3.5.3), but also significant research efforts have already focused on the interface design. Best efficiencies often have been achieved combining high-quality films with improved interfaces to promote charge transport and minimize recombination.

Heterojunction Interface. See Section 2.3 for a general introduction to semiconductor junctions. Due to free carriers, a Fermi energy level is formed, which results in a band bending, leading to a depletion region. We find a built-in electric field at the junction. Charge recombination can be found in the bulk and at the interfaces. Usually grain boundaries in the bulk can trap charges, which then act as recombination centers. At the interface, interfacial defects can lead to increased recombination. Especially at heterojunctions such defects are impossible to avoid as lattice mismatch, uncoordinated electrons, and thermal vibrations will exist at such interfaces [170].

For perovskite solar cells the basic principles of semiconductor physics apply. However, one difference is that the diffusion length of the minority carriers is comparable to the total thickness of the cell [139, 171].

Currently it is not clear how the lattice mismatch between the top and bottom contact to the perovskite film influences the defect structure. In most perovskite solar cell architectures, free charges are mainly extracted through the n-selective contact (in many cases a perovskite/TiO_2 contact). However, little is known about the exact atomic structure and interactions at this interface. It can be expected that the lattice mismatch will lead to undercoordinated electrons and dislocations. Roiati et al. observed an oriented permanent dipole at this interface, indicating an ordered perovskite layer covering the TiO_2. This ordering might be induced by specific local interactions and might be a reason for the very efficient carrier transport within the perovskite layer [172]. Two possible configurations are proposed by Shi et al., which is illustrated in Figure 3.71 [173].

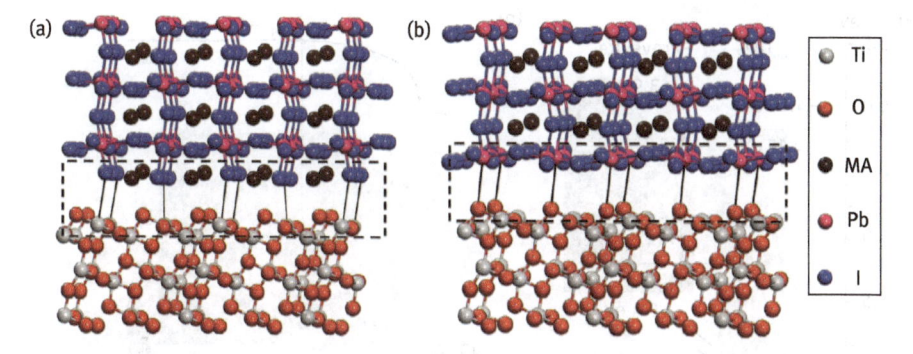

Fig. 3.71: Possible interfacial atomic structure at the perovskite CH$_3$NH$_3$PbI$_3$/TiO$_2$ interface where (a) iodine atoms are coordinated with titanium atoms or (b) lead atoms are coordinated with oxygen atoms. Figure adapted from [173] with friendly permission by Wiley and Sons (2015).

In anatase TiO$_2$ the major exposed plane is (101) with oxygen atoms at the surface and undercoordinated titanium atoms exposed below. Iodine atoms may have a strong and dominating interaction with the titanium atoms and might induce a strong charge displacement at this interface, leading to an interfacial polarization and the observed static dipole moment. However, another possibility might be that lead atoms coordinate with oxygen atoms as strong binding between lead and oxygen has been observed in PbI$_2$ systems [174].

It is also not clear how the mixed halide perovskite with Cl will change the binding and how this will influence the charge transfer.

Not all perovskite solar cells contain the TiO$_2$ layer. In some cases this electron-selective layer has been replaced by ZnO [175] or an organic layer based on fullerenes. In all cases the impact of the binding between the perovskite and the n-type layer has not yet been studied in detail. It can be expected that this will affect not only the charge transfer and recombination at this interface but also the chemical stability and degradation effects based on chemical reactions.

On the other side of the perovskite layer usually a hole-transport layer (HTM) is deposited. As this will be deposited on top of the perovskite layer and usually has an amorphous phase such as the standard HTM Spiro-OMeTAD, it will not influence the growth of the perovskite layer. However, the interaction of perovskite layer and HTM is expected to have a strong influence on the charge transfer. It has been investigated how methoxybenzene (CH$_3$OC$_6$H$_5$) anchors on the perovskite surface. Simulations suggest that adsorption can take place on the (001) surface terminated on PbI$_6$ octahedra as a methoxy group finds there a stable minimum in the interstice of four corner-sharing octahedra. The flat PbI$_2$ (001) surface, however, does not offer such stable minimum for adsorption of methoxybenzene [176].

Interfacial Electric Field. To measure the interfacial electric field, kelvin probe force microscopy (KPFM) has been used. Results demonstrate a continuous electric potential across the cell [177, 178].

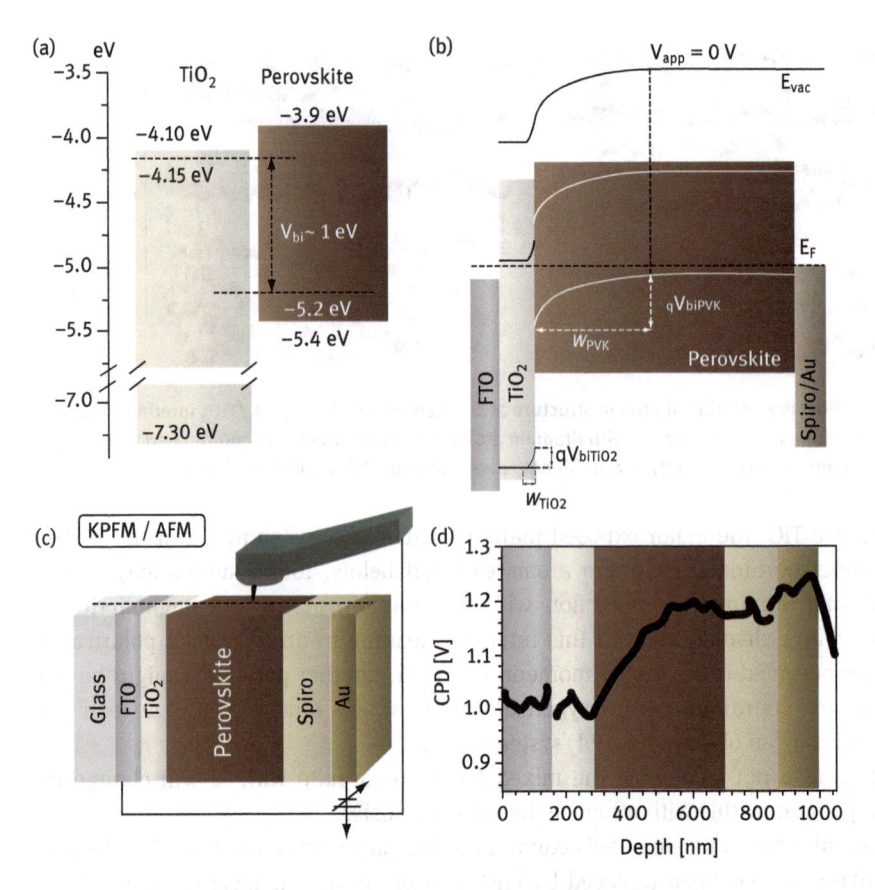

Fig. 3.72: Illustrative diagram of a perovskite solar cell: (a) Energy diagram, a built-in voltage of approximately 1 eV is expected, (b) band diagram in equilibrium showing the formation of a p–n-type heterojunction, (c) drawing of the kelvin probe force microscopy setup indicating the lateral path over which contact potential difference (CPD) is monitored, and (d) the average KPFM signal over several tracks. Reprinted with permission from [178]. Copyright (2014) AIP Publishing LLC.

Clearly visible is a large drop in the contact potential difference (CPC) at the TiO$_2$–perovskite interface (Figure 3.72), leading to an effective electric field with width of several hundreds of nanometers, which supports the fast electron extraction at this interface. On the other side, the perovskite/HTM shows a much weaker electric field. Electron-beam-induced current extraction (EBIC) is another useful tool to map the electron extraction in the different layers of a perovskite solar cell. These measurements show a similar result with two regions of high charge separation and collection, located at the two interfaces of the perovskite layer with the hole-blocking contact and the electron-blocking contact, respectively. This is the typical characteristic of a p–i–n solar cell with a layer of low-doped, high-electronic-quality semiconductor in between a p- and an n-layer. In accordance with KPFM measurements at the interface with the hole-blocking layer (usually TiO$_2$), the extraction capability is more pronounced than at the interface with the HTM. When replacing the electron blocker with a gold contact, only one heterojunction remains and we have a characteristic of p–n-type solar cells [179]. This shows that perovskite cells can act as p–n-junction cells or more as p–i–n-type solar cells depending on the exact nature of the contacts.

Interface Engineering. Controlled manipulation of the interface is an important aspect to improve the performance and stability of perovskite solar cells. Lowering the interfacial energy barriers for charge transport allows to enhance charge transport and suppress recombination. Next to the perovskite layer, interface engineering is one of the key issues for high efficiencies in perovskite solar cells. Many different interfacial modifications have been introduced and we list here only a small fraction of proposed interfacial improvements for perovskite solar cells. In many cases the TiO_2 layer has been modified, either with an additional coating layer or through doping of the TiO_2. In both cases improvement of the device performance has been reported, which is usually explained by the passivation of trap states, suppressed recombination, and improved electron transfer. Also the improved formation of the perovskite layer on top of the modified TiO_2 surface is in some cases mentioned as a reason for the observed improved performance. In some cases it has been demonstrated that the interface modification also improved the device stability. Table 3.3 lists some examples for modifications of the TiO_2 layer.

Similar results that have been observed with TiO_2 layers have been reported for ZnO [191]. Not all perovskite solar cells are using metal oxide layers such as TiO_2 or ZnO as electron extraction layers. Other architectures use fullerene molecules, which are deposited on top of the perovskite layer and can also passivate surface traps in the perovskite film [192, 193]. It is expected that undercoordinated iodine ions within the perovskite structure are responsible for surface traps. By using supramolecular halogen bond complexation or a Lewis base, it is possible to successfully passivate these crystal sites [194, 195].

Tab. 3.3: Interface engineering: examples of TiO_2 modifications in perovskite solar cells

Modification	Finding	References
Ultrathin graphene quantum dots	Strong quenching of perovskite PL at 760 observed due to faster electron extraction	180
Fullerene monolayers (C_{60}SAM)	Faster electron transfer to TiO_2 (p–i–n configuration)	181, 182
HOCO-R-NH_3 anchor groups	Retardation of charge recombination, better crystal growth of perovskite	183
Sb_2S_3 layer	in combination with inorganic CuSCN HTM, degradation strongly suppressed	184
MgO-coated TiO_2	Retarded recombination times (three times slower than without MgO layer)	185

(continued)

Tab. 3.3 (continued)

Modification	Finding	References
Cesium carbonate–coated TiO_2	Retarded recombination times	186
Al_2O_3 and Y_2O_3 passivation layer	Removal of surface traps, retardation of charge recombination	187
Al_2O_3 doping	Passivation of bulk and surface traps, increased conductivity of TiO_2 layer, stability enhancement	188
Lead-doped TiO_2 nanofibers	Rapid electron collection, no HTM	189
Yttrium-doped TiO_2	In combination with additional interface engineering: improved morphology of perovskite layer, improved electron transfer	190

The most common electron blocking contact is the HTM Spiro-OMeTAD, also commonly used in solid-state dye-sensitized solar cells. P3HT has also been used in many cases. However, it seems that Spiro-OMeTAD gives in most cases the higher performance. We will need to know the exact nature of the interface between the perovskite layer and the HTM to understand how it is influencing the charge transfer and recombination dynamics, which might be able to explain the different performances. Currently it does not seem to be possible to predict the best-performing HTM structure, but it is expected that Spiro-OMeTAD is not the ideal material as it provides relatively low hole mobility in the range of 10^{-4} $cm^2V^{-1}s^{-1}$. So other classes of HTMs are under investigation, such as triphenyl-based and carbazole-based materials, among others [196]. Usually the HTMs are modified to increase the charge carrier mobility. The solution of Spiro-OMeTAD is mixed with small amounts of Li-TFSI. It gives similar effects to Spiro-OMeTAD as a p-dopant by moving the Fermi level of the molecular film closer to the HOMO level [197]. There are a number of HTMs and doping combinations. One example is doped poly(triaryl amine) (PTAA) by 2,3,5,6-tetrafluoro-7,7,8,8-tetracyanoquinodimethane (F4-TCNQ), which reduces device series resistance by a factor of 3, increasing the device fill factor to 74% in perovskite solar cells [198]. However, as a solvent for F4-TCNQ, usually acetonitrile is used, which seems to have negative effects on the perovskite layer and reduces its stability. Therefore, care has to be taken when choosing certain additives and solvents to avoid unwanted changes in the perovskite film.

Instead of a polymeric/organic HTM, other materials such as graphene oxide [199, 200], NiO [193, 201, 202], and carbon [203–209] also in the form of carbon nanotubes [210] have been used as contacting electrode materials. In most of these cases an inverted architecture has been chosen to collect the holes in the bottom electrode and have a PCBM layer as an electron-extracting layer on top. Carbon electrodes are very attractive, as they can replace the top metal electrode completely and allow simple solution-based deposition such as printing also on top of the perovskite layer. In HTM-free solar cells a thin Al_2O_3 blocking layer help in increasing the performances of the cell [211, 212].

One additional important aspect, which can be tuned by the interfaces, is the stability of the cell. In Table 3.3 we saw that improved stability has been achieved by using a thin Sb_2S_3 layer at the TiO_2–perovskite interface. A decomposition of the perovskite layer might occur at this interface as TiO_2 has a strong ability to extract electrons from organic materials, which can be observed when used as photocatalysts. It can also extract electrons from iodide (I^-) as observed in dye-sensitized solar cells, which might also happen at this interface and trigger the decomposition of the perovskite layer [184]. Similar to Sb_2S_3 layer at the TiO_2–perovskite interface, an Al_2O_3 layer at the perovskite–HTM layer can improve the stability of the perovskite film [213]. The reason for the improved stability in this case is seen in the encapsulation effect through the Al_2O_3 layer, which hinders the permeation of humidity to the perovskite film, which is very sensitive to moisture. By using a polymer-functionalized single-walled carbon nanotubes embedded in an insulating polymer matrix (PMMA), a strong retardation in thermal degradation has been observed as also here the resistance to water ingress has been enhanced [214].

3.5.4.6 Ferroelectricity

Most inorganic perovskite structures have ferroelectric properties with $BaTiO_3$ as the first discovered ceramics with ferroelectric properties. The reason for ferroelectric behavior is the breaking of the crystal center symmetry. Also in the hybrid perovskite solar cells such ferroelectric behavior is expected arising from the asymmetry of the organic cation. Frost et al. modeled the molecular orientation disorder in $CH_3NH_3PbI_3$ [215]. They observed twinned molecular dipoles whose rotation le to a complex ferroelectric behavior.

The ferroelectric domains are expected to vary with temperature and applied electric field. Kutes et al. found experimental evidence for ferroelectric domains using piezoforce microscopy. The domain size is found to be of ~100 nm grains, which can be switched by a DC bias [216]. Juarez-Perez et al. found a giant dielectric constant in the perovskite with a low-frequency dielectric constant in the dark of $\epsilon_0 = 1,000$, which further increases by a factor of 1,000 under illumination of 1 sun. As this behavior is observed in samples with different architectures and morphologies, they conclude that this effect is an intrinsic property of the $CH_3NH_3PbX_3$ under investigation and might be induced by structural fluctuations. The photoinduced carriers might modify the local unit cell equilibrium and change the polarizability as the methylammonium ion can freely rotate [164]. It is expected that the ferroelectric behavior can be tuned by, e.g. changing the molecular dipole using an organic cation with different electronegativity [215].

The role of ferroelectricity in the perovskite solar cells is not yet clear. However, it might have a strong influence on charge separation as a built-in potential is expected to improve the charge separation. Ferroelectric domains with their charged domain walls can serve as segregated channels for the motion of charge carriers improving the charge carrier lifetime. Patterning of these domain walls might lead to further improvement in perovskite solar cells [215, 217]. Above-bandgap open-circuit voltages (V_{OC}s) have been reported for ferroelectric solar cells due to charge separation based on the electrostatic potential in ferroelectric domain structures [218].

3.5.5 Stability

Even though the first perovskite solar cells were reported in 2009 [16], these solar cells did not receive a lot of attention in the beginning. The reason was the extremely short lifetime of these liquid electrolyte-based solar cells and consecutively a relatively low reported efficiency. The perovskite ($CH_3NH_3PbBr_3$) was used as a sensitizer in a dye-sensitzied solar cell architecture and corroded extremely fast in the iodide/triiodide redox couple. Therefore, the measured efficiency was 3.8%, significantly lower than in dye-sensitized solar cells with common dye. Replacing the liquid electrolyte with an organic hole transporter Spiro-OMeTAD has then led to the breakthrough. Still, the solar cells have only limited lifetime and are especially prone to degradation when exposed to humidity. The reason for this can be seen in the chemical equation describing the formation of methylammonium lead–iodide perovskite:

$$PbI_2(s) + CH_3NH_3I(aq) \leftrightarrow CH_3NH_3PbI_3(s).$$

As summarized by Niu et al., decomposition of the $CH_3NH_3PbI_3$ can happen when the reaction in the equation is driven to the left direction. This is the case when either PbI_2 or CH_3NH_3I reacts with other components. The degradation is strongly driven by the environmental conditions, such as moisture, oxygen, UV light, or solvents or additives still present from the processing and temperature [213]. $CH_3NH_3PbI_3$ tends to hydrolyze in the presence of moisture, leading to a dissolution of the CH_3NH_3I into CH_3NH_2 and HI, which is further decomposed to I_2 and H_2O or directly into H_2 and I_2 in the presence of oxygen. As a result the HI is consumed, which drives the degradation process further. Therefore, the most stable and reproducible perovskite solar cells are fully or at least partially prepared in inert atmosphere inside a glove box to avoid the exposure to moisture and oxygen while fabricating the solar cell. One extreme indication for the degradation is the color change of the perovskite layer from dark brownish to pale yellow as observed, e.g. by Noh et al. at a humidity of 55% [125]. Interestingly, controlled humidity might be beneficial for the fabrication process as reported by Zhou et al. [219]. The moisture might induce a reconstruction mechanism, accelerating the transport of the chemicals in the film. This gave a very high initial efficiency of 19.3%. However, the devices might be less stable.

Therefore, one important issue for long-term stability is the encapsulation of the device to protect it against moisture. Habisreutinger et al. proposed P3HT-functionalized single-wall carbon nanotubes embedded in an insulating PMMA layer as electrode material [214]. Perovskite cells with this layer showed remarkably improved moisture and thermal stability compared with Spiro-OMeTAD-based cells. The hydrophobic PMMA successfully avoided any penetration of moisture into the cell. They therefore concluded that non-hygroscopic, ideally hydrophobic HTM layers are needed to improve the stability of the cells. Hygroscopic layers, such as Spiro-OMeTAD doped with Li-TFSI, should be avoided. The Li-TFSI with its hygroscopic nature seems to accelerate the degradation of the perovskite cell [214].

Another degradation mechanism is related to the used electrode materials. For instance TiO_2, the most common anode material, is known for its photocatalytic activity. As such it strongly oxidizes organic materials. It can extract electrons from iodide (I^-) as evidenced in dye-sensitized solar cells. It might do the same with the iodide in the perovskite, destroying its structure, as suggested by Ito et al. [184]. They used a Sb_2S_3 underlayer to avoid this unwanted reaction and to significantly improve the device stability.

One important issue is the solvents present in the solution for the deposition of the HTM. In many cases Spiro-OMeTAD is used with some additive, such as Li-bis (trifluoromethanesulfonyl)imide (Li-TFSI) and 4-*tert*-butylpyridine (tBP). However, these additives seem to have strong effects on the perovskite layer. tBP corrodes the perovskite layer and needs to be replaced [220]. Li-TFSI is usually dissolved in acetonitrile, which also corrodes the perovskite layer [221]. Therefore, the solvent and additives used for depositing the HTM on top of the perovskite layer have to be carefully chosen. There might be also solvent residues that can lead to degradation of the perovskite layer. Impurities in the solvent can have strong effects on the device stability.

Other aspects of stability, such as thermal stability, also need to be considered when looking at the lifetime of perovskite solar cells. The use of organic layers in the device, e.g. the stability of the HTM layer, will also influence the final device stability. Before successful commercialization of perovskite solar cells, more research is needed to investigate the interplay between the different layers in detail to find the ideal combination in terms of efficiency and stability.

3.5.6 Outlook

In the last few years perovskite solar cells have become an important focal point of photovoltaic research. Rapidly their efficiencies increased dramatically to values of now over 20%. Still, why this material performs so well is only rudimentary understood. Many important questions remain. So it is not really understood whether the ferroelectric effect plays an important role in the high efficiency, or what exactly the role of Pb is, and whether it will be possible to replace it successfully with a nontoxic atom. Other questions concern stability. Is it possible to get intrinsically stable perovskite solar cells or will the huge advantage of the "easy" crystallization of perovskite films at low temperatures always also allow easy degradation of this structure? Does the hybrid organic/inorganic structure play a significant role toward the film-forming properties? Finally the questions remains, whether this organic–inorganic hybrid halide perovskite is unique in its properties or whether other materials might also allow to fabricate high-efficiency solar cells from solution. Even though the efficiencies are not expected to raise much further since they are already close to the Shockley–Queisser limit, the research on perovskite solar cells will certainly remain very exciting and we can expect many fundamental questions to be answered in future.

4 Characterization Techniques

In this chapter we will discuss characterization techniques commonly used in organic and hybrid solar cell research. Understanding how these techniques work is essential when reading the literature since the basics of widely used measurements are not always outlined in papers. This overview of characterization techniques should also be helpful for researchers in the field.

We will first discuss characterization techniques that are useful to analyze the properties of individual solar cell compounds, such as their absorption and emission behavior and their conductivity. After that we will take a look at the basic device characterization, which allows to measure, for instance, the power conversion efficiency or the quantum efficiency spectrum of solar cells. The last two sections of the chapter discuss more on advanced steady-state and time-resolved measurement techniques that will give deeper insights into the fundamental mechanisms involved in charge generation, transport, and recombination.

4.1 Characterization of Solar Cell Components

In order to research and understand a solar cell as a whole, it is essential to first analyze its components. This includes the optoelectronic properties of the photoactive materials as well as the blocking layers and the external electrodes, particularly the transparent conductive front electrode. In the following we will discuss four basic measurement techniques, namely UV-vis absorption spectroscopy, photoluminescence and electroluminescence spectroscopy, ellipsometry, and the most commonly used characterization procedure for transparent conductive substrates.

4.1.1 Absorption Spectra of Photoactive Materials

The most important property of any solar cell material is its ability to absorb photons. As we saw in Section 3.1.2, considering the context of the Shockley–Queisser limit, high-efficiency materials have to be tailored so that they absorb photons mostly in the UV, visible, and near-IR regions. Therefore, one of the most fundamental analysis techniques is the UV-vis absorption spectroscopy, which probes the light absorption of a sample typically in the range of 300–1200 nm (this range can be smaller or broader depending on the experimental setup).

The easiest setup for an absorption measurement is a transmission-mode UV-vis, as schematically shown in Figure 4.1. In a typical configuration, two lamps are used as the excitation source, a tungsten lamp that emits mostly in the visible and near-IR and a deuterium lamp for UV excitation. Excitation light is passed through the sample, most commonly either a solution in a cuvette or a film on a transparent substrate, and the transmitted intensity I_1 is compared to the incident intensity I_0, which is determined in a reference measurement without sample. Behind the sample chamber the light hits a grating and the diffracted light is analyzed using a diode array. Such a setup allows faster measurements than a monochromator with a photodiode and allows to use white light excitation. However, more precise measurements are usually done by monochromatic scanning of the transmission at different wavelengths.

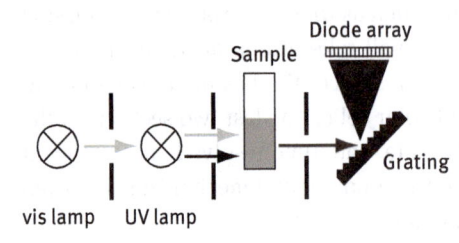

Fig. 4.1: Schematic of a setup for UV-vis absorption experiments. In order to yield sufficiently high illumination intensities, a combination of a UV lamp and a lamp emitting in the visible spectrum is commonly used as the light source. The light is passed through the absorbing sample, commonly either a dilute solution of the absorber in a cuvette or a thin film on a substrate. The transmitted light then hits a grating, and the spectrum and intensity of the light is analyzed with a diode array.

For an analysis of the absorption properties of the samples, the fraction I_1/I_0 has to be taken into account. This fraction is used to define the wavelength-dependent absorbance *A*

$$A(\lambda) = -\log_{10}\left(\frac{I_1(\lambda)}{I_0(\lambda)}\right), \tag{4.1}$$

which can be used to calculate the *absorption* Abs as

$$\text{Abs}(\lambda) = 1 - 10^{-A(\lambda)}. \tag{4.2}$$

Logarithmic evaluation of the fraction I_1/I_0 results in a linear dependency of the absorbance on the sample thickness; i.e. if for instance light passes through twice as thick a sample (film thickness, cuvette thickness, or doubled solution concentration), twice the absorbance is observed. This is the case since light extinction in a medium where interference effects do not play a role (as for instance in very thin films) can be described using Beer–Lambert's law

$$I(z) = I_0 \exp(-\alpha z), \tag{4.3}$$

where I_0 is the incident light intensity, I is the intensity of the transmitted light, z is the sample thickness, and α is the so-called *attenuation coefficient* (this term is often used synonymously to *absorption coefficient*). α is given in units of mol^{-1}cm^{-1} for solutions, i.e. absorbance per mole of absorbing species and per centimeter light is passing through the sample. For solids the unit is simply cm^{-1}, i.e. absorbance per film thickness that is passed by the excitation light.

At this point we have to mention two major issues that have to be taken into account in such transmission UV-vis experiments. First, I_1 is not only a function of light that is absorbed in the sample, but can also be reduced by reflection of the sample or indirect transmission, i.e. scattering of light out of the pathway into the detector. Using appropriate reference samples, for instance empty cuvettes or cuvettes filled only with solvent, can help reduce measurement errors, rendering the method ideal for the quantitative analysis of absorbing species on solutions. Nevertheless, transmission-mode UV-vis spectroscopy is not suitable for strongly scattering or reflective samples. In this case an integrating sphere is used to accommodate the sample, which has its interior surface covered with a diffuse white reflective coating, with small holes for light entrance and exit port for light detection. Multiple diffuse back reflections of scattered light in this sphere lead to a homogeneous light distribution, which is detected. This allows to measure the optical absorption also in highly scattered and reflecting films, as backscattered light will still reach the detector as described in more detail in Section 4.2.1 about total absorption measurements. Second, sample absorption can be significantly over- or underestimated in the case of thin films. If the film thickness is in the order of the used wavelengths, interference effects will dominate light propagation through the sample and Beer–Lambert's law will not be applicable anymore. A prominent example of this is the presence of a nanometer-thin gold coating on a thin polymer film on top of a glass substrate, which renders the structure a one-dimensional photonic crystal and can enhance transmission (and thus reduce the measured absorption in a transmission-mode UV-vis experiment) compared to the case without the gold film.

Despite these limitations, transmission-mode UV-vis spectroscopy is the most commonly used method for absorption measurements, which is often applied to thin films, especially in the organic solar cell community. This is the case since these measurements are conceptually simple and fast to perform and also because the equipment is comparably inexpensive. Spectra are then given in arbitrary units since interference effects do not allow complete quantitative measurements. Nevertheless, peak positions and the general shape of the absorption spectrum can be investigated, which provides already essential insights into the suitability of a material for solar cell applications.

4.1.2 Emission Spectra of Photoactive Materials

As we have seen in Chapter 2, every organic semiconductor has two characteristic optical spectra, an absorption spectrum and an emission spectrum. The spectra not only give insights into how suitable a material is for application in a solar cell, but are also related to the electronic properties. Furthermore, since these spectra change on the basis of the aggregation of molecules, they can be used to characterize the morphology and crystallinity of films of organic semiconductors. In Section 4.1.1, we discussed the basic experimental setup for absorption measurements. In this section, we will take a look at a typical setup for photoluminescence experiments, which give access to the emission spectra of the sample upon excitation with photons.

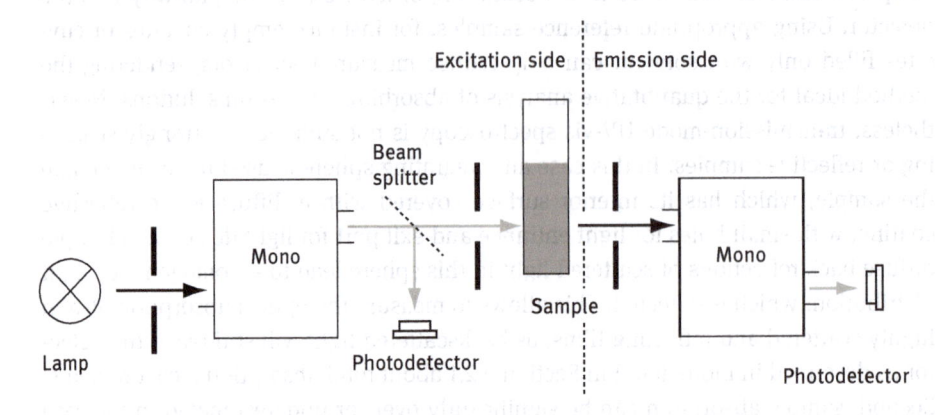

Fig. 4.2: Schematic of a photoluminescence setup. The sample is excited by monochromatic light via a lamp with a broad spectrum like a tungsten or a Xe lamp going through a monochromator. A semitransparent mirror focuses a small fraction of the excitation light onto a photodetector in order to allow normalization of the data to the excitation photon flux. Upon excitation there is emission of photons from singlet and triplet transitions in the sample. In order to analyze this emission spectrally, the light from the sample is guided through a second monochromator and its intensity is probed with a photodiode. Typical samples are either dilute solutions in a cuvette or thin films on a supporting substrate.

A schematic of such an experimental setup is shown in Figure 4.2. Monochromatic excitation light is provided by the combination of a white light lamp (typically a high-power Xe lamp) and a monochromator or by lasers, if only single excitation wavelength is of interest. After the monochromator, the light passes through a semi-transparent mirror so that a small fraction of the light can be measured with a photo-detector in order to account for the wavelength dependence of the excitation light. The larger fraction is used for excitation of the sample, which most commonly is either a dilute solution of the emitting material or a thin film. In the case of a film, two configurations are possible, either a transmission mode configuration, where the emission is detected on the substrate side opposing to the illuminated side, or a reflection mode configuration, for which also nontransparent substrate can be used (provided they do not show a photoluminescence at the used wavelengths, which could possibly eclipse the signal from the sample). The photoluminescence is then typically collected with a concave mirror on the emission side of the setup and focused through a second monochromator onto a photodetector in order to acquire the emission spectrum.

When evaluating the raw data of the setup, the signals of both photodetectors have to be re-normalized with respect to their specific quantum efficiency spectra. Furthermore, appropriate band-pass filters have to be used in the light path. First, the excitation light is typically blocked before the emission monochromator so that only emission from the sample is detected. Furthermore, band-pass filters can be used to block higher order wavelengths that can also pass through the most commonly installed Czerny–Turner monochromators, which rely on turning optical gratings.

Fundamentally, two types of photoluminescence spectra can be acquired with such a setup, as will be discussed next.

Photoluminescence Emission Spectra. The more commonly used spectra are *photoluminescence emission spectra*, which are acquired by exciting the sample at a fixed wavelength and scanning through different wavelengths with the emission monochromator. Such measurements allow us to determine the emission properties of a sample if a certain species is excited. Photoluminescence emission spectra are also commonly measured in setups with a slightly different configuration, where the lamp–monochromator combination is replaced by one or a few continuous wave lasers. Even though such setups are limited to certain excitation wavelengths, they exhibit the advantage of a simpler excitation side and allow excitation intensity–dependent measurements since typically much higher monochromatic intensities can be realized.

Photoluminescence Excitation Spectra. The second type of spectrum that is often of interest is the so-called *photoluminescence excitation spectrum*. Such a spectrum is found when detecting at a fixed wavelenght on the emission side while scanning through a range of excitation wavelengths. Such a measurement gives access to the information how different species in the sample contribute to the emission at a certain wavelength. For instance, photoluminescence excitation spectra are well-suited in order to identify energy transfer processes, when emission of the energy acceptor is observed not only upon direct excitation of the acceptor but also upon excitation of the energy donor.

Since most photoluminescence spectrometers are automated setups, it is also possible to combine the two spectra types, i.e. measure one emission spectrum each for a series of excitation wavelengths. Such a combined measurement yields the full picture of the photoluminescence properties of a sample, and single lines of the resulting three-dimensional plot can be extracted to investigate certain emission or excitation spectra.

Electroluminescence. As mentioned in Section 3.3.1.2, organic donor–acceptor combinations typically exhibit emission feature in addition to the separate emission of donor and acceptor due to the presence of charge transfer states at donor–acceptor interfaces. For many systems these states can be directly observed in photoluminescence experiments. Additionally, charge transfer state emission is electronically accessible in a sample that allows to apply an electric field to the active layer. The same is true for the emission of single materials in case suitable contacts are used, which allow electron injection into the LUMO and hole injection into the HOMO of the material, from where they recombine radiatively. Such an experiment is called an *electroluminescence* measurement and yields interesting insights into the quantum efficiency for radiative recombination processes upon charge injection. As derived by Rau, the quantum efficiency for electroluminescence EQE_{EL} of a solar cell can be related to its external quantum efficiency EQE_{PV} as

$$EQE_{EL}(\lambda, V) = EQE_{PV}(\lambda) I_{BB}(\lambda) \left(\exp\left(\frac{qV}{k_B T}\right) - 1 \right), \qquad (4.4)$$

where V is the applied voltage, λ the wavelength, and I_{BB} the black body spectrum at temperature T. For a solar cell that is described by the ideal diode equation, Equation (2.140), and has an ideality factor of $n = 1$, this expression can be written as

$$J_0 EQE_{EL}(\lambda) = q EQE_{PV}(\lambda) I_{BB}(\lambda), \qquad (4.5)$$

where J_0 is the dark saturation current.

Electroluminescence experiments can be performed with a setup similar to the one schematically shown in Figure 4.2, where only the emission side is used. Samples are typically realized on transparent conductive glass substrates, similar to thin-film solar cells, and a constant voltage or current source connected to the external contacts provides electrical excitation.

4.1.3 Ellipsometry

Besides absorption and emission spectra the fundamental interaction of light with the different materials in a photovoltaic device has to be understood in order to investigate light coupling into and out of the device, and light intensity distribution in its different layers. The material properties that determine the nature of this light–matter interaction are contained in the complex refractive index $\tilde{n}$ of the material, which is given as

$$\tilde{n}(\lambda) = n(\lambda) + ik(\lambda), \qquad (4.6)$$

where n is the refractive index and k the extinction coefficient. Both refractive index and extinction coefficient are wavelength-dependent and determine how light is propagating and being absorbed inside the material, as well as how it is reflected and transmitted at interfaces between materials with different values of $\tilde{n}$. Alternatively, the *complex dielectric function* or *permittivity*

$$\tilde{\varepsilon} = \varepsilon' + i\varepsilon'' \text{ with } \tilde{\varepsilon} = \tilde{n}^2 \qquad (4.7)$$

is used instead of $\tilde{n}$, which, however, is completely equivalent.

In the context of refractive index, light is described as electromagnetic wave with oscillating electric and magnetic fields, which are perpendicular to each other and the direction of light propagation. For most materials the coupling between the material and the electric field **E** of a light wave is much stronger than the coupling to the magnetic component, which is why for ellipsometry only the electric field of a light wave has to be considered. For a given direction of light propagation (we call it z), the orientation of this field is unambiguously described by two components, E_x and E_y, since $E_z = 0$. This orientation of **E** is called the *polarization* of the light. For random orientations of **E** the light is *unpolarized*, whereas polarized light is given for fixed relation between E_x and E_z. Three categories of polarization are commonly used, namely (1) linear polarization, where **E** oscillates along one direction; (2) circular polarized light, where E_x and E_y oscillate at a constant phase of 90°; and (3) elliptically polarized light, where E_x and E_y oscillate with a constant phase ≠ 90°. The index n determines how fast **E** oscillates in space, with faster oscillations being present in optically thicker media. This is directly apparent from the wavelength λ in a material, which is given as

$$\lambda = \frac{\lambda_{vacuum}}{n}, \tag{4.8}$$

where λ_{vacuum} is the wavelength in vacuum. Accordingly, also the phase velocity of light changes with n as

$$v = \frac{c}{n}, \tag{4.9}$$

where c is the vacuum speed of light.

The extinction coefficient k determines the absorption properties of the material, and there is a direct relation between k and the absorption coefficient α from the Beer–Lambert law in Equation (4.3):

$$\alpha(\lambda) = \frac{4\pi k(\lambda)}{\lambda}, \tag{4.10}$$

where λ is the wavelength. According to the Beer–Lambert law, a wave propagating over a distance z through a medium with extinction coefficient k is damped with a factor $\exp(-\alpha z)$, i.e. **E** continues to oscillate, but the amplitude of the oscillation decays exponentially.

According to these, a considerable knowledge of the individual complex refractive indices $\tilde{n}$ directly allows the computation of light intensities throughout a stack of layers of different materials, as present, e.g. in a thin-film solar cell. A common approach for simulations of light intensities inside thin-film solar cells is the use of a transfer matrix algorithm as briefly mentioned in Section 4.2.1. The complex refractive index of a material is directly accessible via *ellipsometry*, which takes advantage of the *Fresnel equations* to gain access to n and k.

Before we look at the experimental setup that will be used to perform ellipsometry, we briefly introduce the Fresnel equations, which describe how polarized light incident under an angle onto an interface between two different materials is reflected and transmitted. First, we need *Snell's law*, which gives the angle with respect to the normal ϕ_2 under which light is transmitted when incident under an angle ϕ_1 through a material with index n_1 onto an interface toward a material with index n_2:

$$n_1 \sin(\phi_1) = n_2 \sin(\phi_2). \tag{4.11}$$

These two angles, ϕ_1 and ϕ_2, along with the indices, n_1 and n_2, are the parameters used in the Fresnel equations given below. These set of equations describe the fractions of light reflected and transmitted at an interface between two materials (with indices n_1 and n_2) as a function of the angle of incidence ϕ_1 and the polarization of light. The two treated cases are s-polarized light and p-polarized light, for which the reflection R_s and R_p, and transmission T_s and T_p, respectively, are given as

$$R_s = \frac{n_1 \cos(\phi_1) - n_2 \cos(\phi_2)}{n_1 \cos(\phi_1) + n_2 \cos(\phi_2)} \tag{4.12}$$

$$R_p = \frac{n_2 \cos(\phi_1) - n_1 \cos(\phi_2)}{n_1 \cos(\phi_2) + n_2 \cos(\phi_1)} \tag{4.13}$$

$$T_s = \frac{2 n_1 \cos(\phi_1)}{n_1 \cos(\phi_1) + n_2 \cos(\phi_2)} \tag{4.14}$$

$$T_p = \frac{2 n_1 \cos(\phi_1)}{n_1 \cos(\phi_2) + n_2 \cos(\phi_1)}. \tag{4.15}$$

For linearly polarized light the convention is that *p-polarized* light means the case that the polarization direction lies in the plane of incidence, i.e. the plane described by the surface normal and the incidence vector, whereas *s-polarized* light refers to the case of the polarization direction being perpendicular to the plane of incidence.

An ellipsometry measurement now uses the Fresnel equations to compute the refractive index of the tested material from polarization-dependent reflection results. A schematic of the experimental principle is shown in Figure 4.3. Linearly polarized light is incident under an angle Φ onto the sample, which commonly is a thin film on a comparably thick silicon substrate with known optical properties. This choice of sample geometry not only allows us to determine how much light is reflected at the sample–air interface but also assures that some light is reflected at the material–silicon interface after being attenuated due to the *k*-value of the material. Thus, ellipsometry allows to measure both *n* and *k*, i.e. the whole complex refractive index $\tilde{n}$. The reflected light is measured using a polarization analyzer, so that for a given polarization of the incident beam the intensity distribution of the reflected beam can be determined as a function of the polarization. This measurement can then be performed for a series of wavelengths, polarizations of the incident light, and angles of incidence.

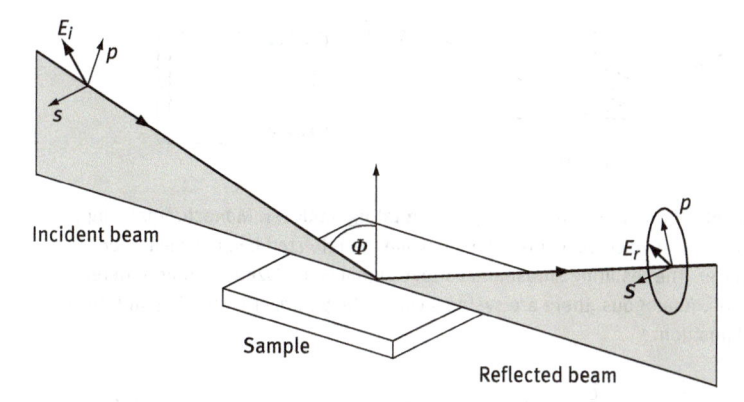

Fig. 4.3: Working principle of an ellipsometry experiment. Linearly polarized light is incident on a sample at an angle Φ close to the Brewster angle of the sample. The polarization of the reflected beam changes according to the Fresnel equations, leading to a change in p-polarized and s-polarized components with respect to the incident beam. Analysis of the polarization change allows us to determine the dielectric function of the sample.

Most ellipsometers use an experimental setup as depicted in Figure 4.4. The light source is a combination of a white light lamp and a monochromator, where the light is focused through a polarizer onto the sample. This allows us to determine n and k as a function of the wavelength. Often a photodetector behind a rotating polarizer is used as a detection unit, resulting in an oscillating signal. This signal is then analyzed using a lock-in amplifier in phase with the rotation frequency in order to yield good signal-to-noise ratios.

Analyzing the raw data is a complicated process, including data modeling, which requires numerical optimization. In brief, a model consisting of the different layers of the sample is constructed and appropriate start values for parameters like film thickness and optical constants are chosen. Then, a number of Lorentz oscillators are chosen in order to model the response of the sample in different wavelength regions. Numerical iteration then yields best-fit parameters for n and k as a function of the wavelength.

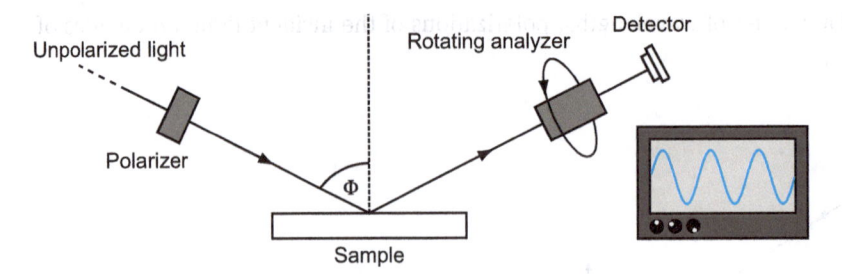

Fig. 4.4: Working principle of an ellipsometry setup with rotating analyzer. Monochromatic light that goes through a polarizer is used as incident beam, while the reflected beam is measured behind a rotating analyzer. The resulting sinusoidal voltage signal is analyzed in order to determine the reflected polarization. Analogous, there are systems where the polarizer is rotating and the analyzer is kept at a fixed position.

Besides using ellipsometry to determine the optical constants of a material, it can also be used for very accurate thickness measurements, in particular for nanometer thin films, which cannot be precisely measured with many other techniques. For thickness measurements the optical constants have to be known and the thickness is found as only free parameter in appropriate data modeling.

4.1.4 Quality of Transparent Contacts

Besides the properties of the active components of a solar cell, the quality of the transparent contact is often most decisive for the efficiency of the device. This is the case since two properties of the contact play a crucial role for solar cell operation, namely (1) the conductivity of the contact, which is responsible for the bigger part of the series resistance of the device, and (2) the transparency of the contact, which determines how much of incident light can be coupled into the active layer. Accordingly, both properties have to be characterized and *Haacke's figure of merit* has been introduced as a measure that weights both the transparency and the conductivity of a transparent conductive substrate, which can be used to describe its quality. The following three sections discuss how transparent conductors are commonly characterized and how their quality is evaluated.

4.1.4.1 Sheet Resistance

The first task in an operating optoelectronic device, which is performed by a transparent contact, is the transport of charge carriers. In light-emitting diodes, charges are transported toward the active film, where radiative recombination is responsible for the lighting of the device. In solar cells, photogenerated charge carriers are extracted from the active film and transported toward the external connection of the device, from where it can operate a load. Since transparent contacts are commonly realized as thin coatings (typically between 0.1 and 1 μm) of conducting materials on glass, the parameter of interest is the so-called *sheet resistance* of the film. This is the case since charge transport occurs mostly in lateral direction, i.e. parallel to the substrate. In a film with thickness t and electrical resistivity ρ, the sheet resistance is given as

$$R_{\text{sheet}} = \frac{\rho}{t} = \frac{1}{\sigma t} = \frac{1}{q(n\mu_n(n) + p\mu_p(p))t}. \quad (4.16)$$

As we see the sheet resistance can be equivalently expressed as a function of the electrical conductivity σ, which is the reciprocal of ρ. The last expression gives the sheet resistance as a function of the electron and hole densities n and p, and their mobilities $\mu_n(n)$ and $\mu_p(p)$, respectively. Note that the mobilities commonly show an implicit dependence on the charge densities. This dependence is of great importance for the design of high-performance transparent contact materials. In order to maximize the conductivity of the film and with this minimize R_{sheet} both the mobilities and the charge densities have to be high. However, high charge densities result in reduced transparency in the red and near-IR, as will be discussed in Section 4.1.4.3. Therefore, the charge mobility plays a crucial role in the efficient operation of a transparent conductor.

R_{sheet} has the unit Ω, but is commonly written as $\Omega\,\square^{-1}$. The $\square$ is not a unit but is only used to differentiate a sheet resistance from a bulk resistance. The origin of this unit is as follows: the electrical resistance of a block of material with cross-sectional area A and electrical resistivity ρ, through which a current runs perpendicular to the cross section over a distance l, is given as

$$R = \rho \frac{l}{A}. \quad (4.17)$$

Accordingly, the electrical resistivity

$$\rho = R \frac{A}{l} \quad (4.18)$$

has the unit Ωm. Since the sheet resistance is the resistivity per film thickness (see Equation (4.16)), the unit of R_{sheet} accordingly is Ω. Thus, when only looking at the units, a sheet resistance can be easily confused with a bulk resistance, which is the reason for the common use of the unit $\Omega\,\square^{-1}$.

R_{sheet} can be directly measured using a simple four-point probe experiment. Four terminal sensing uses two separate circuits for the sensing of voltage and current, which makes it more sensitive compared to a conventional two-terminal measurement. In particular, it eliminates influences of contact resistances and is therefore well-suited for measurements of highly conductive materials.

Even though four-point probe experiments can be performed using random positions of the four contacts using the *van der Pauw method*, the measurement is simplest using a linear configuration with equal distances between all four contacts pins as illustrated in Figure 4.5. In such a configuration the transparent conducting film is contacted from top. A voltage V is applied between the two pins in the center and the current I is measured between the outer two pins. The sheet resistance is then found as

$$R_{\text{sheet}} = \frac{\pi}{\ln(2)} \cdot \frac{V}{I}, \qquad (4.19)$$

where the prefactor has the simple form $\pi/\ln(2)$. For all other configurations of voltage and current electrodes a non-analytical prefactor, which depends on the exact positions of the four pins, has to be numerically estimated.

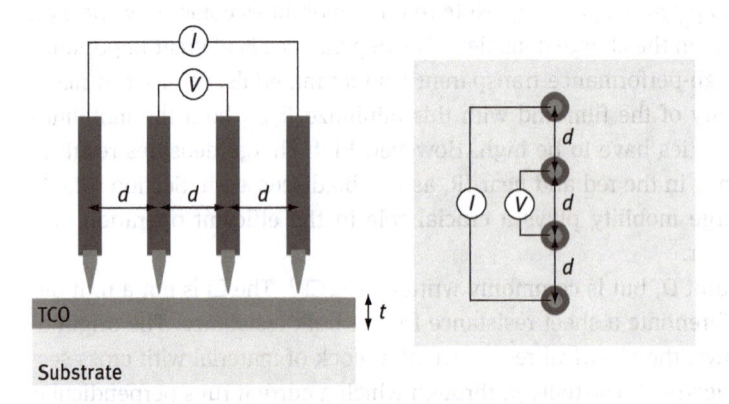

Fig. 4.5: Schematic of a typical four-point probe experiment (side and top views) used to determine sheet resistances of conducting films. Four contact pins are arranged in a straight line at identical distances and a current is applied between the middle two. The voltage is then measured across the outer contacts and the sheet resistance is found via Ohm's law.

4.1.4.2 Transmission

Besides the transport of charges the transparent contact has to be designed so that light incoupling into the active film is maximized (for a light-emitting diode the outcoupling is of importance). Although the exact amount of light that can couple into the active film depends not only on the substrate and the transparent conductor but also on interference effects due to the thickness of the active film and the potential presence of a reflecting top contact, it is useful to characterize the transmission through the bare transparent conductor film since it contains important information about the reflectivity of the substrate and the absorption behavior of the material.

Fundamentally, the magnitude of interest is the ratio of incident and transmitted intensity, I/I_0. Apparently, the ideal transparent conducting substrate exhibits $I/I_0 = 1$, i.e. all incident light passes through the substrate. Real transparent conductors show lower ratios due to (1) photon absorption in the substrate,

(2) photon absorption in the conducting film, (3) light back-reflection at one of the three present interfaces (air–substrate, substrate–conducting film, conducting film–air), and (4) light scattering. The latter is typically very weak in good films and light is either passing through the substrate directly, reflected back, or absorbed. Furthermore, absorption by the substrate plays only a minor role since it typically occurs in the UV range, where the solar light intensity is comparably low. However, reflection and absorption of the conducting film have to be taken into account when maximizing the transmission through a transparent contact.

In general, the transmission

$$T = \frac{I}{I_0} \tag{4.20}$$

through a slab of material that is significantly thicker than the wavelength of incident light in the medium is given as [222]

$$T = \frac{(1-R)^2 + 4R\sin^2(\psi)}{e^{\alpha d} + R^2 e^{-\alpha d} - 2R\cos\left(2(\phi+\psi)\right)}, \tag{4.21}$$

where d is the film thickness, R the reflectivity, and

$$\phi = \frac{4\pi n d}{\lambda}, \; \psi = \tan^{-1}\left(\frac{2k}{n^2 + k^2 - 1}\right). \tag{4.22}$$

Here, n is the refractive index of the material, and the absorption coefficient α depends on the wavelength λ and the extinction coefficient k as

$$\alpha = \frac{4\pi k}{\lambda}. \tag{4.23}$$

Neglecting interference in the film, which is described by the term $2R\cos\left(2(\phi+\psi)\right)$, and for small values of ψ, which are present in thin films since k is typically much smaller than 1, this expression reduces to

$$T = \frac{(1-R)^2}{e^{\alpha d} + R^2 e^{-\alpha d}}. \tag{4.24}$$

Accordingly, the goal is to (1) find materials that show only weak absorption in the wavelength region of interest (for solar cells mostly the visible and near-IR) and (2) reduce the reflectivity of the transparent contact. The latter is possible via index matching between the conducting film and the substrate and by using anti-reflection coatings on the front side of the substrate. The former can be generally addressed by using thinner films, which, however, typically come at the price of higher sheet resistance. The balanced optimization of these two aspects is a requirement when fabricating high-efficiency transparent contacts, as will be discussed in the context of Haacke's figure of merit in Section 4.1.4.3.

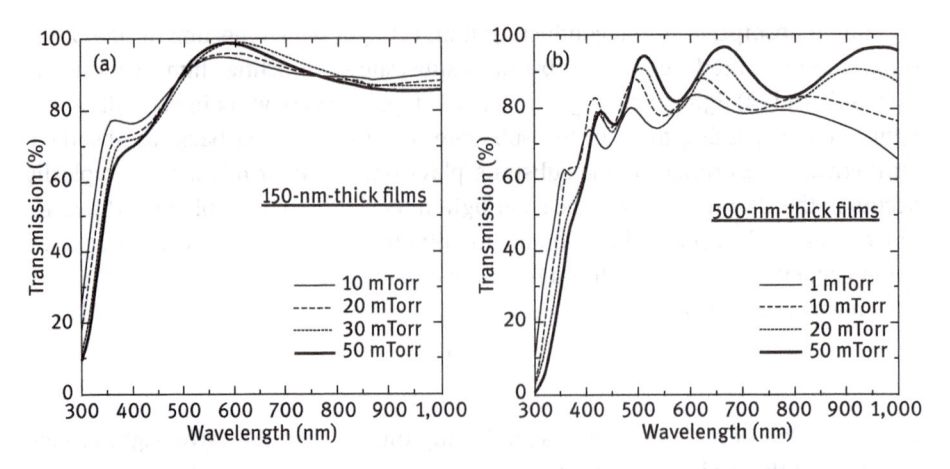

Fig. 4.6: **Transmission Spectra for ITO thin films on glass substrates processed at different sputtering pressures.** (a) Spectra for 150-nm-thick films. (b) Spectra for 500-nm-thick films. Reprinted with permission from [223]. Copyright (1999) AIP Publishing LLC.

In the case of a transparent conductor, thicknesses in the range of the wavelength interference effects start to significantly influence the transmittance of the film. Depending on the refractive index of the material, its absorption, and the film thickness, the transmission spectrum shows resonances at specific wavelengths. This is illustrated in Figure 4.6 for indium–tin oxide films processed at different sputtering pressures (which changes the crystallinity of the films and with this the refractive index). Both cases show the characteristic reduced transmission at wavelengths below 350 nm, owing to photon absorption across the bandgap of indium–tin oxide. However, the transmission spectra does not have a simple Beer–Lambert shape, but exhibits resonances that shift with film thickness and material composition. Since these resonances originate from Fabry–Pérot-type interference due to reflection at the glass–conductor and conductor–ambient interfaces, the transmission spectra are also influenced by the optical properties of the material that is coated on top of the transparent conductor. Therefore, transmission measurements of the bare transparent conductor film on a glass substrate in air only give indications of its suitability for implementation in devices, while the real transmission has to be determined when looking at the fully assembled device, e.g. in simulations based on the optical constants and film thicknesses of the involved materials.

4.1.4.3 Haacke's Figure of Merit for Transparent Contacts

As mentioned in the previous sections, a good transparent conductor has to be designed so that it exhibits both high transmission and low sheet resistance. In Section 4.1.4.2, we saw that while the film thickness and index determine the interference profile in the film, the main loss in transmission is due to absorption across the bandgap of the material. Thus, mostly the bandgap of the transparent conductor determines its transparency. On the other hand, the sheet resistance depends on the concentration and mobilities of charge carriers, as discussed in Section 4.1.4.1.

In this context it has to be taken into account that for high doping concentrations the plasma resonance of free charges comes into play, since it shifts from the far IR closer to the visible. The frequency of this plasma resonance for free electrons in a dielectric medium with relative permittivity ε_r is given as

$$\omega_p^2 = \frac{n_e q^2}{\varepsilon_r \varepsilon_0 m_e}, \qquad (4.25)$$

where n_e is the electron density and m_e the electron mass. Apparently, this frequency increases for increasing charge densities, i.e. the medium becomes non-transparent in the IR and near-IR for high doping concentrations. This is the first trade-off between transparency and conductivity that has to be considered for the design of transparent conductors. Furthermore, while the sheet resistance can be directly reduced for increased layer thickness, the absorption in the film also increases, i.e. the transparency decreases, which is the second trade-off.

In order to find a single number that helps to characterize the quality of a transparent conductor while taking the interplay of transparency and sheer resistance into account, several figures of merit have been proposed. There is no universal figure of merit that covers all applications equally well, as in different applications the importance of low sheet resistance versus transmittance will be different. This will also be briefly discussed in Section 5.1.1.2 in the context of the fabrication of transparent conductors. The most commonly used criterion for application of the transparent conducting oxides (TCOs) for solar cells is *Haacke's figure of merit*, shown in Equation (4.26), which was proposed by Haacke in 1976 [270]:

$$\Phi_{TC} = \frac{T^{10}}{R_{sheet}}, \qquad (4.26)$$

It weights the transparency T with a power of 10 in comparison to the sheet resistance R_{sheet} since light intensity losses are generally considered more detrimental for efficient solar cell generation than an increased series resistance (even though contact resistances cannot be completely neglected, the series resistance of a thin-film solar cell is mostly determined by the sheet resistance of the transparent conductor). For simplicity the monochromatic transparency at a wavelength of 550 nm is taken into account, since this wavelength lies in the center of the spectral range, which is typically converted in photovoltaics. While this convention reduces the experimental effort to determine Haacke's figure of merit, it also renders this number to be only an indicator, which must not be understood in absolute terms. For instance, in the case of interference patterns in the transmission spectrum the transmission at a specific wavelength can be higher compared to a film of different optical properties, while the integrated transmission is lower. Again, this issue can be addressed only when taking the whole device configuration into account, e.g. in optical simulation of light transmission and reflection over the whole spectrum.

4.2 Steady-State Device Characterization

After looking at the individual components of a thin-film photovoltaic device and their properties, we will now take a look on the characterization of the fully assembled device. While understanding the properties of the components helps to understand the fundamental mechanisms in the solar cell, its complete function and the suitability of its design can be assessed only in the fully assembled device. In the following we will focus on the three most important device characterization techniques, which give access to the total absorption of the device, its current density–voltage curve, and its external quantum efficiency spectrum.

4.2.1 Total Absorption Measurements

The power conversion efficiency of every photovoltaic device depends on two basic and very different efficiencies, namely the efficiency at which light is absorbed and the efficiency at which the energy of an individual photon is converted into electric energy. In Sections 4.1.1 and 4.1.4, we saw how the (beneficial) absorption behavior of photoactive materials can be determined in transmission-mode UV-vis spectroscopy and how the (parasitic) absorption in the transparent conducting substrate can be characterized. In the fully assembled cell these properties are combined and in thin-film devices especially additional effects come into play due to interference effects. Essentially, light incident on a complete photovoltaic device can be (1) reflected; (2) absorbed in parts of the device, which are not photo-active; and (3) absorbed in the active film of the device. Obviously, (1) and (2) are loss mechanisms, as the involved photons cannot take part in the energy conversion process, and only (3) leads to photovoltaic function. Therefore, it is essential to monitor absorption and reflection in order to minimize losses and thus optimize the device efficiency.

A suitable tool to perform such an analysis is a UV-vis absorption spectrometer with an integrating sphere detector, as schematically shown in Figure 4.7. The light source is a combination of suitable lamps providing sufficient output in the UV, visible, and near-IR, in combination with a monochromator. The detector is a combination of an integrating sphere with a photodetector, which is placed on one side of the sphere so that no direct light is impinging on it. The integrating sphere is covered with a white scatterer diffuser like *Spectralon* (sintered Teflon) or $BaSO_4$, which exhibit wide bandgaps and behave approximately like Lambertian scatterers. Since light is incoupled only through a small entrance slit, all light inside the sphere is eventually detected by the photodetector due to a series of diffuse scattering events on the inside wall of the sphere.

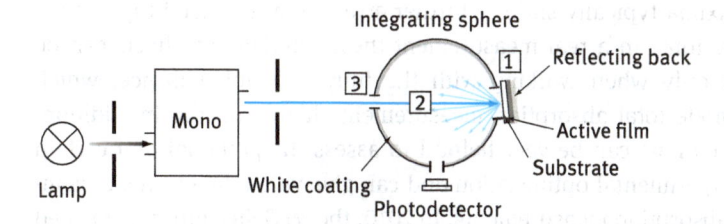

Fig. 4.7: UV-vis absorption setup with an integrating sphere used for total absorption measurements. Depending on the type and positioning of the sample, such a setup can be used to measure (1) total absorption in reflection, (2) total absorption for an arbitrary sample, and (3) direct and diffuse transmission.

Commonly, three different sample positions are used in integrating sphere absorption spectroscopy as indicated by (1), (2), and (3) in Figure 4.7. For total absorption measurements on fully assembled devices with reflecting (metal) top contacts, the sample is placed on the back port of the sphere, at position (1). The sample is illuminated under a small angle on the order of 10° in order to avoid direct light reflection out of the sphere. This configuration measures the absorption in the complete device and takes direct as well as diffuse scattering of the substrate taken into account. Total absorption of samples without reflecting top contacts are acquired at position (2), inside the sphere, also with the sample at a small angle with respect to the incident light. In order to measure direct and diffusive transmission the sample is placed at position (3), in front of the sphere. For all measurements typically two baselines have to be measured, one with the front port being closed for zero light intensity in the sphere and one with a diffuse reflecting sample on the back port for maximum light intensity in the sphere.

Especially, measurements of fully assembled devices at the back port of the integrating sphere are essential to investigate the real light distribution and absorption inside a thin-film solar cell. As mentioned before, interference effects dominate the light intensity distribution inside the device for active layer thicknesses in the order of, or thinner than, the wavelength. Therefore, only a measurement that takes the standing wave inside the solar cell into account reflects the processes occurring during solar cell operation, whereas simple characterization of the transmission of the active layer might lead to wrong estimates of its absorption properties. Examples for such standing waves are shown in Figure 4.8 for the case of an organic bilayer solar cell. Pettersson and co-workers were one of the first ones to show such simulations based on a transfer-matrix algorithm for an organic bilayer solar cell [224]. These examples illustrate how differences in the active layer thickness result in different electric field intensity distributions within the devices. The electric field intensity distribution in the device determines how much light is absorbed in the active layer and thus is decisive for the quantum efficiency of the solar cell. Note that the curves shown in Figure 4.8(b) also depend on the wavelength of incident light, with the minima and maxima typically shifting further away from the metal top contact for lower energy photons. In a real measurement these interference effects can be taken into account only when working with the fully assembled device, which renders reflection-mode total absorption measurements highly useful. In addition, transfer-matrix simulations can be very helpful to assess the potential of a certain cell design before experimental optimization and can give valuable insights in combination with total absorption measurements. In 2010, the McGehee group published a paper on the implementation of a transfer-matrix algorithm for use in the context of bulk heterojunction solar cells [225]. The authors further published the *Matlab* script used in this publication, which can be downloaded from their Website[1] and easily adjusted for the simulation of various solar cell architectures.

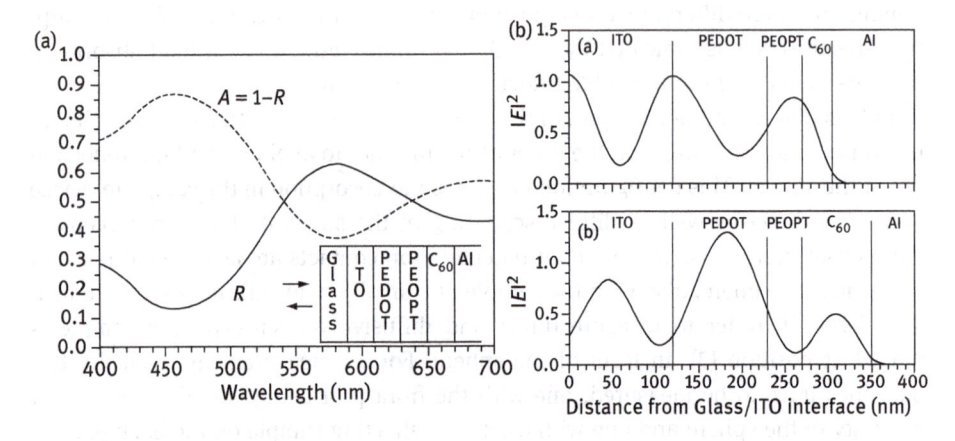

Fig. 4.8: Simulated absorption and emission of a layered solar cell with reflecting top contact along with the distribution of electric field intensity inside the device. (a) Wavelength-dependent total absorption and reflection for linearly polarized light incident through the glass. (b) Electric field intensity distribution for light with a wavelength of 460 nm inside two devices with different acceptor layer (C_{60}) thicknesses). Reprinted with permission from [224]. Copyright (1999) AIP Publishing LLC.

Estimates for light absorption and reflection in different layers of the device can also be found purely based on experiments using an integrating sphere via successive measurements during a layer-by-layer assembly of a solar cell. These measurements can be used to reconstruct reflection and absorption due to the substrate, conducting contact, and active layer, as shown in Figure 4.9. However, since the complete picture of light interference in a thin device is gained only by measuring the presence of the metal top contact, such a reconstruction should always be compared to estimates based on simulated spectra and are typically more accurate for thicker active layers.

---

[1] http://web.stanford.edu/group/mcgehee/transfermatrix/

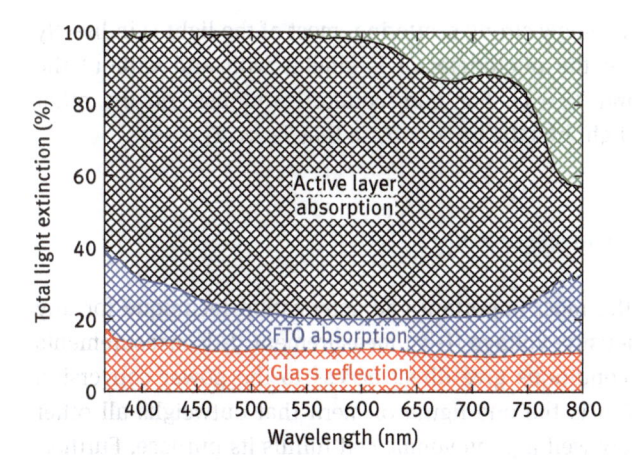

Fig. 4.9: Example of a total absorption curve for a thin-film solar cell. Absorption and reflection of the different layers can be estimated from transmission and reflection measurements at different positions in a UV-vis with an integrating sphere. Figure reproduced from [226] with permission of The Royal Society of Chemistry.

Besides total absorption measurements a UV-vis absorption setup with an integrating sphere can also be used to determine how strongly a sample is scattering light, its so-called *haze*. Diffuse light scattering is a popular way to increase the optical path length through an active film, which enhances the absorption while keeping the film thickness small and which potentially has favorable implications like superior electronic properties and lower fabrication costs. In thin-film devices this is often implemented by using inscattering or backscattering nanostructures in front of or behind the active layer. In order to characterize such films (among others), haze measurements are often performed, which summarize the scattering properties of the film in one number.

The haze of a sample can be determined from a sequence of four measurements, two with sample and two without the sample. For each case the direct and indirect transmission is determined, i.e. the two measurements without sample serve as reference. This sequence of measurements yields four transmission spectra

T_1: no sample, white standard on back port
T_2: sample, white standard on back port
T_3: no sample, back port open (i.e. perfectly black)
T_4: sample, back port open (i.e. perfectly black)

from which the haze can be calculated as

$$\text{Haze} = \frac{T_4}{T_2} - \frac{T_3}{T_1}. \qquad (4.27)$$

While T_3/T_1 is a reference term, the most interesting part T_4/T_2 relates the transmitted light intensity without white reflector on the back port to the transmission with reflector. For samples that show strong scattering, most of the light is indirectly transmitted, i.e. the signal on the photodetector is similar with and without the reflector at the back port and $T_4 \approx T_2$. For non-scattering samples, on the other hand, almost all transmitted light hits the back port, so that $T_4 \ll T_2$.

4.2.2 Current Density–Voltage Measurements

We now come to probably the most important characterization technique for any solar cell device, a current density–voltage (J–V) measurement. J–V measurements under defined illumination conditions give direct access to the power conversion efficiency of the device, which is the one figure of merit that outweighs all other since it ultimately decides how well a photovoltaic cell fulfills its purpose. Furthermore, J–V measurements in the dark are performed very commonly as well, since they allow to assess the diode quality of the solar cell, which already gives many insights into present recombination losses.

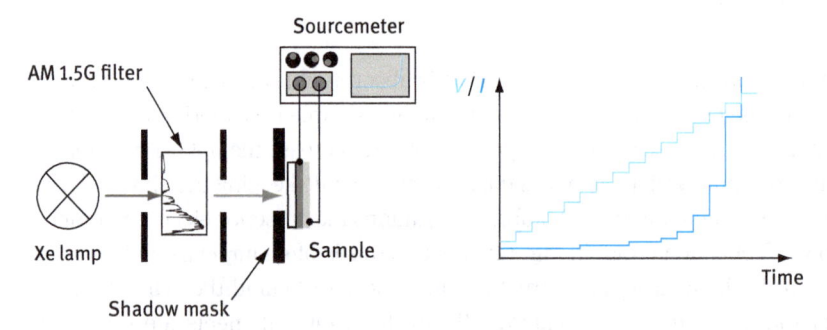

Fig. 4.10: Setup for measurements of current density–voltage curves under simulated solar light. Commonly, a Xe lamp is used as a light source and the spectrum is modified with appropriate filters to simulate AM 1.5G solar light. The solar cell sample is then illuminated through a shadow mask defining the active area and a voltage step function is applied to the device while measuring the current response.

A schematic drawing of a J–V setup is shown in Figure 4.10. A solar cell sample is placed behind an appropriate shadow mask, which defines the active area. The issue of proper masking will be discussed later in this section. Most commonly, the device is illuminated with a combination of a Xe lamp with an appropriate filter so that the spectrum of light impinging on the solar cell equals that of AM 1.5G sunlight (such a combination is referred to as a *solar simulator*). The convention is it to use an intensity of 100 mW cm^{-2} as *1 sun* condition and a temperature of 25°C (see Table 3.1, Section 3.1.1 for standard measurement conditions), which then allows to calculate the solar cell efficiency (see Section 3.1.5, Equation 3.14). However, there are many instances where J–V curves are published for intensities slightly lower than 100 mW cm^{-2}, mostly between 60 and 80 mW cm^{-2}. The electronic measurement is then performed by a sourcemeter unit, which applies a voltage between the front and back contacts of the solar cell and measures the current. Parameters for such a measurement involve the voltage range, the voltage steps, and the integration time per voltage step, which determines the overall sweep time for the measurement. A simplified picture of the raw data of a J–V measurement is shown on the right-hand side of Figure 4.10. Since the current changes exponentially with the applied voltage, according to the diode equation that was given in Equation (2.139), sufficiently small voltage steps have to be chosen to yield a smooth current response curve. Furthermore, as will be discussed later, the sweep time has to be long enough to allow the system to reach a dynamic equilibrium at each applied voltage, since the device performance can otherwise be systematically over- or underestimated.

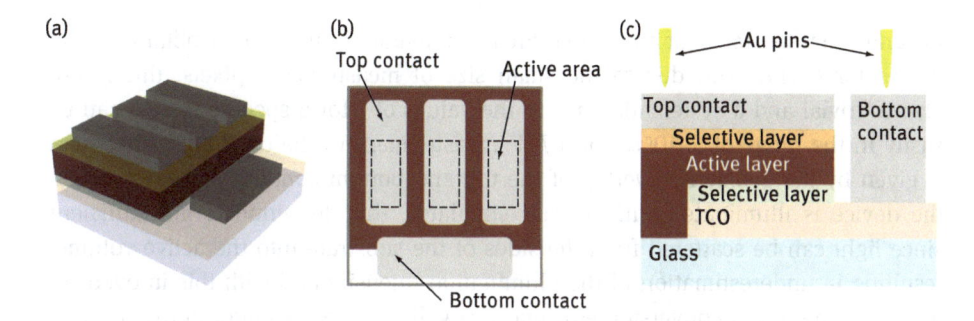

Fig. 4.11: Geometry of a typical lab solar cell used for different organic and hybrid devices. **(a)** Three-dimensional overview of the layered structure of the solar cell. **(b)** Top view of the device, showing the bottom contact on the TCO substrate and the three top contacts that define one individual solar cell each. The active area that is illuminated during a measurement is commonly smaller than the contact in order to avoid errors due to edge illumination. **(c)** Side view of the device showing gold pins for contacting the solar cell during electrical measurements. The TCO substrate is removed on one side of the substrate in order to avoid short cuts in the cell due to piercing through the active film on the TCO by the gold pins. The active layer is etched away on the opposing side in order to allow direct contact to the TCO electrode. In the laboratories, different device layouts are used. Another common layout uses a circular instead of rectangular shape for the individual solar cells on a device.

The particular geometry of organic and hybrid solar cell devices differs significantly among different laboratories and depending on the scientific question under investigation. Many solid-state devices, however, in principle, share the experimental geometry shown in Figure 4.11, yet with a different number, shape, and size of the individual contacts. The illustrated device geometry is shown for an organic or hybrid solar cell assembled on a TCO substrate with a size on the order of 1–4 cm². The TCO is covered by an electron- or hole-selective contact, on which the photo-active film is deposited. Often on top of the active film another selective layer is placed, which renders the metallic top contact specific for the opposite type of charge carriers than the front contact. For the electronic measurement the two electrodes of the solar cell are contacted and connected to the sourcemeter of the measurement setup. The active layer is removed on one side of the device in order to allow direct electrical contacting of the TCO, while the TCO is removed on the other side in order to avoid shorting of the device when contacting the back side. This is necessary since typical active films thicknesses of organic and hybrid solar cells are only on the order of 0.1–10 µm, so that contacts, such as small pins, easily pierce through the whole active film.

Since a measurement setup, as shown in Figure 4.10, yields current–voltage curves, one is typically more interested in current density–voltage curves, and the active area A of the device has to be known in order to calculate the current density J from the absolute current I as

$$J = \frac{I}{A}. \qquad (4.28)$$

Accurate measurement of the active area is imperative in order to obtain correct values for J. However, due to the small size of measurement pixels, this is not always trivial and a systematic error in the values of J for a specific system can be easily in the range of 5–10%. For a J–V measurement in the dark, the active area is given by the geometric overlap of the top and bottom contacts. However, when the device is illuminated with a solar simulator, this definition is not sufficient since light can be scattered from the sides of the substrate into the active volume, resulting in underestimation of the illumination intensity and with this in overestimation of the power conversion efficiency, as will be discussed later in this section. Therefore, the device is illuminated through a shadow mask with an area smaller than or equal to the electronic active area, as indicated in Figure 4.11(b). In the case of smaller illuminated area compared to the electrically active area, the resulting J–V curve slightly underestimates the performance of the device due to the presence of additional dark current from the dark areas of the pixel. However, these deviations are small since the dark saturation current is typically small compared to the photocurrent. Nevertheless, an ideal J–V setup includes a cell geometry where the deviation between the illuminated area and electrically active area is minimized.

Proper masking of the solar cell and illumination through a well-defined shadow mask is imperative to obtain realistic efficiency values. As shown in Figure 4.12, a good mask completely covers the back and the sides of the device so that illumination is possible only through a defined hole in the device holder. Furthermore, incident light has to be parallel and must not be focused through the opening of the mask since proper light intensity calibration and calculation of the current density from the current raw data are impossible. Furthermore, when using a mask design with several individual solar cell pixel, as exemplarily shown in Figure 4.12, care has to be taken that the pixels are electrically insulated from each other. This is imperative since during the measurement of one pixel, only this pixel is contacted, while the other pixels are often also illuminated. In case of electrical contact between the pixels, e.g. due to highly conductive charge-selective layers like PEDOT:PSS, pixels other than the illuminated also contribute photocurrent, leading to overestimation of J_{SC} and possibly power conversion efficiency [227–229].

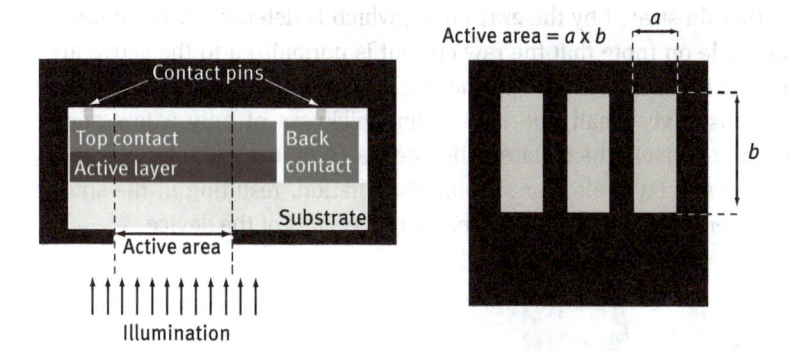

Fig. 4.12: Schematic of a solar cell measurement holder. The device is placed inside a light-tight box and the top and back electrodes are connected to the measurement setup. Illumination is provided through a shadow mask, which defines the active area of the device. This illuminated area has to be smaller than the electrically connected area of the cell in order to avoid measurement errors. In case highly conductive materials are used in or on top of the active film, only one pixel must be illuminated at the same time, whereas several pixels can be illuminated simultaneously if only one is contacted and the pixels are electronically not connected.

The importance of proper masking when measuring thin-film solar cells has been outlined in a number of publications [229–231]. One of the most impressive examples of the impact of wrong masking is shown in Figure 4.13 for the case of a dye-sensitized solar cell. The active film in such cells is commonly assembled between two conducting glass substrates that are glued together. The maximum active area is given by the area of the dye-sensitized film, which is commonly a circle in the center of the device, as shown in Figure 4.13(a). Illuminating the whole device and calculating the current density using the area of this circle result in the black curve shown in Figure 4.13(b), which overestimates the efficiency by approximately 50%. This is the case since light can enter the active area from the sides, e.g. scatter at the edges of the substrate. Furthermore, the active film is typically inhomogeneous toward the edges and accurate measurement of the circle dimensions is difficult. A well-defined active area can be obtained by using a shadow mask as shown in Figure 4.13(b), which can be crafted with very accurate dimensions. However, using only such a mask in front of the device without covering the edges can result in an even more pronounced overestimation of the device performance (red curve). Light can still be scattered into the active film from the sides of the substrates, but the effect is even stronger due to the smaller active area that is used for current density calculation. In the given example this leads to an efficiency estimate that is roughly 100%

higher than the efficiency for a proper measurement (green curve). A proper measurement requires not only masking of the front in order to yield a well-defined active area but also covering all edges and the back side of the sample in order to avoid any unwanted illumination of the active film. The effect of light scattering into the device is further illustrated by the gray curve, which is determined when illuminating the device side-on (note that the raw current is normalized to the active area for the case of front illumination and not the perpendicular projection of the active area, which is vanishingly small due to the film thickness of only a few micrometers). Due to the relatively thick glass substrates, a significant amount of light is funneled into the active layer also for side-on illumination, resulting in the shown measurement problems when not properly masking the sides of the device.

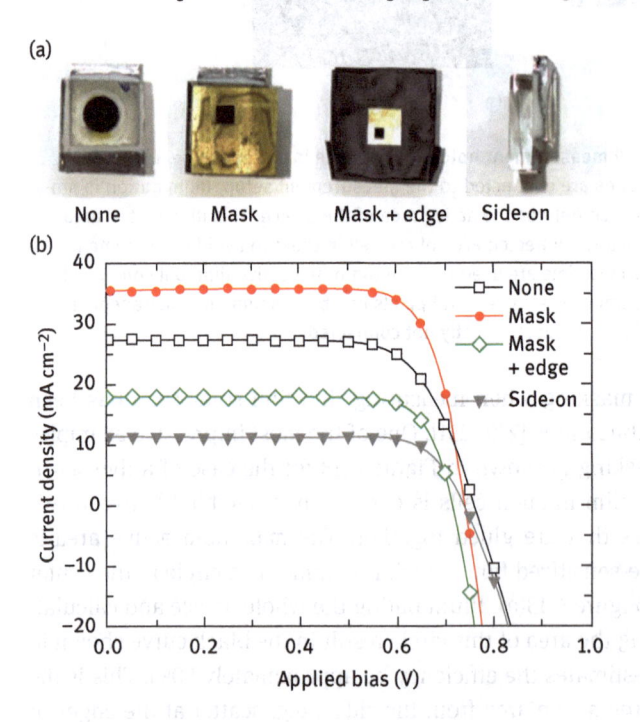

Fig. 4.13: Impact of different shadow maskings on the J–V curves of dye-sensitized solar cells. For the case without mask the active area is defined by the geometrical overlap of front and back electrodes. For the side-on experiment the active area to which the current is normalized is set to the value of the shadow masks. Figure reproduced from [228] with permission of The Royal Society of Chemistry.

Besides these effects of incomplete masking, the active area plays an important role for the accuracy of J–V measurements. Especially for very small active areas on the order of 1 mm², the active area might not be determined without a certain error. For instance, for the case of a circular pixel of 1 mm diameter, which is crafted 50 μm too large, the resulting current density is overestimated by 10.3%. Apparently, the same absolute inaccuracy for a diameter of 50 µm results in an error of only 1% for a diameter of 1 cm, which is why institutes like NREL or Fraunhofer demand pixel sizes of at least 1 cm².

Another important parameter in J–V measurements is the sweep rate, i.e. the wait time between voltage steps during the measurements, and additionally the measurement direction in case the sweep rate is chosen too high. As illustrated in Figure 4.14, faster sweeping results in more pronounced deviations for both upward and downward sweeps since the system is not given sufficient time to reach the new equilibrium current upon application of a different voltage.

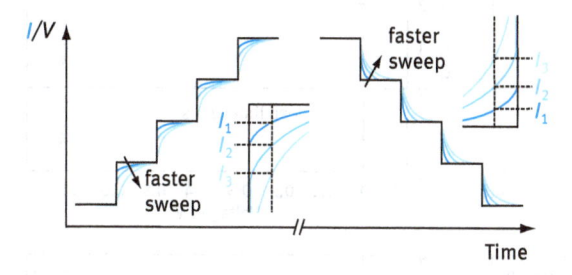

Fig. 4.14: Illustration of current under- and overestimation due to fast sweep times. If time between voltage steps is shorter than the time the system requires to adapt to the new condition, the measurement over- or underestimates the current in the system. For rising voltages fast sweeping leads to underestimation of the currents, whereas for falling voltages the current is overestimated.

For complete J–V measurements this results in drastic deviations between curves measured from J_{SC} toward V_{OC} and reversed as shown in Figure 4.15 for a dye-sensitized solar cell with an ionic liquid electrolyte [229]. When measuring from forward bias toward reverse bias too quickly, the current is generally over estimated, potentially resulting in current overshooting as well as in overestimation of the V_{OC} as shown in the upper left J–V example. In contrast measuring from reverse bias toward forward bias results in underestimation of the current and potentially the V_{OC}. Only for sufficiently slow sweep rates both forward and reverse curves superpose each other since the equilibrium current can be reached for every applied voltage.

The effect of too rapid J–V measurements plays a particularly important role in liquid electrolyte, and also in perovskite solar cells (see Section 3.5.4.1), where slow charge transport or polarization processes result in retarded current response. While researchers address this issue in their publications by showing both forward and reverse curves, and sometimes also giving average curves, the only appropriate method to obtain completely realistic J–V curves is sufficient reduction of the sweep rate.

4.2.3 External Quantum Efficiency Measurements

Besides current–voltage characterization, the most fundamental measurement of a solar cell is the determination of its external quantum efficiency EQE, which was introduced in Equation (3.4). The EQE is the quantum efficiency for conversion of an incident photon to an electron by the photovoltaic device and is in general a

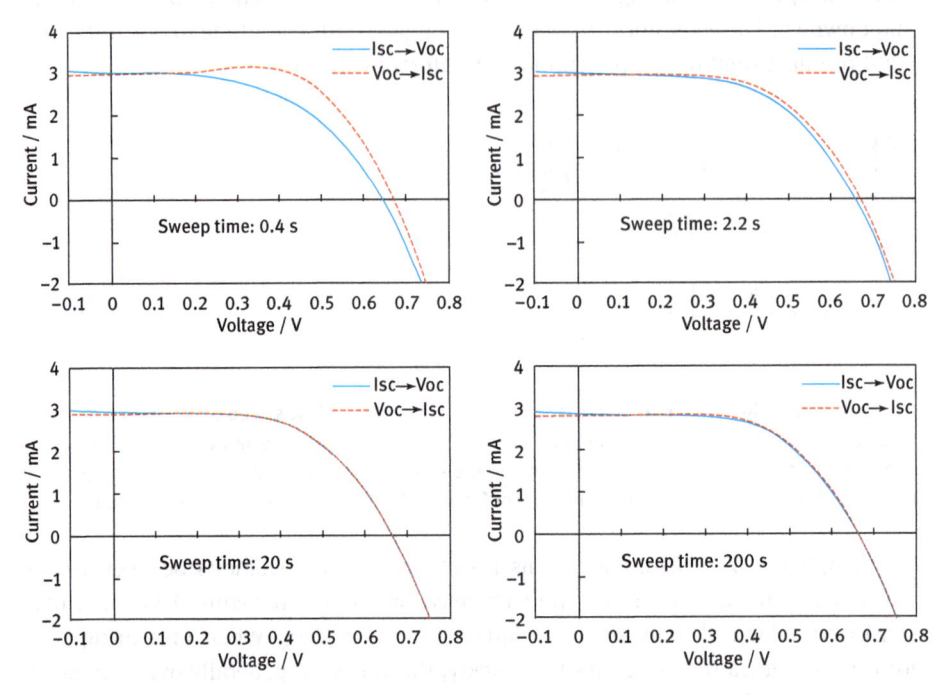

Fig. 4.15: Effect of sweep time and sweep direction on *J–V* curve measurements. Data are shown for a dye-sensitized solar cell with an ionic liquid electrolyte on the forward sweep from J_{sc} to V_{oc} and reverse direction with sweep times between 0.4 s and 200 s for the sweep range from −0.1 V to +0.8V. This figure was published in [229], Copyright (2012) Elsevier.

function of the photon wavelength λ, i.e. EQE = EQE (λ). Knowledge of the EQE (λ) spectrum reveals which spectral regions contribute to photocurrent generation, which is imperative to analyze when optimizing a solar cell. Furthermore, in combination with measurements of the total absorption spectrum $A(\lambda)$ (see Section 4.2.1), it is possible to calculate the internal quantum efficiency IQE (λ) as

$$\text{IQE} = \frac{\text{EQE}(\lambda)}{A(\lambda)}, \tag{4.29}$$

i.e. the quantum efficiency of conversion of an *absorbed* photon to an electron.

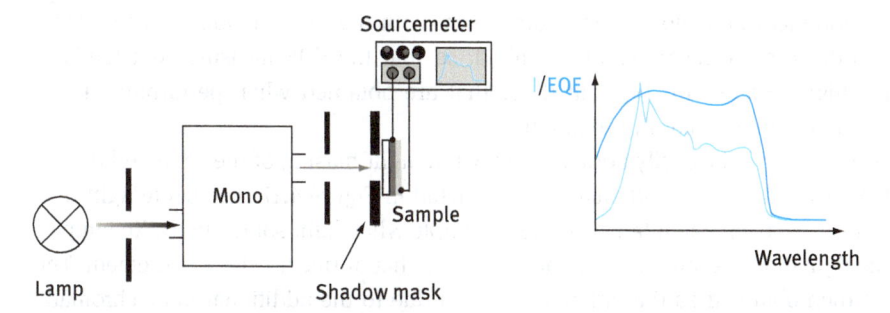

Fig. 4.16: Experimental setup for external quantum efficiency measurements. The solar cell sample is illuminated with monochromatic light, typically from a combination of a Xe lamp with a monochromator. The photocurrent of the solar cell is then acquired as a function of the wavelength leading to raw current data that is a convolution of the lamp spectrum and the EQE of the device.

Experimentally, the EQE can be measured with a setup schematically depicted in Figure 4.16. A monochromatic light source, for instance a combination of a Xe lamp and a monochromator, is used to illuminate a solar cell sample through an appropriate shadow mask. The device is kept at short circuit conditions and the current through the device is measured with a sourcemeter for each wavelength position of the monochromator. The raw current data have to be normalized to the number of incident photons, which is found by calibration of the setup with a reference solar cell with known spectral response for each wavelength. It is then possible to directly determine the EQE at a given wavelength as

$$\text{EQE} = \frac{\text{number of photogenerated electrons}}{\text{number of incident photons}}. \tag{4.30}$$

Since the EQE is the spectral response efficiency of the solar cell, the EQE spectrum can be used to predict the short circuit current density $J_{\text{SC,pred}}$ of the device under illumination with an arbitrary spectrum of light $S(\lambda)$ as

$$J_{\text{SC,pred}} = \int q \text{EQE}(\lambda) S(\lambda) \, d\lambda, \tag{4.31}$$

where $S(\lambda)$ is given in photons per second. Accordingly, $J_{\text{SC,pred}}$ calculated for an AM 1.5G solar spectrum should be identical to the measured J_{SC} under illumination with a solar simulator. Comparing these two values is an important and easy check for the reliability of J–V and EQE measurements and avoids measurement errors due to wrong or missing calibrations [231].

Since a typical setup for EQE measurements, as shown in Figure 4.16, uses a Xe lamp in combination with a monochromator as light source, light intensities at a given wavelength are small compared to the standard 1 sun intensity of 100 mW cm^{-2} used in solar simulator J–V measurements. Accordingly, the charge density in the device is very different, giving rise to deviations of J_{SC} and $J_{\text{SC,pred}}$ for solar cells with a nonlinear behavior of the photocurrent with the illumination intensity. Most of such devices are recombination-limited, so that the EQE measurement tends to predict higher values for J_{SC} than those that are obtained when performing a real measurement under solar illumination.

This issue can be easily addressed by white light-biasing of the solar cell during the EQE using a setup as schematically depicted in Figure 4.17. The white light-bias is provided by a solar simulator (or often a simple white light source like a Xe lamp or a white light diode) at an intensity comparable to that during a $J-V$ measurement. The EQE is then obtained as the differential signal due to the additional monochromatic illumination. Since the monochromatic intensity is small compared to the background, the measurement is performed using a chopper in the light path of the monochromatic light and a lock-in amplifier. Thus, the background current signal is subtracted and only the additional current due to the monochromatic illumination is measured, while the solar cell is kept at realistic (1 sun) conditions, i.e. with much higher charge densities present than in the case of a setup without white light bias.

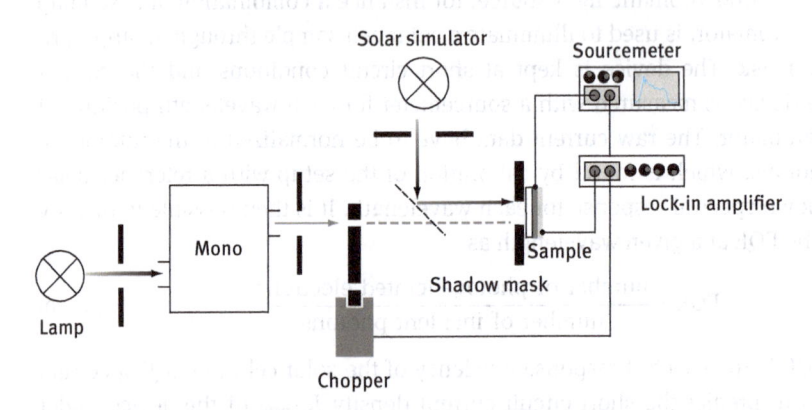

Fig. 4.17: Experimental setup for external quantum efficiency measurements using a lock-in technique. Especially organic and hybrid solar cells sometimes show a nonlinear behavior of the EQE with the light intensity. In order to account for this a white light or solar spectrum, illumination is focused on the solar cell while measuring the EQE. The monochromatic light is then applied as chopped modulation and the differential photocurrent response is measured using a lock-in amplifier.

In general, the chopping frequency has to be chosen so that it is slower than the inverse current response time of the device. Otherwise, the current does not reach its equilibrium during the off-state of the chopper and the next illumination event finds the device slightly higher light biased than for the equilibrated case under white light illumination. This is of particular importance for solar cells such as liquid-electrolyte dye-sensitized solar cells or metal-organic perovskite solar cells, where slow current responses are observed.

For many types of solar cells both types of EQE setups yield comparable results, in particular, if the device operates linearly over a broad range of light intensities.

However, white light-biasing is the only reliable method to obtain correct EQE values for a device under solar illumination, especially for solar cells with pronounced charge density–dependent charge–carrier recombination at short circuit. Therefore, only white light-biased EQE setups are considered state of the art today.

4.2.4 Light Intensity–Dependent Current–Voltage Measurements

While J–V measurements under simulated solar cells allow to directly measure the most important parameters of the device, like J_{SC}, V_{OC}, FF, and PCE, a lot of additional information about the underlying mechanisms of device operation can be gained by performing J–V measurements over a broad range of illumination intensities. For instance, such measurements allow to monitor the behavior of J_{SC} or V_{OC} as a function of light intensity, which gives insights into the recombination mechanisms limiting the device performance.

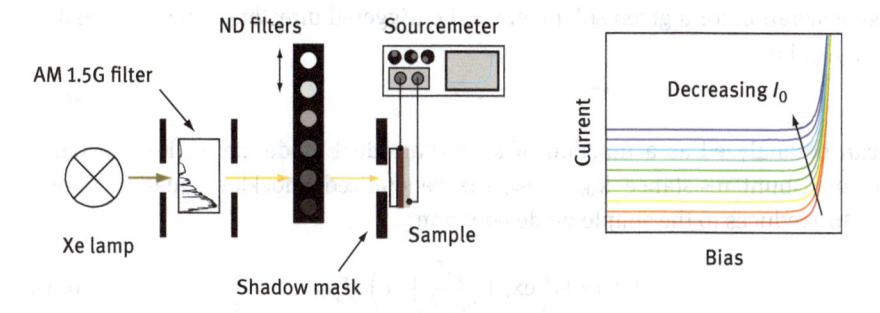

Fig. 4.18: Experimental setup for light intensity–dependent current density–voltage measurements. A regular J–V setup can be upgraded using an array of suitable neutral density filters so that measurements can be performed for various illumination intensities. Other methods for reduction of light intensity involve the use of metal meshes, or a combination of lenses and apertures.

One possibility to realize a setup for light intensity–dependent J–V measurements is schematically shown in Figure 4.18. A regular J–V setup including a solar simulator, a sourcemeter, and appropriate shadow masks is upgraded using a series of neutral density filters, which reduce the illumination intensity while keeping the spectrum (approximately) unchanged. Alternatively, metal grids of different porosity can be used or a combination of apertures and appropriate optics. Furthermore, if the illumination spectrum does not have to exactly match the solar spectrum, a combination of different light-emitting diodes can be operated at various intensities. Ideally, such a setup is fully automated so that J–V curves at a series of different intensities are recorded by a computer.

We will now have a closer look at how light intensity–dependent J–V measurements can be used to characterize charge–carrier recombination in organic and hybrid solar cells. We start with a general discussion of recombination mechanisms. First, note that geminate recombination, i.e. recombination of excited states before separation into free charges, even though being a loss mechanism that directly reduces the photocurrent, typically does not affect the J–V curve since geminate electron–hole pairs (excitons) are electrically neutral. This is true at least in the case of a weak electric field–dependent exciton splitting. In addition, Auger recombination can be neglected in organic and hybrid solar cells due to the short lifetime of excitons. Typically, the mechanism of formation of charge transfer excitons and their recombination has no effect on the electrical behavior of the device at illumination intensities lower than approximately 2.5 suns [232, 233]. We therefore limit the following discussion on recombination of already separated charge carriers, which can be monitored by electrical measurements.

Light Intensity Dependence of the V_{OC}. One of the most important parameters for recombination analysis is the V_{OC} of the solar cell, which describes the point where charge generation rate G and charge recombination rate R are equal. Since the charge generation for a given solar cell can be triggered directly by the illumination intensity I_0, i.e.

$$G \propto I_0, \tag{4.32}$$

V_{OC} can be analyzed as a function of I_0. For an ideal diode, i.e. series resistance $R_s = 0$ and shunt resistance $R_{sh} \to \infty$, the generalized Shockley equation (Equation (3.16)) reduces to the simple diode equation

$$J(V) = J_0 \left(\exp\left(\frac{qV}{nk_BT}\right) - 1 \right) - J_{ph}. \tag{4.33}$$

Furthermore, the photocurrent J_{ph} scales linearly with the illumination intensity, i.e. $J_{ph} \propto I_0$. The slope of the V_{OC} as a function of $\log(I_0)$ in units of k_BT/q thus is the ideality factor n. For good diodes, i.e. $R_{sh} \gg R_s$, the exponential forward part of the J–V curve is only slightly affected by R_s and R_{sh}, and n can be estimated from the slope of the dark J–V curve, without performing light intensity–dependent measurements [234]. However, especially high R_s can lead to an overestimation of n, which is then better approximated using the slope of the V_{OC} with I_0.

As discussed in Section 2.3.1 in the context of the current through a p–n junction, the ideality factor can be identified with the dominant type of recombination, where $n = 1$ represents the case of purely bimolecular recombination (of free charges), whereas $n = 2$ is present for monomolecular (trap state–assisted) recombination. We will now derive how these two recombination mechanisms can be identified from their footprints in light intensity–dependent J–V measurements.

In order to clarify the nature of charge–carrier recombination that determines the dependence of V_{OC} on I_0, we first consider purely Langevin-type (bimolecular) recombination. In this case, both electrons and holes migrate freely through the active layer and recombine with a certain probability if the Coulomb attraction is strong enough to enable interaction between two opposite charges. In this case the recombination rate R_L depends on both charge carrier densities and can be written as [235]

$$R_L = y\,(np - n_{int}p_{int}), \tag{4.34}$$

where y is the Langevin coefficient, n and p are electron and hole density, and n_{int} and p_{int} denote the intrinsic carrier concentrations of electrons and holes, respectively. With $n_{int}p_{int} = n_{int}^2$ for non-doped materials (like organic semiconductors) those are determined by the temperature T, the energy gap E_{gap}, and the effective density of states N_C as

$$n_{int} = N_C \exp\left(\frac{qE_{gap}}{2k_B T}\right). \tag{4.35}$$

For pristine materials, y depends on both electron and hole mobility, μ_n and μ_p, respectively, as

$$y = \frac{q}{\varepsilon}(\mu_n + \mu_p), \tag{4.36}$$

where ε is the relative permittivity of the material. In contrast, for a donor–acceptor blend, Langevin recombination is dominated by the slower charge carrier since the faster carrier effectively is always available for recombination at the interface. In this case, the recombination constant is [236]

$$y = \frac{q}{\varepsilon}\min(\mu_n, \mu_p). \tag{4.37}$$

For generation rate G and charge separation probability P, the rate equation for the one-dimensional current (in z-direction, i.e. toward the external electrodes) is [234]

$$\frac{1}{q}\frac{\partial}{\partial z}J(z) = PG - (1-P), \tag{4.38}$$

which simplifies to

$$G = \frac{1-P}{P}R \tag{4.39}$$

at V_{OC}, where the current is 0 and the cell is in thermal equilibrium. Here P can be used to describe geminate recombination of excitons, while G denotes the generation rate of excited states. For purely Langevin-type recombination, R_L from Equation (4.34) can be used and further simplified to

$$R_L = y n p, \tag{4.40}$$

since $np \gg n_{int}p_{int}$. Assuming that quasi-Fermi levels for electrons and holes are constant at open circuit (which is the case since drift and diffusion current densities cancel out and are effectively zero), the quasi-Fermi levels for electrons and holes (Φ_n and Φ_p, respectively) are equal to the potentials at the external contacts, i.e. their difference is the V_{OC}, and constant throughout the device [237]. This can be used to find an expression for the product of electron and hole density np as a function of the V_{OC}:

$$np = n_{int}^2 \exp\left(\frac{q(\Phi_p - \Phi_n)}{k_B T}\right) = n_{int}^2 \exp\left(\frac{qV_{OC}}{k_B T}\right). \tag{4.41}$$

With this, the Langevin recombination rate can then be written as

$$R_L = y n_{int}^2 \exp\left(\frac{q(\Phi_p - \Phi_n)}{k_B T}\right) = y n_{int}^2 \exp\left(\frac{qV_{OC}}{k_B T}\right). \tag{4.42}$$

Thus, by solving Equation (4.42) to V_{OC}, a logarithmic dependence of V_{OC} on the generation rate, i.e. the illumination intensity, is found as

$$V_{OC} = \frac{E_{gap}}{q} - \frac{k_B T}{q} \log\left(\frac{(1-P) y N_C^2}{PG}\right) \tag{4.43}$$

with effective density of states N_C and bandgap energy E_{gap}, which in this case denotes the energy difference between the HOMO of e.g. P3HT and the LUMO of $PC_{60}BM$ [234, 238]. Accordingly, for a device operation dominated by Langevin recombination, an approximate dependence like

$$V_{OC} \propto S \frac{k_B T}{q} \log(I_0), \tag{4.44}$$

with a slope of $S = 1$ is expected in light intensity–dependent J–V experiments, while material-specific parameters like E_{gap}, y, N_C, and P do not affect the dependence of V_{OC} on I_0.

In the case of trap-assisted charge recombination, the solar cell operation can be described by the Shockley–Read–Hall (SRH) equation and the recombination rate R_T is [239, 240]

$$R_T = \frac{C_n C_p N_C (pn - p_{int} n_{int})}{C_n(n + n_{int}) + C_p(p + p_{int})}, \tag{4.45}$$

where C_n and C_p denote the density of electron and hole traps, respectively. For a combination of Langevin and trap-assisted recombination, the overall recombination rate then becomes

$$R = R_L + R_T, \tag{4.46}$$

and Equation (4.44) cannot be solved analytically. However, the dependence of V_{OC} on G is still approximately logarithmic, only with a slope $S \neq 1$, i.e.

$$V_{OC} \propto S \frac{k_B T}{q} \log(G) \tag{4.47}$$

with larger values of S indicating a stronger impact of trap-assisted recombination [241, 242].

Light Intensity Dependence of the J_{SC}. When measuring light intensity–dependent J–V curves, it is also useful to investigate the behavior of the short circuit current J_{SC} with the illumination intensity. Since $J_{SC} \approx J_{ph}$ for good solar cells, i.e. devices with field-independent charge generation and negligible impact of series and shunt resistance, a linear behavior of J_{SC} with I_0, or more precisely the generation rate G, is expected for a device operating according to the Shockley diode equation due to

$$J_{ph} \propto I_0 \propto G. \tag{4.48}$$

Experimentally, this is not found to be always the case, but a sublinear behavior is often observed [234, 243]. However, satisfying data fitting usually can be obtained using a power law

$$J_{SC} \propto I_0^{\alpha}, \tag{4.49}$$

where α typically ranges between 0.75 and 1. As outlined by Koster et al., $\alpha = 0.75$ is observed in devices [234] with highly unbalanced electron and hole mobilities, which suffer from the built-up of space charge. Such a situation is also apparent when observing increasing values of α for the light intensity behavior of the measured photocurrent at increasing reverse bias.

Besides, $\alpha \neq 1$ is also often associated with monomolecular (trap-assisted) recombination, which reduces the photocurrent of the device compared to the case of purely bimolecular recombination, which dominates when approaching the V_{OC} [244].

Charge Extraction Probability. Besides plotting V_{OC} and J_{SC} as a function of I_0, another very useful representation of light intensity–dependent J–V data is the interpretation of the obtained curves as bias-dependent charge extraction probability $P(V, I_0)$. This probability can be obtained from the J–V curves by subtraction of the dark J–V curve and subsequent normalization of the curves to an estimate for the photocurrent J_{ph} at a given intensity I_0. As estimated, the dark curve-corrected current at moderate reverse bias (for instance -0.5 V) can be used in order to account for potential space charging, which can be overcome by an extracting bias. This leads to the expression

$$P(V, I_0) = \left| \frac{J(V, I_0) - J(V, 0)}{J(-0.5V, I_0) - J(-0.5V, 0)} \right|, \tag{4.50}$$

where $J(V, I_0)$ are the light intensity–dependent J–V curves [232, 245].

One example for such an estimate of $P(V, I_0)$ from light intensity–dependent J–V curves is shown in Figure 4.19. While analyzing the raw data plotted together does not directly reveal much about the device operation, monomolecular and bimolecular recombination losses can be identified by clear footprints in the $P(V, I_0)$ curves. Analogous to the considerations concerning the light intensity dependence of J_{SC} and V_{OC}, monomolecular recombination dominates at small bias voltages around the J_{SC}. Therefore, a device that is limited by monomolecular recombination will show a light intensity dependence of the $P(V, I_0)$ curves in this bias region. On the other hand, bimolecular recombination comes into play especially when approaching the V_{OC}, where high charge densities are present in the device. Pronounced bimolecular recombination therefore induces a strong light intensity dependence of the $P(V, I_0)$ curves approaching the V_{OC}.

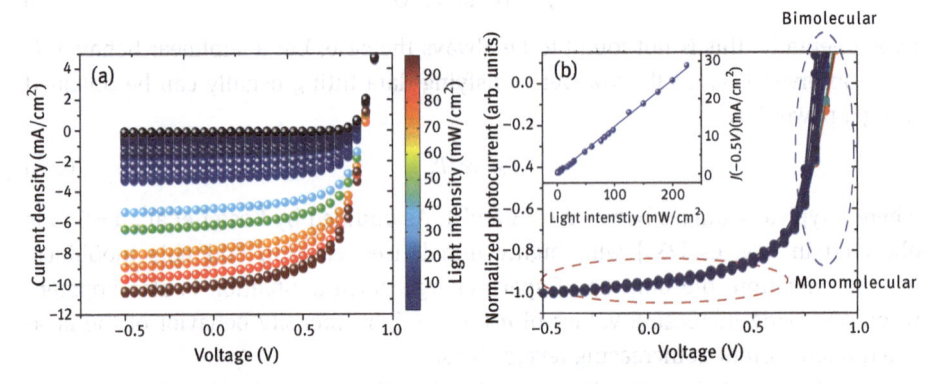

Fig. 4.19: Example for light intensity–dependent *J–V* results, which can be used to calculate the charge collection probability. Normalization of each *J–V* curve to the photocurrent allows to directly identify the role of monomolecular and bimolecular recombination in the device at different illumination conditions. Figure reprinted with permission from [245]. Copyright (2010) the American Physical Society.

4.3 Time-Resolved Characterization Techniques

Besides our discussion of steady-state techniques, we also want to highlight the most important time-resolved characterization techniques, which are used to monitor the dynamics of excitons and polarons in organic and hybrid solar cells. We will first discuss three electronic experimental techniques, namely transient photovoltage/photocurrent decay measurements, photo-charge extraction measurements, and impedance spectroscopy. These will be followed by a brief description of two fs-laser-based techniques, which are transient absorption spectroscopy and time-resolved photoluminescence spectroscopy.

4.3.1 Transient Photocurrent and Photovoltage Decay Measurements

A widely used measurement technique to monitor polaron dynamics is a combination of transient photovoltage and photocurrent decay experiments (TPV and TPC) [69, 246–249]. These experiments can be performed either as a light-off experiment, where the time evolution of the photovoltage or photocurrent is monitored after the illumination source is switched off (with sufficiently short switching times), or using background illumination and a small perturbation caused by a fast LED or a laser. The latter is more common for organic and hybrid solar cells and will be discussed in this section.

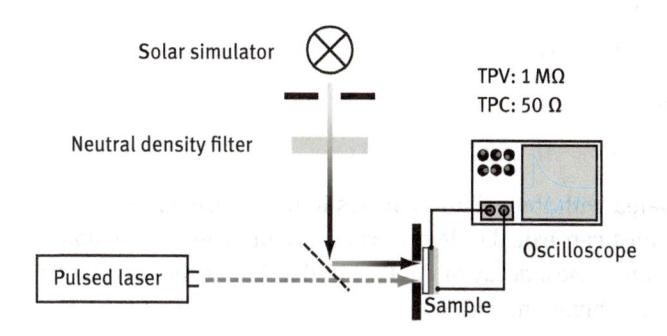

Fig. 4.20: Experimental setup for transient photovoltage and photocurrent decay measurements (TPV and TPC). A solar cell sample is white light-biased using a solar simulator with defined light intensity adjusted using appropriate neutral density filters. A small perturbation is generated using a pulsed laser or a fast LED with pulse lengths on the order of nanoseconds. The decay back to equilibrium can then be monitored using an oscilloscope with termination resistance 1 MΩ for TPV and 50 Ω for TPC.

Figure 4.20 shows a schematic of an experimental setup, which can be used for TPV/TPC measurements. A solar cell sample is illuminated with a solar simulator or a suitable background illumination source, often in combination with filters that allow to adjust the illumination intensity over a few orders of magnitude. Additionally, a small perturbation is caused by short light pulses, e.g. emitted by a pulsed laser. The intensity of these pulses is small compared to the background so that the system response to the perturbation is approximately linear. Performing TPV/TPC experiments in this small-perturbation regime is a prerequisite for straightforward yet meaningful data analysis since it allows to model the detected voltage and current decays with a simple exponential function. Time-dependent photovoltage and photocurrent are measured using an oscilloscope, where for TPV a quasi-V_{OC} situation is achieved using a termination resistance $R_{termination} = 1\,\text{M}\Omega$ in the oscilloscope, while TPC is performed at quasi-J_{SC} measuring the potential drop over a termination resistance $R_{termination} = 50\,\Omega$ and using Ohm's law

$$J(t) = \frac{V(t)}{R_{termination}}. \tag{4.51}$$

During a TPV/TPC experiment, the short light pulse temporarily changes the light intensity at which the device is operated by a small amount, so that V_{OC} or J_{SC} slightly increases. Since the perturbation is only a short pulse, the device is out of equilibrium after light-off and decays back to the equilibrium value of V_{OC} or J_{SC} at the background illumination intensity. As mentioned before, the time-dependent voltage and current for experiments in the small-perturbation regime are described by a monoexponential decay on top of the equilibrium offset voltage and current $V_0 = V_{OC, \text{equilibrium}}$ and $J_0 = J_{SC, \text{equilibrium}}$, respectively. Thus, there are characteristic decay rates k_{TPV} for the voltage decay

$$V(t) = \Delta V_0 \exp(-k_{TPV} t) + V_0, \qquad (4.52)$$

and k_{TPC} for the current decay

$$I(t) = \Delta I_0 \exp(-k_{TPC} t) + I_0. \qquad (4.53)$$

These rates can be associated with the polaron kinetics in the device. Since at open circuit conditions no charges can exit the device, charge–carrier recombination is the only mechanism that can cause a decay of the V_{OC}, so that k_{TPV} is the characteristic rate of charge carrier recombination, and

$$\tau_{TPV} = k_{TPV}^{-1} \qquad (4.54)$$

can be interpreted as charge carrier lifetime. At short circuit, charges can easily flow out of the device and charge–carrier recombination can be neglected, so that k_{TPC} is a good estimate for the charge carrier extraction lifetime, since charge extraction is the predominant mechanism causing a decay of the J_{SC}. Analogous to the TPV case a characteristic charge extraction time can be found as

$$\tau_{TPC} = k_{TPC}^{-1}. \qquad (4.55)$$

As outlined by Shuttle and co-workers, TPV and TPC experiments can be further used to measure the charge carrier density dependency of the recombination rate using the so-called *differential charging method* [250]. This technique relies on a series of TPV and TPC measurements at increasing background illumination intensity, which allows to calculate the differential capacitance of the device at a given background intensity from the differential voltage ΔV_0 measured at open circuit and the differential charge ΔQ measured at short circuit as

$$\Delta C = \frac{\Delta Q}{\Delta V_0}, \qquad (4.56)$$

where ΔQ is the integral over the current decay, i.e.

$$\Delta Q = \int \Delta I_0 \exp(-k_{TPC} t)\, dt, \qquad (4.57)$$

at a given background light intensity. Integrating all differential capacitances up to V_{OC} at the highest background illumination level then yields the overall capacitance of the device at this illumination condition. This integral together with the dimensions of the device can then be used to calculate the charge density n as

$$n = \frac{1}{qAd} \int_0^{V_{OC}} \Delta C dV, \qquad (4.58)$$

where A is the active area of the device, d the film thickness, and q the elementary charge.

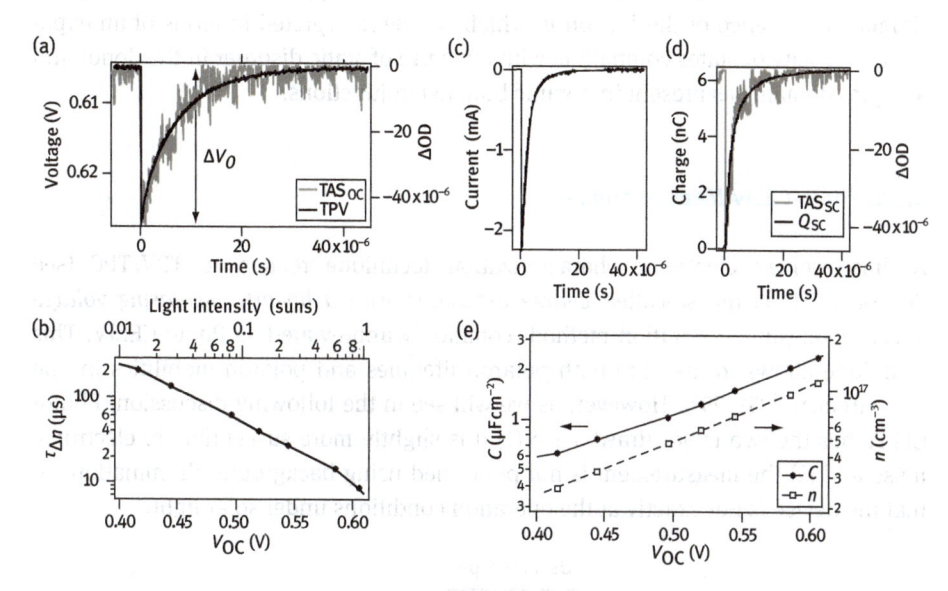

Fig. 4.21: **Example for transient photovoltage (TPV) and photocurrent (TPC) decay experiments.**
(a) TPV decays nicely coincide with decays from transient absorption experiments, outlining that charge dynamics can be directly monitored using this technique. (b) TPV experiments at different background illumination intensities allow to determine the charge carrier lifetime as a function of the device V_{OC}. TPC shown in (c) can be time-integrated in order to obtain a transient capacity curve as shown in (d). (e) Integrating these capacities over different background intensities can then be used to find an estimate for the charge carrier density as a function of the V_{OC} using the differential charging approximation discussed in the text. Figure adapted from [250]. Copyright (2008) American Institute of Physics.

An example for such a combined TPV/TPC experiment on conventional P3HT:PCBM bulk heterojunction solar cells at varying background illumination is shown in Figure 4.21 [250]. In Figure 4.21(a), a typical TPV decay shown as a black line is compared to the dynamic photoinduced absorption of polarons in the polymer, shown as a gray line, which is monitored using transient absorption spectroscopy (see Section 4.3.4) on a fully assembled solar cell kept at open circuit. Both decays show the same time evolution, indicating that both methods measure the same property of the device, the polaron lifetime. A plot of the obtained TPV lifetime as a function of the light intensity and with this the V_{OC} of the device, as presented in Figure 4.21(b), shows an exponential behavior of the polaron lifetime with the V_{OC}. Figure 4.21(c) shows a typical current decay, which can be converted to a charge decay, as shown in Figure 4.21(d), where this decay is compared to transient absorption results obtained by keeping the device at short circuit. Apparently, also for TPC the decay is in good accordance with a transient absorption-based polaron measurement, outlining its validity in the context of the differential charging approach. Combined TPV and TPC results then allow to plot the relationship of the V_{OC} and the charge carrier density n as discussed above. From Figure 4.21(e), it is apparent that there is a logarithmic dependence of the V_{OC} on n, which can be interpreted in terms of an exponential density of states compatible with a picture of static disorder in the donor and acceptor domains as present in organic bulk heterojunctions.

4.3.2 Photo-CELIV Measurements

A time-resolved electronic characterization technique related to TPV/TPC (see Section 4.3.1) is the so-called *charge extraction with a linearly increasing voltage sweep* upon photo-excitation method, commonly abbreviated as Photo-CELIV. This technique allows to measure both polaron lifetimes and polaron mobilities in one measurement [251, 253]. However, as we will see in the following discussion, Photo-CELIV has the two major drawbacks: (1) it is slightly more susceptible to electronic noise and (2) the measurement is not performed using background illumination, so that the device is not exactly at the operation conditions under solar light.

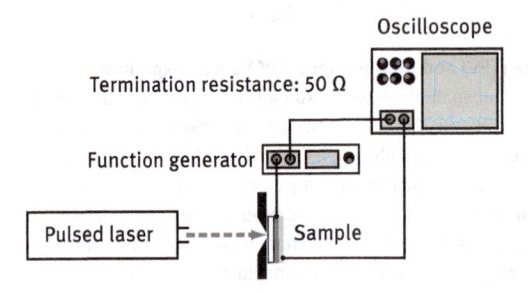

Fig. 4.22: Schematic of the experimental setup used for photoinduced charge extraction with a linearly increasing voltage sweep (Photo-CELIV) measurements. A photovoltaic sample is connected in series with a function generator and the low-resistance termination of an oscilloscope. The sample is kept at flat-band conditions and illuminated with a pulsed laser. The laser triggers an extraction voltage sweep after a defined delay time, and the resulting current signal is measured as a potential drop with the oscilloscope.

The experimental setup for a Photo-CELIV measurement is shown in Figure 4.22. A solar cell sample is kept in the dark and illuminated with short pulses, either with a pulsed LED or with a laser. The charge extraction sweep is provided by a function generator, which is connected in series to the solar cell. The current transients are then obtained via measurement of the potential drop over the 50 Ω termination resistance of an oscilloscope.

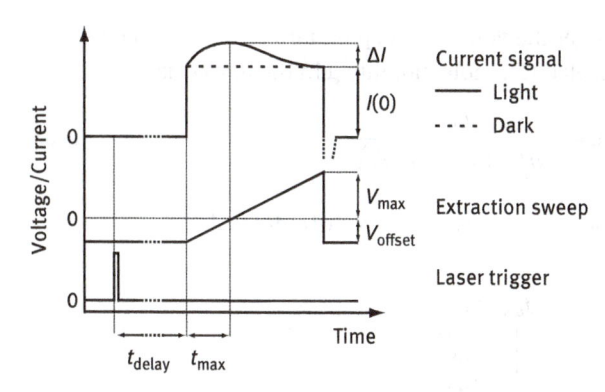

Fig. 4.23: Signals used and detected during a Photo-CELIV experiment. A pulsed laser is triggered at t_0 to generate a single shot. The device is kept at flat band conditions during a certain delay time after the pulse, before a linearly increasing voltage sweeps out the remaining photo-generated charge carriers. Analysis of the resulting CELIV signal allows to determine charge mobilities and recombination lifetimes.

A typical Photo-CELIV signal along with the timing of the laser pulse and the voltage sweep is shown in Figure 4.23. A short laser pulse excites the sample at time $t = 0$, while the device is kept at flat band conditions, i.e. at some forward bias V_{offset}. After a wait time t_{delay} the voltage is quickly swept from flat band to strong reverse bias, with typical sweep times being on the order of 10 μs and maximum voltages V_{max} around 1–5 V (reverse bias). Since every thin-film hybrid and organic solar cell exhibits a constant geometrical capacitance C due to its metal–semiconductor–metal structure the response in the dark (i.e. in the absence of photocarriers) to the linearly increasing voltage sweep is a constant current. This is the case since for voltage V and charge Q with

$$C = \frac{Q}{V} = \text{const.} \quad (4.59)$$

and for current I

$$I = \frac{dQ}{dt} \quad (4.60)$$

one finds

$$I = \frac{d}{dt}CV = C\frac{dV}{dt}, \quad (4.61)$$

i.e. a constant dark current $I = I(0)$ for a linear voltage sweep. Due to the excitation pulse there are also photocarriers in the tested solar cell sample, which do not leave the device during the delay time since the cell is kept at flat band conditions. During the voltage sweep, however, these carriers are extracted from the active layer, resulting in an additional current signal with peak value ΔI on top of the rectangular current response. The area under this additional current signal gives the amount of charge remaining after a wait time t_{delay}, during which charge carrier recombination occurs. Therefore, this charge can be evaluated as a function of the delay time in order to monitor charge carrier recombination. Furthermore, the time at which the maximum current is observed after the voltage sweep is started (t_{max}) can be used to calculate the effective averaged electron–hole mobility µ in the device as

$$\mu = \frac{2d^2}{3Rt_{max}^2\left(1 + 0.36\frac{\Delta I}{I(0)}\right)} \tag{4.62}$$

with active layer thickness d and voltage sweep rate $R = dU/dt$ [245].

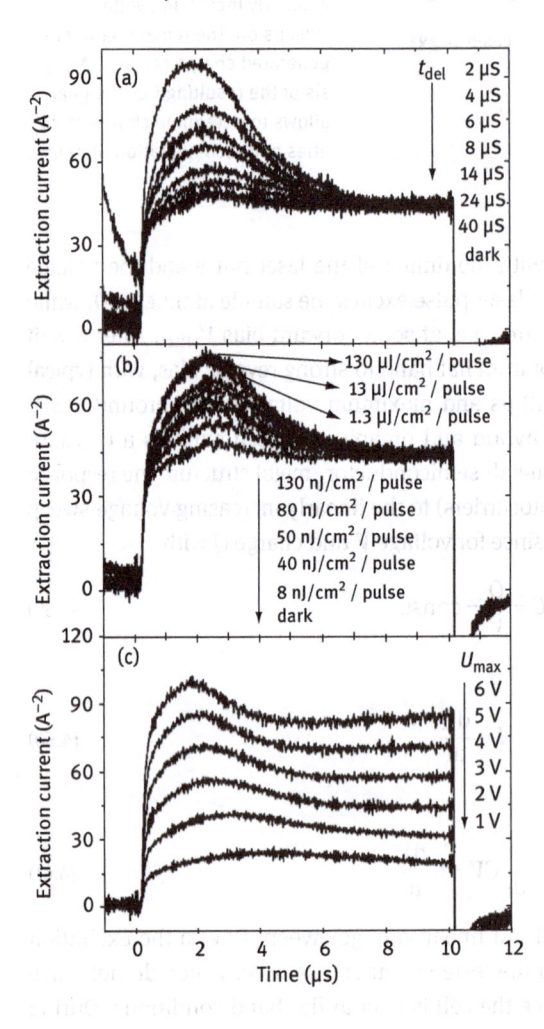

Fig. 4.24: Measured Photo-CELIV curves at **(a)** various delay times (t_{del}) between the light pulse and the voltage pulse; **(b)** various incoming light intensity at fixed $t_{del} = 5$ µs; and **(c)** various U_{max}. Figure adapted from [254]. Copyright (2005) American Institute of Physics.

Figure 4.24 gives an overview of the impact of various measurement parameters on the shape of the Photo-CELIV signal [254]. In Figure 4.24(a) the delay time is varied in order to monitor charge carrier recombination. For increasing t_{delay} more charges recombine before they are swept out of the device, resulting in decreasing photocharge on top of the rectangular capacitive current response. Figure 4.24(b) shows measurements for different laser intensities for one fixed delay time, i.e. different amounts of initially generated photocarriers. While more charge is observed for higher light intensities, t_{max} is constant, indicating that the increased charge density in the device has no impact on the charge carrier mobility over the range of tested laser intensities. The impact of different values of V_{max} on a Photo-CELIV measurement is illustrated in Figure 4.24(c). Since the sweep is always started at the flat band potential, increasing V_{max} means increasing sweep rates, which results in decreasing t_{max}. Typically, voltages around 5 V are used, allowing to clearly identify the peak value. Besides, fast sweeping guarantees that all photocarriers detected during the sweep as indicated by the current signal dropping back to its rectangular capacitive dark current behavior. However, some sensitive devices may be damaged by such a high reverse bias, so that in some cases lower V_{max} has to be used.

4.3.3 Impedance Spectroscopy

In this section we give a short introduction of impedance spectroscopy, one of the most powerful characterization techniques for thin-film solar cells (and other devices), which allows insights into static as well as dynamic electronic processes in the device. The fundamental principle of this technique is a measurement of the frequency-dependent complex impedance $Z(\omega) = Z'(\omega) + iZ''(\omega)$ of the sample and subsequent modeling of the data with appropriate equivalent circuits, which in turn allow to attribute certain physical mechanisms in the device to components in the equivalent circuit, e.g. resistances and capacities. Note that ω denotes the radial frequency, which is related to the frequency f as

$$\omega = 2\pi f. \tag{4.63}$$

Experimentally, the complex impedance is determined using Ohm's law

$$Z(t) = \frac{V(t)}{I(t)} \tag{4.64}$$

after applying an ac voltage $V(t)$ and measuring the current response $I(t)$ of the sample. Figure 4.25 illustrates a typical measurement setup on the left-hand side and the fundamental principle of the measurement on the right-hand side. The setup is performed using a potentiostat, which can apply ac voltages with frequencies typically ranging between mHz and MHz (i.e. 6–9 orders of magnitude) on a solar cell sample, which can also be illuminated in order to allow characterization of the device under working conditions. The device is kept at a fixed dc offset voltage V_{offset}, defining the point on the J–V curve where the measurement is performed. On top of this offset voltage there is a small sinusoidal ac voltage

$$\hat{V}(t) = V_0 \sin(\omega t), \tag{4.65}$$

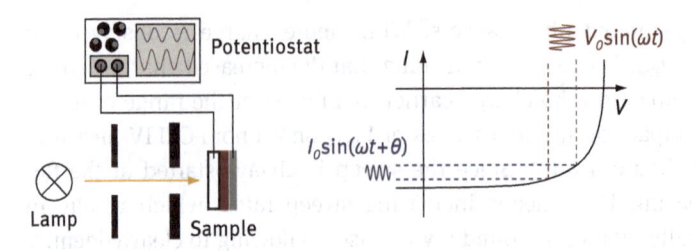

Fig. 4.25: Schematic illustration of an impedance spectroscopy setup. A solar cell sample is light-biased if required and connected to a potentiostat, which performs the impedance measurement. For this purpose a small alternating voltage is applied in addition to a dc bias voltage, which determines the position on the J–V curve at which the sample is probed. Due to the small amplitude of the sinusoidal voltage the sample response also is sinusoidal (plus a dc offset), shifted by a phase θ. The measurement is performed for a wide range of frequencies, typically between mHz and MHz. Amplitude and phase shift of the current signal are recorded and can be used to directly calculate the frequency-dependent impedance $Z' + iZ''$.

with amplitudes V_0 typically significantly below the thermal voltage (25 mV at room temperature, with $k_B T \approx 25$ meV), e.g. 10 mV in most cases. Since this ac signal is a small perturbation the current response is linear in very good approximation; i.e. there is a constant offset current I_{offset}, and a time-varying small ac current

$$\hat{I}(t) = I_0 \sin(\omega t + \theta), \tag{4.66}$$

with amplitude I_0 and a phase shift θ with respect to the applied voltage. With this the impedance can simply be written as

$$Z(t) = \frac{\hat{V}}{\hat{I}} = \frac{V_0}{I_0} \frac{\sin(\omega t)}{\sin(\omega t + \theta)} =: Z_0 \frac{\sin(\omega t)}{\sin(\omega t + \theta)}. \tag{4.67}$$

Using Euler's formula

$$\exp(i\phi) = \cos(\phi) + i \sin(\phi) \tag{4.68}$$

the ac components of voltage and current can be written as

$$\hat{V}(t) = V_0 \exp(i\omega t) \tag{4.69}$$

and

$$\hat{I}(t) = I_0 \exp\left(i(\omega t + \theta)\right). \tag{4.70}$$

Using Ohm's law from above, the time dependency cancels out and the resulting impedance is intrinsically dependent on ω since the current amplitude $I_0 = I_0(\omega)$ is a function of the frequency of the applied ac voltage:

$$Z(\omega) = \frac{\hat{V}}{\hat{I}} = \frac{V_0}{I_0(\omega)} \exp(i\theta) = Z_0(\omega)\left(\cos(\theta) + i \sin(\theta)\right) = Z'(\omega) + iZ''(\omega). \tag{4.71}$$

Since $Z(\omega)$ can be determined directly from the amplitude and phase shift of the current response to an applied small ac voltage perturbation, impedance spectroscopy can be performed by measuring these two quantities in order to obtain the frequency dependency of the total impedance of the device. The frequency dependence found can then be interpreted using equivalent circuits with a certain total impedance, which shows the same frequency behavior, so that in turn the elements of the equivalent circuit can be interpreted as representatives of physical properties of the device. However, care has to be taken to chose a model that is not overly complicated in order to make sure that the interpretation is meaningful (while a model with 100 circuit elements might yield a better fit to the frequency-dependent data than a model with only three elements; a complex model makes a straightforward interpretation impossible since it is unclear which mechanism in the device is associated with which element in the equivalent circuit).

In order to construct such equivalent circuits it is necessary to know the complex impedances of certain circuit elements. Then, the total impedance of a circuit can simply be calculated using Kirchhoff's circuit laws. Table 4.1 gives an overview of the four most important circuit elements used for impedance spectroscopy modeling. While a resistance has a purely real impedance, capacitance and inductance have purely imaginary impedances. The constant phase element is a special case of a non-ideal capacitor, which is useful to model effects like electrical double layers.

Tab. 4.1: Overview of the most important circuit elements used in impedance spectroscopy. The dc quantity is given along with the complex impedance that has to be taken into account for ac circuits.

Circuit element	dc quantity	Impedance	Symbol
Resistance	R	R	
Capacitance	C	$(i\omega C)^{-1}$	
Constant phase element	Q	$(i\omega Q)^{-n}$, $0 < n \leq 1$	
Inductance	L	$i\omega L$	

As an example we now take a closer look at the simplest equivalent circuit that can be used to model a photovoltaic device: a combination of a series resistance R_s and a parallel combination of a resistance R_p and a capacitance C. This circuit is shown in Figure 4.26 on the upper left-hand side. Such a circuit is suitable since every electronic device features a certain series resistance and the active layer of the solar cell can be approximately modeled with the parallel R–C combination. The capacitance originates from the geometry of a thin-film solar cell (a low-conductive active material between two metal electrodes), while the resistance allows some current flow through the active film. Note that for an ideal solar cell the geometrical capacitance does not change for different applied bias voltages, while the resistance decreases exponentially in forward direction (and thus allows an exponentially increasing diode current).

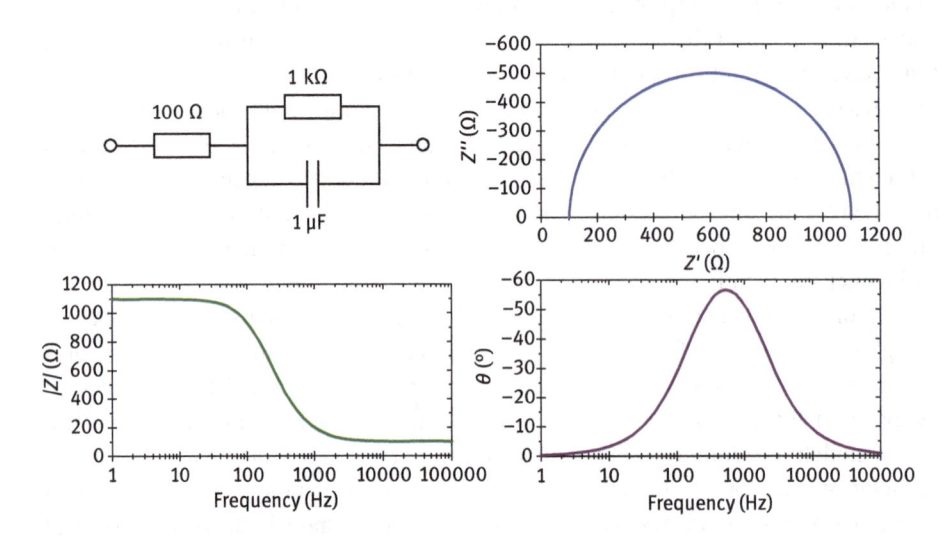

Fig. 4.26: Example of the impedance spectra of a simple circuit with a parallel combination of a 1 kΩ resistance and a 1 µF capacitor in series with a 100 Ω resistance. The figure summarizes the most essential plots that can be obtained using impedance spectroscopy data, namely (1) a Z'–Z" plot, a so-called Nyquist plot, (2) the Bode plot of the absolute value of the impedance |Z| as a function of frequency f, and (3) the Bode plot of the phase θ between applied ac voltage and the resulting current as a function of frequency f.

According to Kirchhoff's circuit laws, the total impedance of the equivalent circuit in Figure 4.26 is found from the impedances of the three components as

$$Z_{\text{total}} = Z_{R_s} + \frac{1}{Z_{R_p}^{-1} + Z_C^{-1}} = R_s + \frac{1}{R_p^{-1} + i\omega C} = \\ = R_s + \frac{R_p}{1 + \omega^2 C^2} - i\frac{R_p^2 \omega C}{1 + \omega^2 C^2},$$ (4.72)

i.e. we found a simple analytical expression for the real and imaginary part of Z, Z', and Z'', respectively. Experimentally one now investigates the frequency dependence of Z, which is outlined by simulations of our equivalent circuit presented in Figure 4.26. The upper right graph is a so-called *Nyquist plot*, which represents the behavior of Z in the complex Z'–Z'' plane, where by convention the y-axis is plotted as $-Z''$. The left-hand side of the arc corresponds to the high-frequency case, where current can pass the capacitor, so that R_p is effectively bypassed by C. For decreasing frequencies the total impedance has both real and imaginary part and current can flow through both the parallel resistance and the capacitor. This is also apparent from the Bode plot of the absolute value of the total impedance, which drops in the mid-frequency range. For low frequencies, i.e. toward the dc case, the resistance of the capacitor becomes infinite and all current flows through R_p. Since the Z'–Z'' curve is shifted along the real axis by the impedance of the series resistance R_s the total resistance in this case is then $R_s + R_p$. The lower right plot additionally shows the Bode plot of the phase shift between applied ac voltage and the current response, which clearly indicates the resonance frequency of the circuit. Note that the axes of the Nyquist plot are isometric, which allows to directly characterize the impedance behavior in the Nyquist plot as circular (for the case of an R–C parallel combination) or elliptical (which is the case if the capacitor is replaced by a constant phase element with $n < 1$).

In order to further illustrate the nature of impedance spectroscopy data we show a three-dimensional representation of the frequency-dependent real and imaginary part of Z in Figure 4.27. The black dots show the measured data points, whereas the red, green, and blue curves are the projections in certain directions. While the Nyquist plot gives information about the relationship of Z' and Z'', the frequency information is lost. Both Z' and Z'' can be plotted versus the frequency in order to illustrate the behavior of the device, but often $|Z|$ is plotted instead along with a Nyquist plot as shown in Figure 4.26.

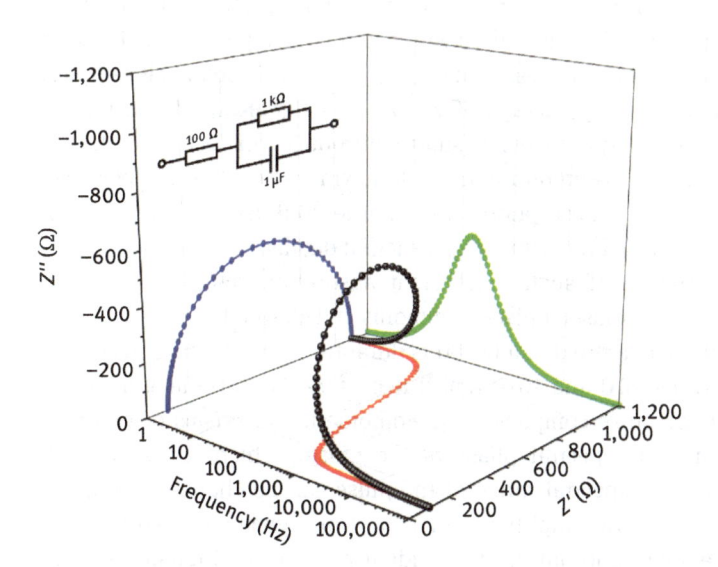

Fig. 4.27: Three-dimensional representation of the impedance spectrum of the R–RC circuit from the previous figure. The graph shows the frequency-dependent real and imaginary part of the total impedance, Z' and Z'', respectively. Accordingly, one projection yields a Nyquist plot (Z'–Z''), while the other two are Bode plots of Z' and Z''.

As mentioned before, impedance spectra are typically recorded for a set of different bias voltages and possibly also illumination intensities. Subsequently, a reasonable model, i.e. equivalent circuit diagram, has to be found, which ideally allows to model all recorded curves, but contains as few elements as possible. The voltage and intensity behavior of the cell is then reflected in changing values of the circuit elements, e.g. decreasing R_p for increasing bias voltage. While data fitting can be easily done using commercially available software packages, finding an appropriate equivalent circuit is highly challenging when performing impedance spectroscopy analyses. Here, we do not want to go into any detail, but limit our discussion to the fundamental principles of impedance spectroscopy presented so far. The interested reader is referred to a number of excellent review articles and publications [255–258].

4.3.4 Transient Absorption Spectroscopy

Many processes in organic and hybrid solar cells occur on femtosecond timescales and are therefore impossible to monitor using electronic measurement techniques. This is particularly the case for exciton dynamics and charge transfer processes, both of which are essential to understand in detail how an excitonic device could be pushed toward its performance limit. One of the most powerful methods to monitor processes on such short timescales is transient absorption spectroscopy, a pump–probe technique that uses two fs laser pulses with a defined temporal offset in the fs range in order to probe dynamic processes in organic and hybrid solar cell samples. Transient absorption spectroscopy setups of various designs are used and related techniques like pump–push experiments as shortly introduced in Section 3.3.1.3 are frequently found throughout the literature. In the following we discuss one possible, more or less standard transient absorption setup in order to illustrate its fundamental working mechanism without the actual experimental details.

A schematic illustration of such a transient absorption setup is shown in Figure 4.28. As source for fs laser pulses commonly a Ti:Sa laser is used, which allows to generate pulses of approximately 100 fs duration with a center wavelength that can be tuned between 670 and 1070 nm. If necessary the pulse length can be drastically reduced using pulse compressors (a combination of prisms or gratings, which allows to control the spectral phase of the pulse), which allow to reach timescales below 10 fs. The spectral range of the pulse can be changed using, for instance, an optical parametric amplifier, which basically generates two photons with widely tunable energy ratio out of one incident photon. Additionally, second harmonics generation can be used in order to double and triple the energy of photons in the pulse (of course at the cost of intensity). In our exemplary setup we employ only a Ti:Sa laser as the source for fs pulses and a second harmonics generator (SHG) for simplicity.

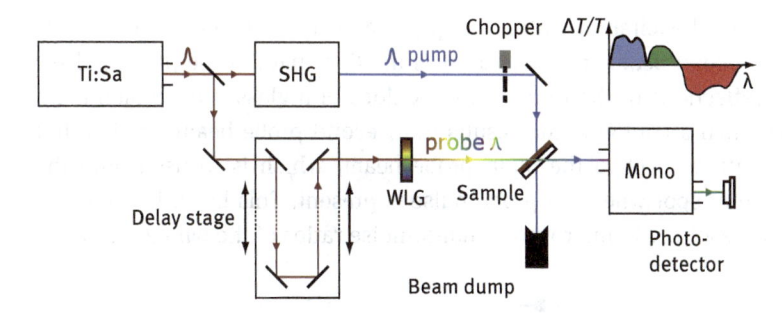

Fig. 4.28: Schematic experimental setup for transient absorption spectroscopy experiments.
A pulsed fs laser like a titanium:sapphire oscillator with appropriate amplifiers is used as a light source, generating pulses centered around 800 nm. A beam splitter guides a portion of the laser light toward the sample, which serves as a pump. Often, a second or third harmonic generation, or an optical parametric oscillator, is used to change the wavelength of the pump to allow excitation in the UV or visible. The rest of the laser light is running over a delay line to allow tuning of the time between pump and probe arriving at the sample, and converted into a white light pulse using supercontinuum white light generation in a nonlinear medium like sapphire. Since the pump signal is chopped, the setup allows to measure the difference in sample transmission with and without pump, i.e. $\Delta T/T$.

The excitation pulses are first guided on a beam splitter, which defines pump and probe light paths. In our example the pump beam is generated using second harmonics generation, which puts the spectral range of the excitation light in the blue or green visible spectral region. It is then focused onto the sample and chopped with a mechanical chopper, typically at a frequency around 1 kHz. Behind the sample the pump beam is guided into a beam dump in order to avoid reflections at components of the setup, which could cause unwanted signals.

The probe beam is guided over a mechanical delay stage, a combination of aligned mirrors that allows direct control over the path length that the probe light has to travel. This path length can be adjusted by changing the distance between the two pairs of mirrors, mechanically via a translation state or using piezo motors behind the mirrors for µm control. The optical delay results in a time delay between a given pair of pump and probe beam, where a path length difference of 1 mm corresponds to a time difference of roughly 3.3 ps. The probe beam is further focused into a medium that allows supercontinuum white light generation, most commonly a thin slice of sapphire glass. The resulting broadband pulse allows to probe the sample across the whole UV-visible spectrum plus the near-IR. After passing through the sample at the same spot as the pump pulse the probe pulse is detected, typically using a combination of a monochromator and a photodiode. The detected signal is modulated with the chopping frequency of the pump beam and acquired using lock-in amplification, so that the differential transmission signal $\Delta T/T$ for the cases with and without pump pulse can be measured. Accordingly, a transient absorption setup measures the time-dependent differential transmission spectrum of a sample upon optical excitation with a pump pulse of defined center wavelength. In many cases a second beam splitter is used in the light path of the probe pulses, often using the reflections of the front and back sides of a glass slide, which is hit under an angle. Such a configuration results in a second probe beam parallel, but slightly shifted with respect to the main probe beam, which is focused onto the sample at a different spot where no probe pulse is present. This beam is used as a second reference, resulting in improved signal-to-noise ratios of the whole setup.

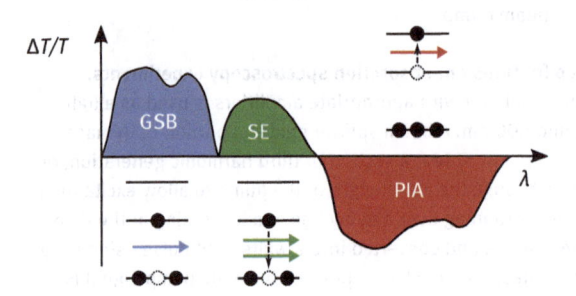

Fig. 4.29: Example of a transient absorption signal. Fundamentally, three features are contained in every transient absorption signal, owing to the processes schematically indicated. Those are ground-state bleach (GSB), stimulated emission (SE), and photoinduced absorption (PIA).

We now take a closer look at a typical transient absorption spectrum obtained for one certain time delay between pump and probe via scanning through a range of energies with the detection monochromator. Figure 4.29 shows a sketch of a transient absorption spectrum showing the most commonly found spectral features. In our example the spectrum is plotted over wavelength, where in many cases it is also displayed over photon energies, i.e. with a reciprocal axis. As mentioned above, the most common representation of the experimental results is to plot the differential transmission $\Delta T/T$, i.e. the change in transmission through the sample at a given wavelength due to previous excitation with the pump pulse. The three most important spectral features, which are typically found in every transient absorption spectrum, are shown in blue, green, and red:

GSB: The first feature is the so-called *ground-state bleach* or *photobleach*, which originates from a reduction of available ground-state chromophores in the sample due to the pump pulse. When the pump light is absorbed by a chromophore it remains in an excited state for a certain time, so that the fundamental S_0–S_1 singlet absorption is not possible on this chromophore. For times shorter than the exciton lifetime there is therefore an increased transmission of the probe pulse in the spectral region of the material's ground-state absorption. The ground-state bleach can be easily identified by comparing the transient absorption spectrum to steady-state absorption measurements.

SE: The second feature is located at lower energies than the ground-state bleach and is commonly referred to as *stimulated emission*. This feature resembles the steady-state photoluminescence spectrum of the probed material and originates from a stimulated S_1–S_0 transition. After a chromophore has been excited by the pump pulse a probe photon of matching energy can lead to the emission of a second coherent photon in a stimulated emission process (the probe photon has to match the energy of the luminescence photon). Since for such a process two photons are detected for the one incident photon stimulating the emission, the net differential transmission is positive; i.e. the stimulated emission peak is located in the red with respect to the ground-state bleach and has the same sign.

PIA: The third commonly observed feature is typically energetically much broader and originates from several processes depending on the sample under investigation. It is called *photoinduced absorption* and originates from absorption events due to the presence of excited state, for instance an S_1–S_2 transition. Especially for donor–acceptor samples like polymer–fullerene blends, further processes can be responsible for photoinduced absorption, e.g. polarons or charge transfer states. While it can be challenging to deconvolve the transient absorption spectrum and in particular the photoinduced absorption, detailed analysis of this spectral feature allows deep insights into the kinetics of charge generation, a key mechanism in organic and hybrid solar cells.

During a transient absorption experiment the sample is typically measured for time delays spanning several orders of magnitude, mostly in the range of fs up to ns (an optical delay length of 30 cm, which can be easily realized experimentally, is equivalent to a time delay of 1 ns). In case significantly longer delay times are of interest the system typically contains two lasers, a pump laser and a probe laser, with the probe laser being triggered by the pump laser with a defined time delay. Due to a certain time jitter this method, however, can be used only for minimum delays on the order of µs since the error in timing between pump and probe is too large for much smaller delays. On the other hand, in a setup like the one shown in Figure 4.28, pump and probe pulse originate from the same laser source, which leads to highly stable delay times down to the fs range (in case environmental conditions like temperature or humidity, which could affect the travel time of the pulses, are kept constant).

As an illustrative example for transient absorption data we show results published by Huang and co-workers in 2013 [259]. The authors present a polymer–fullerene bulk heterojunction (P3HT:PCBM) containing a certain amount of a near-IR dye (SQ), which extends the absorption spectrum of the device and can contribute charge carriers via electron injection from its LUMO into the LUMO of PCBM. Furthermore, it participates in the charge generation process upon photon absorption in the P3HT since there is a fast Förster resonance energy transfer from P3HT to SQ, which competes with the electron injection from P3HT into PCBM. This energy transfer was directly observed using transient absorption spectroscopy as shown in Figure 4.30.

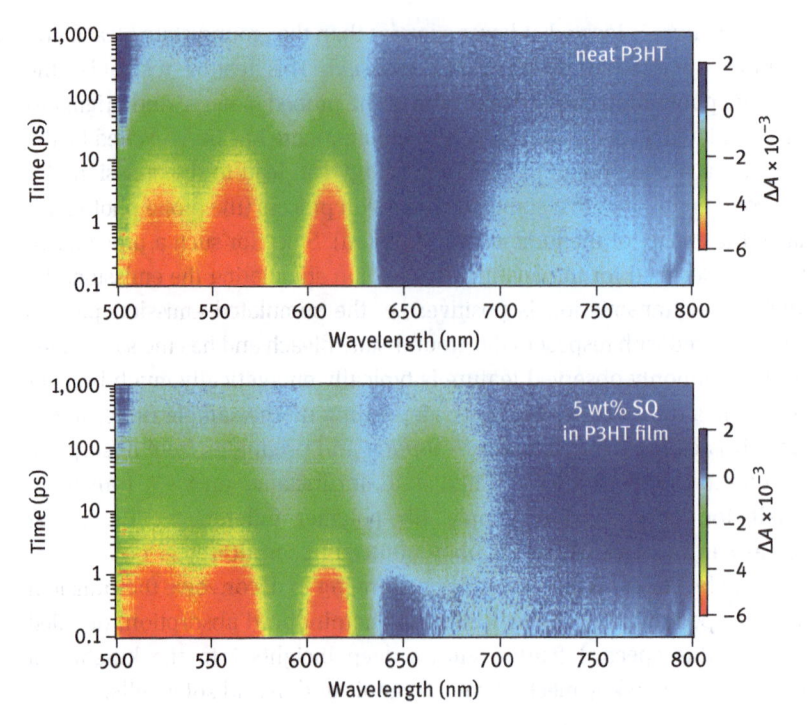

Fig. 4.30: Example for Förster resonance energy transfer from P3HT into a near-IR dye monitored using transient absorption spectroscopy. The transient absorption spectra for the neat P3HT indicate typical photobleach (500–625 nm), photoinduced absorption (650 nm), and stimulated emission (680–750 nm) features. In the presence of the squaraine dye SQ the P3HT emission is only visible at very early times after excitations, while for times exceeding 1 ps the ground-state bleach of the dye appears, indicating that an energy transfer has taken place. Reprinted by permission from Macmillan Publishers Ltd: Nature Photonics [259]. Copyright (2013).

Note that the z-scale is given as ΔA; i.e. it shows the differential absorption instead of the differential transmission so that the signs of the spectral features are reversed. The upper plot shows the time evolution of a pure P3HT film upon excitation at 500 nm, with clearly visible 0–0, 0–1, and 0–2 features in the ground-state bleach and a prominent photoinduced absorption feature at 650 nm, which can be attributed to absorption of S_1 states in the P3HT. In comparison, the lower plot shows time-dependent transient absorption spectra for a mixed film of P3HT and 5 wt% of the energy transfer dye. As can be seen, there is a significantly reduced lifetime of the ground-state bleach of the polymer, i.e. a faster decay of excitonic population. This is caused by a depopulation of excited states in the polymer due to an energy transfer to the SQ dye, which is further apparent from an additional ground-state bleach signal occurring after approximately 2 ps between 650 and 700 nm. This can be directly attributed to excited states in the SQ dye, which are populated after the energy transfer. This example illustrates how transient absorption spectroscopy can be directly used to monitor the kinetics and energetic motion of excited states in complex systems and outlines its relevance for the research field of organic and hybrid solar cells.

4.3.5 Time-Resolved Photoluminescence Spectroscopy

Like all continous wave (cw) measurements steady-state photoluminescence spectroscopy monitors a time average of radiative emission events in a sample and thus does not give any information about differences in emission quantum efficiencies or variations of occupation of excited states with time. However, in order to understand the fundamental mechanisms that occur upon light absorption in an organic semiconductor or an organic semiconductor device it can be essential to monitor the time dependency of the photoluminescence.

This can be done using a setup similar to the one shown in Section 4.1.2 with a pulsed excitation source (for instance a fast diode or a pulsed laser) and a fast photodetector connected to an oscilloscope. Either photoluminescence decays can then be recorded for a defined emission wavelength through a monochromator or the integral of the whole photoluminescence is directly measured, i.e. focused directly onto the photodiode without a monochromator. Alternatively, various commercial photoluminescence spectrometers allow the operation in *single photon counting* mode, which can be performed using a conventional photomultiplier tube as detector. Reduction in the intensity of the pulsed excitation source results in emission of virtually *single photons* per excitation events, so that time resolutions on the order of a few ps can be achieved despite the slow response time of the photomultiplier tube since only the onset of the signal has to be detected (the moment when one photon hits the detector).

A more elegant way to perform time-resolved photoluminescence spectroscopy, which also gives access to even better time resolutions on the order of 1 ps, is the use of an fs laser system as an excitation source in combination with a streak camera as a detector. In combination with a spectrograph, which converts wavelength information into spatial information, a streak camera allows to measure time- and spectrally resolved photoluminescence and thus gives access to a comprehensive analysis of photoluminescence kinetics. The function of a streak camera will be discussed in the next paragraph, while we now first take a look at the whole experimental setup, which is schematically depicted in Figure 4.31. Pulsed laser excitation is supplied by an fs laser system, most commonly featuring a mode-locked Ti:Sa oscillator emitting ~100 fs long pulses of tunable center wavelength (between 670 and 1,070 nm). This laser is often combined with an SHG in order to have access to the doubled photon energies. Furthermore, an optical parametric oscillator (OPO), operated either with the Ti:Sa pulses or with the second harmonics of those, can be used for continuous tuning of the excitation spectrum. A combination of Ti:Sa, SHG, and OPO gives access to the full UV-visible spectrum and is thus a powerful source for fs pulsed excitation of most organic and hybrid solar cell materials.

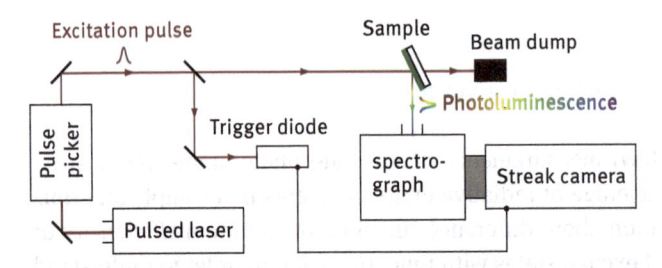

Fig. 4.31: Schematic of a setup for time-resolved photoluminescence experiments using a streak camera as detector. A pulsed laser system is used as an excitation source, typically a titanium:sapphire oscillator with second harmonics generator and/or optical parametric oscillator in combination with an acousto-optic pulse picker to adjust the repetition rate. Light is focused onto the sample, which emits its photoluminescence into a spectrograph. The spectrally resolved signal is then detected with a streak camera, which monitors the signal on a spatial (=wavelength) and a time axis.

Since mode-locked Ti:Sa lasers typically have repetition rates on the order of 80–100 MHz, i.e. the time between two excitation events would be on the order of 10 ns, it can be necessary to reduce the repetition rate of excitation using a pulse picker. Such a pulse picker takes advantage of an acousto-optic modulator, a crystal in which a standing sound wave induced by a piezoelectric transducer induces a grating due to a harmonically varying refractive index. The occurrence of this grating can be switched on and off by an electrical signal on the piezo crystal so that pulses can be "picked" if the grating is on using the first order of light diffracted by the pulse picker crystal. Pulse picking can be necessary, e.g. if certain processes in the sample occur at timescales exceeding these 10 ns since otherwise the system does not completely relax back to its ground state between two pulses. The repetition rate can then be easily reduced down to a few Hertz while maintaining the fs nature of the pulses.

After laser pulse generation and picking, i.e. the fully tunable excitation source, the laser beam is focused onto the sample under investigation in order to generate excited states, which subsequently cause photoluminescence. A tiny fraction of the excitation pulse is guided to a trigger diode in order to have a defined time zero for the transient photoluminescence experiment. The excitation beam transmitted through the sample hits a beam dump in order to avoid a second excitation due to reflection off any surface in the setup. The photoluminescence of the sample is collected by appropriate optics (typically two lenses, with the illuminated spot of the sample being in the focus of the first lens and the second lens being used to focus the photoluminescence into the detection system), and the sample is kept under an angle that avoids direct reflection of the excitation beam into the detector.

The detector unit consists of a spectrograph in front of the streak camera, which measures on two axes, one for time and one for spatial coordinate. The latter axis is used to detect the spectrum of the luminescence signal, which is dispersed parallel to the plane of the optical table by the spectrograph. A small entrance slit (~50 μm) is used in front of the spectrograph in order to yield good spectral resolution even for slightly defocused luminescence.

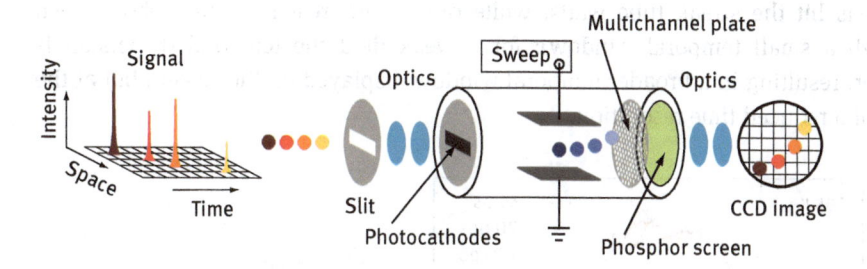

Fig. 4.32: Working principle of a streak camera. After pulsed excitation the sample is emitting time- and wavelength-dependent photoluminescences, where the spectral information is translated into a spatial information by a spectrograph (Figure 4.31). At the entrance of the streak camera the photonic signal is further translated into an electronic signal by a photocathode. Electrodes leave the photocathode on the opposing side and enter the streak tube, where a time-dependent voltage sweep deflects the electrons on their way toward a multichannel plate, which acts as an electronic amplifier. The amplified electron signal is then detected on a phosphor screen using a CCD camera.

The working principle of a streak camera is shown in Figure 4.32 for a time-dependent fluorescence signal that has already passed the spectrograph. In the example the signal contains four fluorescence events centered around four different energies and emitted from the sample at four different times, resulting in the displayed space–time distribution. The luminescence signal is then guided through a small entrance slit (typical slit width ~50 μm parallel to the plane of the optical table) into the streak camera, where it hits a photocathode. Accordingly, signals of different energies excite electrons out of the photocathode at different spots and different times on the photocathode, which subsequently enter the evacuated streak tube. During their flight through the tube the electrons are deflected by a time-varying electric field, which is generated using a voltage sweep on deflection plates parallel to the flight direction; i.e. electrons are deflected in the z-direction, while their position along the axis of the entrance slit remains unchanged. The deflecting voltage sweep is initiated by the trigger signal (compare Figure 4.31) and typically is a linear increase over time, resulting in a linear increase of electron deflection in z-direction with time. After passing the deflection plates the electron signal now contains a temporal information (position in z-direction) and a spectral information (position along the axis of the entrance slit). If necessary, the signal can be emphasized using a multichannel plate with an appropriate gain voltage, before it is detected on a phosphor screen and read by a sensitive CCD camera. Note that the time resolution is independent of the response time of the CCD, which detects only a two-dimensional image already containing temporal and spectral information. As a consequence, the time resolution of a streak camera depends on the resolution of the CCD (or, if smaller, of the multichannel plate) on the one hand and directly on the strength of the deflecting electric field on the other. In case a strong field is used, the time resolution is increased since the electrons are dispersed over a broader range in z-direction (most electrons hit the streak tube walls, while only a few reach the phosphor screen through a small temporal window); for a weak field the temporal dispersion is smaller, resulting in a broader temporal window displayed on the screen, but at the price of a reduced time resolution.

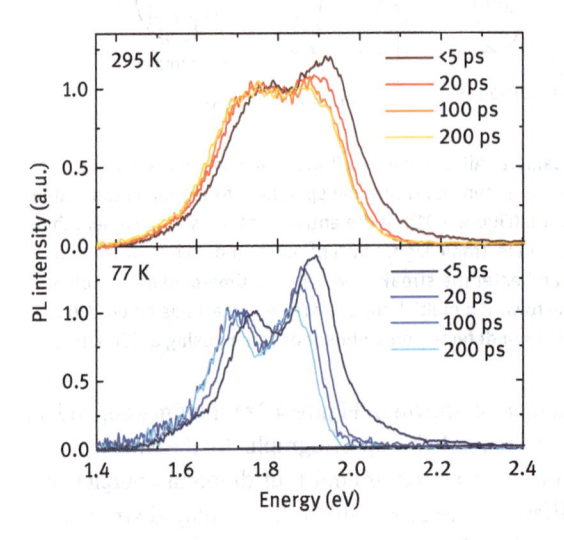

Fig. 4.33: Example for time-resolved photoluminescence spectra acquired using a ps streak camera. The plots show transient photoluminescence spectra of P3HT at room temperature (top) and at 77 K (bottom) for different wait times after excitation. Mostly two differences are apparent when comparing the upper and lower plots, namely (1) a faster energetic relaxation at higher temperatures and (2) overall red-shifted and more distinct spectral features at 77 K.

In addition to the so-far-described detection method using a triggered linear deflection sweep (so-called *single sweep mode*), which allows only repetition rates of up to ~1 MHz, some streak cameras allow operation in a so-called *synchroscan mode*. The deflection voltage in this operation mode is an ac voltage locked to the oscillation frequency of the mode-locked Ti:Sa laser, i.e. with a frequency of 80–100 MHz. The signal is then detected only around the zero-crossing of the sinusoidal voltage, where the signal is approximately linear so that the resulting CCD image can be easily corrected for the small deviation between the sine and a linear function. Due to the high repetition rate and the pronounced nonlinearity of the sinusoidal voltage further away from its zero-crossing the maximum time window in synchroscan mode is approximately 2 ns. However, since roughly 1 million experiments are performed per second short integration times already yield good signal intensities and fast experiments are possible. Additionally, very high time resolutions on the order of 1 ps can be obtained using synchroscan, which makes it the method of choice when investigating fast processes.

An example of transient photoluminescence spectra is shown in Figure 4.33. The plots illustrate the temperature and time dependence of the photoluminescence of a P3HT thin film for excitation at 440 nm. Consistent with the considerations in Section 2.2.2.3 concerning exciton migration a red shift of the photoluminescence with time is observed at both temperatures. During the first few ps this can also be attributed to the presence of highly amorphous polymer phases, which are quickly depopulated as apparent from the more abrupt high-energy onset for later times. At low temperatures the spectral shift is significantly slowed down, consistent with slower exciton migration. Furthermore, the two prominent vibronic 0–0 and 0–1 features of P3HT are more distinct at low temperatures due to reduced spectral broadening and the luminescence is overall slightly red-shifted due to stronger coupling between neighboring polymer chains induced by less conformational variations in the polymer crystal at reduced temperatures.

5 Fabrication and Device Lifetime

The main motivation to work on organic and hybrid solar cells is their high potential as low-cost solar cells. To fulfill this expectation, not only the device physics, discussed in the previous chapters, but also the fabrication issues and lifetimes of these solar cells need to be considered. In this chapter we give a very brief outlook on what needs to be considered to bring this exciting technology to the market. As this topic alone could fill books considering all aspects concerning commercialization of organic and hybrid solar cells, this chapter gives only a selective and incomplete overview of this field. We refer the interested reader to some already published review articles and books that cover different aspects and have also been used as the basis for the information presented in this chapter [260–267].

The most important aspects concerning commercialization of the solar cells are their efficiency, processing technology, and stability (Figure 5.1). Only a situation where all the three aspects give acceptable values will allow for the fabrication of low-cost solar cells. Currently research is focusing mainly on the efficiency issues. Mechanisms leading to limitations of the efficiencies have been discussed in the previous chapters. In this chapter we briefly introduce processing technologies and their impact on the lifetime and stability of organic and hybrid solar cells.

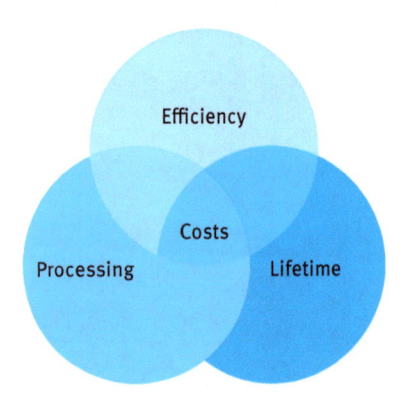

Fig. 5.1: A Venn diagram combining efficiency, processing, and lifetime. For commercialization of organic or hybrid solar cells, the three aspects – efficiency, processing, and long lifetime – have to be combined to achieve low-cost photovoltaic modules.

On a long-term basis the aim is clearly to have organic and hybrid solar cell modules as rooftop grids for power generation. Until this can be achieved, markets for other technologies are very attractive. First of all portable consumer electronics seems to be a promising field, and the first products such as mobile chargers or bags with a flexible solar panel are on the markets. These products require low-weight modules but lifetimes of only 1–2 years with moderate efficiencies below 5%. For indoor consumer electronics, lifetimes of over 5 years seem to be necessary. An interesting thing is the applications such as building-integrated power generation. Here the availability of differently colored solar cells and the possibility to produce semi-transparent devices are clear advantages over conventional solar cells. Window glasses with semitransparent colored modules will need only a moderate efficiency. However, lifetimes should be >10 years. Defects in color or discoloration over time cannot be tolerated for such applications. An attractive implementation of dye-sensitized solar cells from Solaronix in a window facade can be seen at the *SwissTech Convention Center* in Lausanne, Switzerland (Figure 5.2).

Fig. 5.2: Dye-sensitized solar cell facade of the SwissTech Convention Center in Lausanne, Switzerland. Photo by Solaronix, used with friendly permission.

Today, we could see the first prototypes of organic and hybrid solar cells on the market. It seems realistic that soon the mass market will be hit by these emerging solar cell technologies. Similar to organic light-emitting diodes (OLEDs), which are already playing an important and strongly increasing role on the market, it can be expected that organic solar cells will also be very successful. Some predictions see this coming within the next few years. For detailed road maps and market expectations see information from organizations such as *OE-A*, *IDTechEx*, *Plastic Electronics*, and others.

5.1 Processing

As explained earlier, the cost of electricity produced by a solar module strongly depends on its processing cost, efficiency, and lifetime. Increasing the efficiency by a factor of two will decrease the price per power output by the same factor. However, if the lifetime is very short and modules need to be replaced, e.g. every 5 years, then the installation costs will be four times higher than that for the same module if it lasts for

20 years. As installation costs are a significant cost factor, it is generally assumed that solar cells installed for power generation need to have a lifetime of at least 5–10 years. In addition, installation of a rooftop grid-connected solar power generation system has to show significant advances compared to the established silicon-based technology. Here significantly reduced costs are necessary to be able to compete successfully as it is not sure whether this emerging technology will be able to outperform silicon in efficiency or lifetime. Therefore, low-cost processing of organic and hybrid solar cells is the key issue toward successful market implementation. Coating technologies in roll-to-roll fabrication lines with high-throughput processing enable extremely low-cost production, especially if low-temperature, vacuum-free processing is possible.

Currently the cost of an organic solar module can be broken down into the cost of the different layers. Comparably expensive positions are the costs of the transparent conductive oxide (TCO), the donor–acceptor materials, and the metal top contacts. An additional major cost contribution comes from the encapsulation to ensure stable performance of the device over time. Currently extensive research and engineering efforts are going on to reduce costs by maintaining high functionality.

5.1.1 Processing Steps

Most researchers are currently preparing the semiconducting films by spin-coating. This is a laboratory method giving high-quality thin films. During spin-coating, only a very small fraction of the material stays on the substrates and forms the thin film, while most of the material flows over the edge of the substrate and is consequently wasted. At the same time this method enables film formation only on relatively small substrates, up to a few square centimeters, and is therefore not a feasible method for large-scale and low-cost production.

Low-cost processing can be achieved using printing and coating techniques such as roll-to-roll as they allow for the necessary production speed. However, one requirement for roll-to-roll compatible device architectures is that the solar modules need to be processed on flexible substrates. Successive layer deposition needs to be possible. Belt speeds of ≥ 3 m min^{-1} should be possible to allow outputs of more than 100 m² h^{-1}, which is assumed to be economical. The films need to be produced in a single step with even and homogeneous coverage.

In the following section we will briefly go through the different layers forming an organic solar cell and discuss their processing, beginning with the substrate as the first layer.

5.1.1.1 Substrate
The substrate forms the base for all other layers that need to be deposited on top. In laboratories most devices are fabricated on glass substrates. They are stable and

reliable and are also ideal as barrier layers for moisture and oxygen. As glass is highly transparent for visible light, it allows efficient in-coupling of light into the active semiconductor layer. The sample sizes in laboratories are usually small and substrates of ~1–5 cm² are often used. For these sample sizes glass is easy to handle and provides the properties needed. However, as it is not possible to use roll-to-roll processing with glass substrates, flexible substrates are needed for low-cost production of solar cells.

Metal foils can be used, if the light is coupled in from the other side of the device. However, if using the standard cell geometry with illumination from the front side (through the substrate), transparent foils need to be chosen. Such flexible transparent foils need to be thermally stable and should expand little on heating. It is important that they withstand all processing steps including high-temperature deposition and solvents, which are used for the layers successively deposited on top. Generally the foils that withstand higher temperatures are more expensive. Additionally, the foils should have a high surface quality with a smooth and ideally defect-free layer. Often a planarizing coating is applied to improve the smoothness and the surface hardness to prevent easy scratching. The foils need a certain stiffness to withstand the stress induced, e.g. by the deposition of inorganic layers (e.g. TCO electrode). Additionally, roll-to-roll casting needs to be possible without internal deformation of the foil. Finally, the material should not age and get brittle, which possibly induces cracks and can thus lead to device failure.

The most common flexible substrate is polyethylene terephthalate (PET) followed by polyethylene naphthalate (PEN). The chemical structures are depicted in Table 5.1 together with some important property parameters, such as upper processing, glass transition, and melting temperature. As can be seen from this, even though the melting temperatures of these materials are not far from each other, the upper processing temperature for PEN is considerably higher. However, for both materials the maximum processing temperature is far below the limit of 450–500°C usually necessary for sintering TiO_2 particles to a mesoporous TiO_2 film. Therefore, metal-oxide-based devices like dye-sensitized solar cells are usually made of glass substrates.

PET is a very cheap and common material. It degrades by hydrolysis and gets brittle, which leads to cracking. Fortunately, through additives in the formulation of PET the stability over time can be significantly improved. The degradation is mainly attributed to a catalyzing effect of the –COOH ends. A fast degradation can be observed after illumination in the near UV as both materials – PET and PEN – strongly absorb in this wavelength region (≤315 nm). Furthermore, the naphthalene unit in the PEN exhibits an absorption tail of up to 400 nm. Fortunately, PEN forms a ~10 μm oxidized surface layer, which acts as a protective UV barrier [268]. The use of higher molecular weight (and consecutively less –COOH end groups) can improve the stability as well as the use of additives. For UV stabilization often some mixtures of additives are used, most commonly UV absorbers. For more details we refer to the literature (e.g. the chapter by Gugumus in the *Plastic Additives Handbook* [269]).

The additives absorb the UV radiation (~290–400 nm) and convert it into lower energy radiation. The degradation is further accelerated through oxygen, which leads to a stronger photooxidative degradation than just the photodegradation.

As the active layers, which are deposited consecutively on top of the substrate, are also prone to photodegradation and photooxidation, the substrate should act as a barrier avoiding oxygen and moisture diffusion into the active films. In Section 5.2 we address the stability issues of organic solar cells and compare the barrier qualities of different foils for encapsulation of the device.

Tab. 5.1: Chemical structure and properties of PET and PEN foils.

Material	Chemical structure	Upper processing temperature	T_g	T_m
PET		~150°C	78°C	255°C
PEN		~180–220°C	120°C	263°C

5.1.1.2 Transparent Conductive Electrodes

The first layer deposited on top of the substrate is a transparent conducting oxide (TCO). Currently, tin-doped indium oxide (ITO) is used as the state-of-the-art TCO material in many applications, such as LCDs, LEDs, touch screens, and solar cells. ITO coatings generally show an excellent electrical conductivity while exhibiting high optical transparency, making them the standard material of choice despite the high material costs of indium. The standard deposition method in industrial production is DC magnetron sputtering from ceramic ITO targets. As indium is a scarce material, the cost of ITO is expected to rise drastically over the coming years. Therefore there is a high pressure to replace ITO with other materials, which are abundant and show similar performance. To compare the quality of transparent conductors, Haacke defined a figure of merit (see also Section 4.1.4.3):

$$\Phi_{TC} = \frac{T^{10}}{R_{sheet}}, \tag{5.1}$$

with T the optical transmission and R_{sheet} the electrical sheet resistance [270]. Haacke's figure of merit allows a simple and useful comparison for many applications, for instance, transparent conductors in solar cells. Even though this figure of merit seems to be the most common for TCOs for solar application, other figures of merit do exist. For other applications, especially where the transparency is less important than the conductivity, the weighting by the power of 10 in Haacke's figure of merit might not be the best choice.

Currently a number of alternative transparent conductors to ITO exist. Most common are tin- and zinc-based oxides, but stacked metal-oxide/metal/metal-oxide films are also being investigated and show very promising behavior [271]. More recently transparent conducting films based on silver nanowires, graphene or carbon nanotubes, and other nanomaterials have been investigated [272]. Important for application in solar cells is their simple and easy processability on large scale (ideally in a low temperature process from solution). Currently, there are different processing technologies applied. ITO films are usually sputtered, but spray pyrolysis is also possible. Other methods such as chemical vapor deposition, atomic layer deposition, and sol–gel processing are also employed. For roll-to-roll fabrication on a flexible substrate solution-based deposition at low temperatures is desirable. For instance, silver-based pastes are used to form a conductive electrode network. For more information we refer the interested reader to the available literature concerning this topic. One example is the review by Pasquarelli et al. [273] on solution processing of transparent conductors.

Of course the optical properties are very important. Here also the reflection plays a key role, which strongly depends on the refractive indices at the substrate TCO interface. The Fresnel equations can be used to calculate the relative amount of light reflection. We have to consider that the optical properties are wavelength dependent. For many solar cells UV radiation is critical and it might be advantageous to ensure reflection of UV light and allow light transmission only in the visible range. The film properties, such as crystal structure, shape, and roughness, are also important factors, which can enhance or limit the light transmission and optical path length within the active material. Haze of the material can significantly increase the effective absorption in the semiconductor material [274].

For roll-to-roll processing the thin conducting films need to be flexible to withstand the curvature needed. Metal-oxide ceramics are often quite brittle and do suffer upon bending. Here metallic nanowire networks, high-conductive polymer films, and metallic thin films have the advantage of being highly flexible. In a recent review by Angmo and Krebs [275] different ITO alternatives are discussed.

Next to the transparency, conductivity, flexibility, choice of material, and deposition method, the work function of the electrode material plays a role in the application and should be tuned to the used active semiconductor layer in order to guarantee ohmic contacts between the active layer and the external electrodes. The work function describes the energy needed to remove an electron from the conduction band into the vacuum level and differs based on the selected material or material combination.

For glass substrates containing sodium in their formulation, a barrier layer needs to be deposited prior to the transparent oxide to avoid sodium diffusion into the TCO, which will happen at elevated temperatures.

For most applications the conductive electrode also has to be selective to either holes or electrons. Therefore, an additional charge-selective layer is deposited on top of the conductive substrate. PEDOT:PSS is often used as a hole-selective layer. It can be deposited easily as ink by roll-to-roll techniques, such as rotary screen printing (see Section 5.1.2) as it allows the deposition of sufficiently thick layers. As an electron-selective layer, ZnO is a preferred material, as it also allows the formulation as ink and deposition from solution. For many metal oxides annealing to higher temperatures is necessary in order to realize high performances. ZnO also works well as processed and there is no need for such an annealing step.

As the electrodes need to have a certain shape allowing to print modules (see Section 5.1.3), only printing and no coating techniques can be applied for this layer.

5.1.1.3 Active Semiconducting Film

The active semiconducting film should form a donor–acceptor network, which allows fast exciton separation and charge extraction toward the electrodes. In case of dye-sensitized or hybrid solar cells this film consists of an inorganic (usually TiO_2) network, decorated with a dye and infiltrated with an electrolyte redox couple, or an organic hole-transport material. A mesoporous TiO_2 layer can be fabricated via screen printing (see Section 5.1.2), a method that can also be applied to a roll-to-roll procedure. However, since the mesoporous TiO_2 layer needs a sintering step at higher temperatures (400–500°C), efficient dye-sensitized solar cells cannot be currently fabricated on flexible plastic substrates. Therefore, high fabrication speeds as in the case of roll-to-roll technologies are currently not available. Furthermore, owing to its thickness in the order of micrometers the TiO_2 film is prone to crack during bending. Here organic semiconductor materials have a clear advantage, as these materials do not need any high-temperature process and are highly flexible and therefore well compatible with flexible polymeric substrates.

Active semiconducting donor–acceptor blends have been prepared as ink and deposited, e.g. via slot die coating or other coating technologies (see Section 5.1.2). One current disadvantage of standard blend formulations is the use of toxic, organic, often halogenated solvents as ink-base. As these materials are toxic, the search for "green" solvents is an important step toward an environmentally friendly fabrication. This is certainly an issue, which has not been addressed intensively so far and will need further research. Most attempts to replace toxic solvents did not give the same performance, but recent results are promising [276]. As the solubility of the organic semiconducting materials strongly depends on the material design, synthetic efforts are required in order to tailor the solubility of the organic semiconductors for "green" solvents. Also the search for adequate replacements of toxic solvents with nontoxic ones needs to be continued to enable a replacement without compromising on efficiency.

5.1.1.4 Counter Electrode

The counter electrode needs to be deposited on top of other films. In dye-sensitized solar cells the counter electrode is usually a glass substrate with a fluorine-doped tin oxide (FTO) coating, which is covered with a few nanometers of platinum, which is finally laminated on top of the device. Such a technique does not allow roll-to-roll fabrication. For organic solar cells one tries to use highly conductive inks to deposit the counter electrode on top of the device as this is compatible with roll-to-roll fabrication. Often some silver paste formulation is used for this purpose. Similarly as in case of the working electrode, the counter electrode should also be charge specific, which means that charge-selective layers need to be used prior to the deposition of the counter electrode. As the counter electrode does not necessarily need to be transparent (even though this is desired in some applications, especially for building-integrated cells and modules, where the optical appearance plays an important role), metallic thin films can be also used as the counter electrode and additionally act as an encapsulation layer.

Ideally, the printed counter electrode does not overlap with the previously deposited working electrode as such direct overlaying of the electrode structures often leads to shortfalls and malfunctioning of the module. Therefore, such possible contact points need to be minimized. Desirable is a double comb-like structure, with counter and working electrode exactly out of phase. However, due to unpredictable shrinkage of the web during processing, such precision is impossible to achieve. Krebs et al. instead developed a $+5°/-5°$ slanted comb structure where for most cells only one crossing of comb grid lines per unit cells is observed and only few cells with maximum two crossings. Such an optimized geometry of the electrodes turned out to deliver higher reproducibility and minimized failures of cells or modules [262].

5.1.1.5 Encapsulation Layer

An encapsulation layer is needed as the lifetime of a device will depend strongly on the environmental conditions. Under illumination, on being exposed to oxygen and moisture, and at elevated temperatures a solar module will degrade faster than stored in the dark in an N_2 atmosphere at room temperature. To ensure long-term stability encapsulation as also done for OLEDs is obligatory. The encapsulation has the aim to keep moisture and oxygen out, and additional effects, such as UV filtering, can be induced. The important parameters defining the quality of an encapsulation layer are the water vapor transmission rate (WVTR), which is given in $g\,m^{-2}\,d^{-1}$, and the oxygen transmission rate (OTR), which is given in $cm^3\,m^{-2}\,d^{-1}\,bar^{-1}$. Schaer et al. found that for OLEDs water vapor is more destructive than oxygen by a factor of 1,000. Water leads to delamination of the electrodes, whereas oxygen causes oxidation of the electrodes, which is less detrimental to the device performance in comparison [277].

Usually a good water barrier is also a good oxygen barrier film. As WVTR is easier to measure, it is often used as figure of merit for barriers. Reasonable lifetimes seem to be achievable with WVTRs of 10^{-3}–10^{-6} g m^{-2} d^{-1}. Of course, the specific values for a given solar cell strongly depend on its architecture and the stability of the implemented materials, which makes it impossible to give a general rule for the WVTR and leads to a reasonable lifetime [278, 279]. The requirements also depend strongly on the targeted products and the lifetime and efficiency needed.

A list of barrier qualities needed for different applications is given in Figure 5.3. As can be seen, organic light emitting diodes (OLEDs) and organic photovoltaics (OPVs) are on the tip of the triangle, which indicate the highest requirements for these technologies. State-of-the-art packaging films used for food, technical products, and encapsulation of inorganic solar cells are in the range of WVTR of 10^0–10^{-1} g m^{-2} d^{-1}. In general, higher costs can be expected with increasing requirements of the barrier layer.

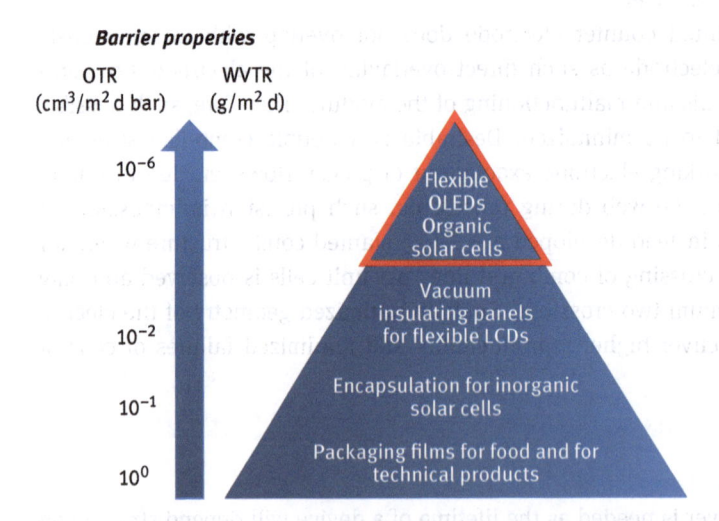

Fig. 5.3: Barrier layer requirements for different technologies. The highest requirements are needed for organic electronic devices. The costs per square meter raise drastically from bottom to top of this triangle. Figure after [280].

Excellent encapsulation is given with glass substrates, which show high barrier properties. The weakest barrier in such glass-laminated cells are the sides, which are glued together. However, glass substrates do not allow roll-to-roll fabrication techniques, have a high weight, and are not flexible, which counteract the advantages of organic and hybrid solar cells. In this respect polymer substrates maintain all the advantages as they are of low cost and flexible. Unfortunately, their barrier properties are weak. Transmission rates for different polymer substrates are plotted in Figure 5.4. As can be seen, none of the films comes in the range of 10^{-3}–10^{-6} g m^{-2} d^{-1} suitable for organic solar cell encapsulation.

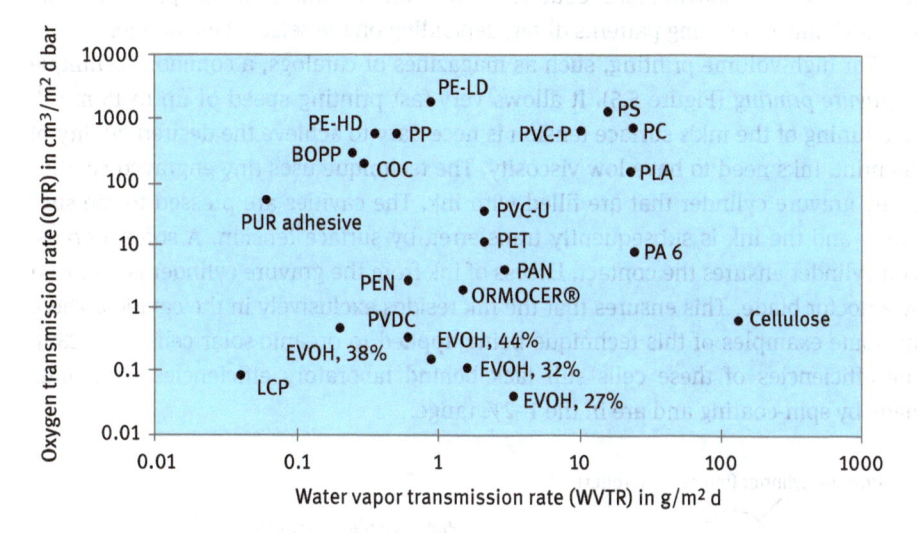

Fig. 5.4: OTR and WVTR for different polymers. None of the polymers fulfills the requirements of a transmission rate between 10^{-3} and 10^{-6} g m^{-2} d^{-1}. Figure adapted from [281] with friendly permission by Wiley and Sons (2003).

The reasons for the poor barrier qualities of polymer foils are small pinholes present in these films, which form pathways for diffusion of oxygen and water. At the position of these pinholes local degradation takes place. To decouple such pinholes a multilayer structure is important. Additional barrier layers such as Al_2O_3, SiO_2, or others will further decrease the diffusion of oxygen and water vapor [282].

Finally, a UV protection layer might further improve the cell stability as highly energetic light contributes strongly to the photodegeneration of the materials. Depending on the substrate type and device architecture UV protection layers can significantly improve the device lifetime. As the UV light usually does not (or only marginally) contribute to the photogenerated current, the device performance does not suffer from shielding off this part of the solar spectrum.

5.1.2 Coating Technologies

The coating technology applied will have a major influence on the quality and forming speed of the films. Printing technologies allow the fabrication of patterned films, which are needed if we want to prepare modules consisting of a number of cells in connection. Many printing technologies have been developed since its invention by Johannes Gutenberg (~1450). They altogether combine the transfer of a wet ink onto a substrate by direct contact. Only ink-jet printing is an exception, as no direct contact to the substrate is made. The different techniques available today have different advantages and disadvantages. Usually the resolution, printing speed, printing costs, and flexibility of printing patterns differ, depending on the selected technique.

For high-volume printing, such as magazines or catalogs, a common technique is *gravure printing* (Figure 5.5). It allows very fast printing speed of up to 15 m s^{-1}. Fine-tuning of the ink's surface tension is necessary to achieve the desired quality of the print. Inks need to have low viscosity. The technique uses tiny engraved cavities in the gravure cylinder that are filled with ink. The cavities are pressed to the substrate, and the ink is subsequently transferred by surface tension. A softer impression cylinder ensures the contact. Excess of ink from the gravure cylinder is removed by a doctor blade. This ensures that the ink resides exclusively in the cavities. There are some examples of this technique being applied to organic solar cells [283, 289]. The efficiencies of these cells still lack behind laboratory efficiencies with films made by spin-coating and are in the 1–2% range.

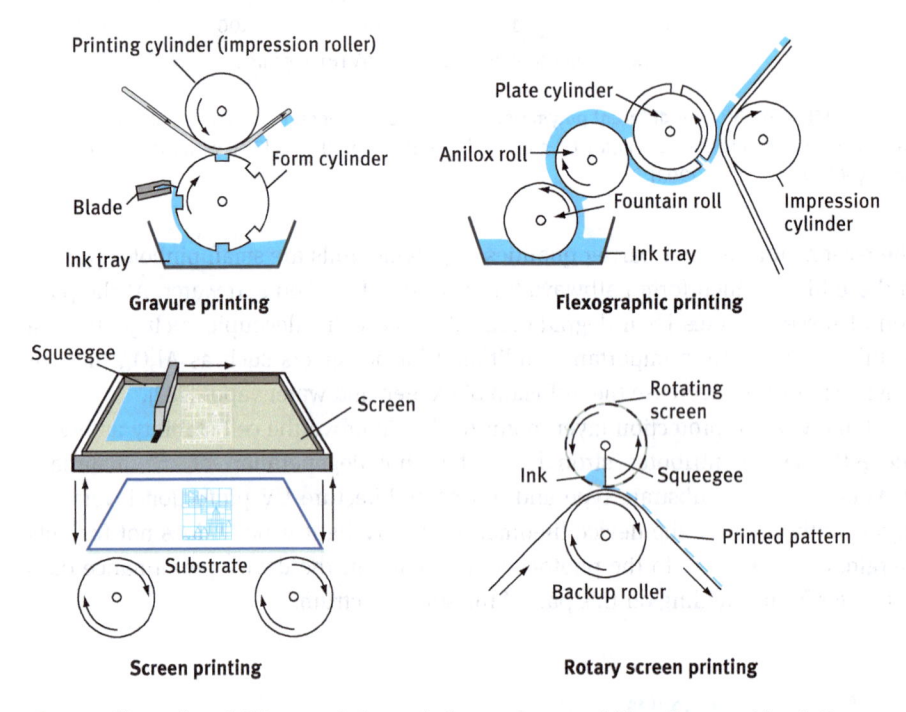

Fig. 5.5: Illustration of different printing techniques: gravure printing, flexographic printing, screen printing, and rotary screen printing. Figure after [261].

Flexographic printing is a technique similar to gravure printing, with the main difference being nonuse of cavities, but the transfer happens over a relief (Figure 5.5). Also this allows for a fast roll-to-roll printing and has been used mainly to deposit PEDOT:PSS layers or conductive electrode grids [286, 290–292].

Another technique especially suitable for electrodes is *screen printing*. It is used as a standard technique in dye-sensitized solar cells to prepare the mesoporous TiO_2 electrode. Here a flat-bed screen print is applied (Figure 5.5). This technique is very useful for laboratory-scale applications, as the screen can be easily exchanged and therefore allows high flexibility. However, for high fabrication speeds a rotary screen print technique needs to be applied (Figure 5.5), where a rotary screen is used, which is significantly more expensive than a flat screen. In organic solar cells the technique is useful for front and back electrodes as it allows the deposition of thick wet layers in the range of 10–500 µm [293–296].

Coating technologies allow the fabrication of thin films but do not allow the transfer of patterns as no direct contact with the substrate is made. Roll-to-roll processing of polymers is possible with knife coating or slot die coating (Figure 5.6). In knife coating the excess of ink is kept ahead of the knife that is in close proximity to the substrate. In slot die coating a small meniscus forms on top of the substrate held by a slot in an ink container.

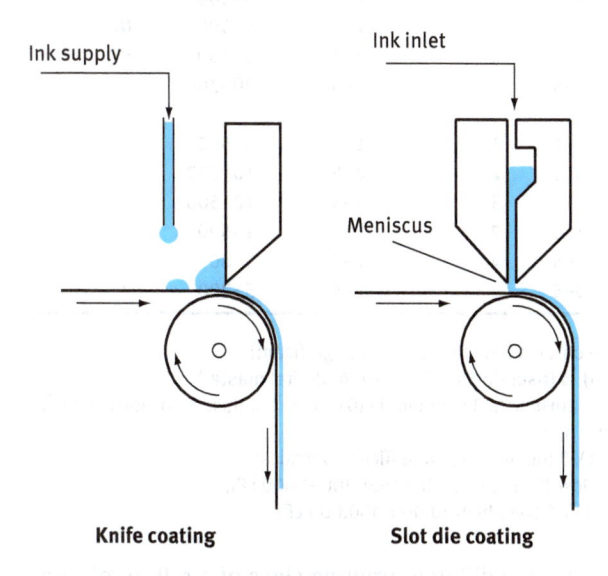

Fig. 5.6: Illustration of knife coating and slot die coating. In knife coating the excess of ink is kept ahead of the knife that is in close proximity to the substrate. In slot die coating a small meniscus forms on top of the substrate held by a slot in an ink container. Figure after [261].

Ink-jet printing is a contactless process allowing pattern transfer. Through precise control over the ink droplets supplied by the ink-jet, their controlled positioning on the substrate is possible. One technique uses piezoelectric injects, which allow the ejection of droplets from a nozzle by demand on the position where the substrate should be covered. Also continuous ink-jets can allow the positioning of the droplets by electronic charging and deflection of the desired droplet onto the substrate and selecting undesired droplets in an ink catcher to be fed back into the ink reservoir.

So far, only small devices have been produced using ink-jet printing and it needs to be seen whether this printing technology is also feasible for fast roll-to-roll fabrication. Certainly, the versatile patterning possibilities make it a highly interesting technique for mass production.

There are a number of other coating techniques including many alterations of the previously described methods. In Table 5.2 a comparison between different techniques is made and their specific properties are summarized.

Tab. 5.2: Comparison of film-forming techniques by printing and coating (data from [297]).

Technique	Ink waste	Pattern	Speed	Ink preparation	Ink viscosity	Wet thickness (µm)	R2R compatible
Spin-coating	5	0	–	1	1	0–100	No
Doctor blade	2	0	–	1	1	0–100	✓
Casting	1	0	–	2	1	5–500	No
Spraying	3	0	1–4	2	2–3	1–500	✓
Knife-over-edge	1	0	2–4	2	3–5	20–700	✓
Meniscus	1	0	3–4	1	1–3	5–500	✓
Slot die	1	1	3–5	2	2–5	10–250	✓
Screen	1	2	1–4	3	3–5	10–500	✓
Ink-jet	1	4	1–3	2	1	1–500	✓
Gravure	1	2	3–5	4	1–3	5–80	✓
Flexo	1	2	3–5	3	1–3	5–200	✓

Ink waste: 1 (none), 2 (little), 3 (some), 4 (considerable), 5 (significant)
Pattern: 0 (0-D), 1 (1-D), 2 (2-D), 3 (pseudo/quasi 2/3-D), 4 (digital master)
Speed: 1 (very slow), 2 (slow < 1 m min^{-1}), 3 (medium 1–10 m min^{-1}), 4 (fast > 10–100 m min^{-1}), 5 (very fast 100–1000 m min^{-1})
Ink preparation: 1 (simple), 2 (moderate), 3 (demanding), 4 (difficult), 5 (critical)
Ink viscosity: 1 (very low < 10 cP), 2 (low 10–100 cP), 3 (medium 100–1,000 cP), 4 (high 1,000–10,000 cP), 5 (very high 10,000–100,000 cP)

In Figure 5.7 we see an example for the different printing steps of a roll-to-roll processed organic solar cell from the group of Prof. Krebs [262].

A preparative method that allows very low-cost production most likely will always be based on roll-to-roll processes. Vacuum processes are not ideally combined with roll-to-roll fabrication. However, as some commonly used materials like the state-of-the-art ITO films are typically sputter-deposited, there are currently no fully vacuum-free processes that allow the fabrication of solar cells with the same performance. Current ink formulations of transparent conducting electrodes (usually grids printed from silver paste) do not yet deliver the same properties as the ITO layers.

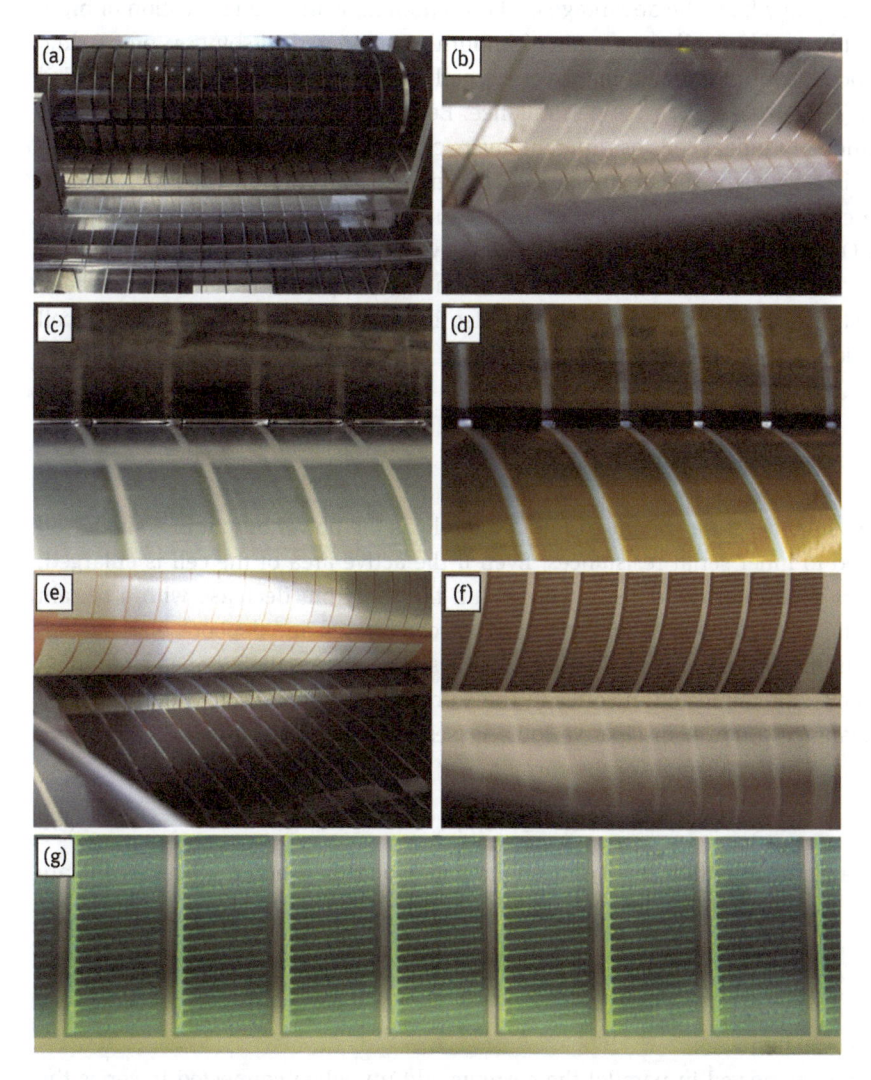

Fig. 5.7: **Roll-to-roll processing of organic solar cells.** Flexographic printing of the front silver grid (a), rotary screen printing of front PEDOT:PSS electrodes (b), slot die coating of ZnO (c), slot die coating of P3HT:PCBM active layer (d), rotary screen printing of back PEDOT:PSS (e), rotary screen printing of back silver electrode (f), and a photograph of the front slanted silver grid with PEDOT:PSS (g). Figure from [262] with friendly permission by Wiley and Sons (2014).

Small-molecule solar cells based on the sublimation of the active semiconducting film need vacuum technology. Only solution-processed films can be made by the above-mentioned coating technologies. Certainly, vacuum processing will be more expensive in terms of fabrication costs when compared to simple roll-to-roll processing in air. However, vacuum-based roll-to-roll processes are also a possible fabrication route. They have the advantage of the vacuum allowing the fabrication of high-quality films with smaller amounts of impurities compared to solution processing. The process control is much higher and small-scale lab processes can be upscaled without large losses in device performance. Besides, this technology allows simple fabrication of tandem cells as different layers can be easily deposited on top of each other without destroying or drastically changing the layers underneath. We see a similar competition between solution-based and evaporated organic semiconductors for the OLEDs. There the vacuum-processed devices seem to outperform the solution-processed ones. Roll-to-roll processes based on vacuum processes have already been demonstrated for OLEDs and have now also been transferred to organic solar cell fabrication.

5.1.3 Module Design

Usually, solar cells are combined into a module as the power output of a single cell is limited to inner serial resistances. Even if the active area of the cell is enlarged, the short circuit current density will remain the same but decrease with a certain size of the cell. The open circuit voltage of the cell is fixed and depends on the materials in use. To overcome this limitation and also to generate higher open circuit voltages, cells are connected in series in the so-called modules. Higher voltages have the advantage of lower dissipative Joule losses P_{loss}, which are proportional to the resistance R and the square of the current I^2:

$$P = R \cdot I^2. \tag{5.2}$$

The disadvantage is that interconnections and spaces between the active cells are lost spaces. Therefore, the geometric fill factor (GFF), defined as

$$\text{GFF} = \frac{\text{area of photoactive regions}}{\text{total module area}}, \tag{5.3}$$

becomes smaller.

When connected in parallel the currents add up, when connected in series the voltages add up.

One possible design for a module of organic solar cells consisting of the layers described earlier is depicted in Figure 5.8. The layers need to be deposited consecutively to allow the desired structuring.

Ideally, the overall efficiency of a module is the product of the efficiency of the single cells and the GFF of the module:

$$\eta_{module} = \eta_{cell} \cdot \text{GFF}. \tag{5.4}$$

In reality, ohmic losses through the interconnection and the current-collecting electrodes will further reduce the overall efficiency of the module. Especially in the TCO we find dissipative power losses due to a certain sheet resistance. Therefore, the dissipative losses within the electrode also depend on the length over which the current has to be transported. For the ideal module all these factors need to be taken into account. Through simulation the optimum balance between GFF, solar cell length, and the distances between the single cells has to be determined, which depends on the material-specific properties. This includes the sheet resistance of back and front contact, which should be as low as possible, the contact resistance between interconnected cells, and the resistance of the photoactive layer.

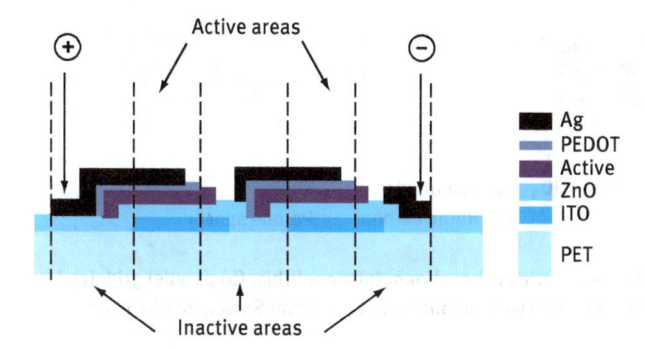

Fig. 5.8: Exploded views of the device structure for a module. The modules are composed of two serially connected cells prepared by knife-over-edge coating of ZnO nanoparticles. The layer sequence on the PET substrate is shown at the right and a schematic cross section (not to scale) of the final device with the overlap of layers, aperture loss, and active area. Illumination is coming from the bottom side passing first the PET layer. Figure after [260].

For dye-sensitized solar cells we find different suggestions for the fabrication of modules: monolithic, parallel grid, and Z- and W-type modules (Figure 5.9). Due to the different geometry of these modules, all of them have some advantages and disadvantages. Often a compromise between simple fabrication method, the appearance of the cell, and efficiency has to be made.

In 1996, the *monolithic design* was invented by Andreas Kay who reported a module efficiency of 5.3% [298]. Monolithic modules are made on a single conductive substrate, which makes the design very cost effective. There is no need for the deposition of a counter electrode, as the commonly used platinum electrode is replaced by a (much cheaper) porous carbon connected to the FTO as counter electrode, which also serves as an electrocatalyst for the I_3^- to I^- reduction and as a current collector. Additionally, a porous (insulating) spacer layer is used to separate working and counter electrode. This setup allows for a relatively simple fabrication, which is compatible with the roll-to-roll processing. The design allows a layer-by-layer fabrication and is not based on the assembly of two electrodes. Therefore, it is also very much tolerant to nonplanarity of the substrate. Disadvantages of this module design are the higher dye uptake due to adsorption on the high surface counter electrode. For building-integrated solar modules this design is not suitable as no semitransparent modules can be made and their appearance will always be very dark.

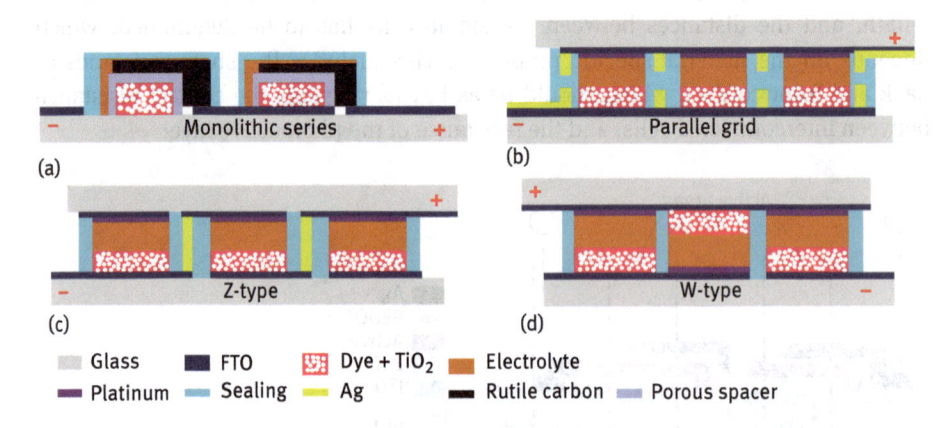

Fig. 5.9: Schematic of different types of series connections: (a) monolithic, (b) parallel grid, (c) Z-type, and (d) W-type. Reproduced from [263] with permission of the Royal Society of Chemistry.

The Z- and W-type modules are named after their resemblance to the letters Z and W, respectively. In the *Z-type module* vertical interconnects between neighboring cells form an in-series connected device. Usually, these vertical interconnects are prepared by a silver contact. The first demonstration was given in 2004 by Toyoda et al. [299]. In this work 64 cells were connected in series, and some outdoor tests were performed. These tests revealed that the performance steadily decreased due to cell degradation. This issue could be improved by a hermetic sealing through glass frit as sealant [300]. As the sealing process requires high temperatures (>500°C), two holes are left for the subsequent dying by introducing the dye solution and filling of a highly viscous ionic electrolyte. One observed disadvantage of this cell geometry is the high series resistance of the photoelectrodes, which makes an optimization of the geometry necessary. Cells need to be in small stripes (<1 cm) to achieve reasonable fill factors [301, 302]. Owing to this there is a large loss in active area through the vertical connections. Furthermore, the vertical silver interconnects are prone to corrosion, especially in combination with corrosive electrolytes.

The *W-type module* design offers a higher active area compared to the Z-design as the vertical metallic interconnects are avoided. In the W-design neighboring cells are of inverted geometry, which allows the alternative bias connection (Figure 5.9). This makes the design simpler and leads generally to higher fill factors as no additional serial interconnections are needed. However, it leads to two types of cells: front- and back-illuminated devices. Due to some light absorption by the iodide/triiodide system as well as the low transmittance of the platinum contact electrode, and also higher losses in the electron collection, the photocurrent density in the back-illuminated device is usually considerably lower than that in the front-illuminated device [303]. As the current has to match, these different cell types need to be adjusted in their width to compensate for this difference ideally under all possible illumination conditions. Unfortunately, perfect compensation is almost impossible under all illumination conditions (direct and diffuse) [304]. Besides, the cells need to be separated from each other to avoid mass transfer of electrolyte between neighboring cells, which can lead to separation of the redox couple [305].

In the *parallel connection design* (Figure 5.9) high short circuit current densities (J_{SC}) can be achieved and the J_{SC}s of the single cells do not need to match as they are just added up. They offer an ease of fabrication as interconnection of working and counter electrodes are not necessary. Charge collection can be enhanced and serial resistance lowered by using metal grids in addition to the standard FTO, which has proven to improve the J_{SC} and fill factor of such cells significantly due to much smaller series resistances [306–310]. As in other types of modules, special care has to be taken to separate the corrosive liquid electrolyte from the interconnections. The active areas of such cells are relatively small due to the extra metallic current collectors. Only if these are wide enough it can be ensured that high currents are collected with minimal voltage drops.

A *combination of series and parallel connected cells* allows high voltage and currents. The output of such a module can be directly tuned to the application where it is intended for. However, due to extra interconnections such cells have a very low active area and special care needs to be taken to optimize the module geometry. Certainly such combined modules suffer otherwise from the same issues addressed above for the different module types. For applications and especially for designing large area panels such combinations will be necessary.

5.2 Stability and Lifetime

In the past years the stability of organic and hybrid solar cells has increased drastically. The reasons for the improved stability are manifold: first of all, the intrinsic photostability of the active polymer materials has improved as well as the understanding and the control of the morphology. Often we find that excellent device performance and stability come together. As the organic materials are especially prone to degradation in their excited or ionized state, fast relaxation into the ground state will lead to more stable cells. Usually, well-performing cells stand out due to fast charge transfer rates allowing such quick relaxation. Furthermore, the inverted solar cell architecture allowing the use of more stable electrode materials (e.g. Ag instead of Al) significantly improved the lifetime without sacrificing on efficiency [311].

The lifetime is defined as the time for which the solar cell maintains at least 80% of its initial efficiency. Up to now the ultimate lifetime of organic and dye-sensitized solar cells is not known. To ensure long lifetimes, the design of the cells should be made to be stable, so no reactive and unstable electrode materials such as calcium, magnesium, and aluminum should be used, the materials should not be very hygroscopic, which is often seen as a problem when using PEDOT:PSS, and in general should show good stability. There are a number of degradation mechanisms decreasing the lifetime. The active organic material should be photochemically stable. UV radiation or highly energetic visible radiation can often induce photochemical processes, such as chain scission, rearrangement, and cross-linking reaction. Electrode materials such as aluminum might oxidize and block charges. Therefore inverted solar cells, using a layered geometry with a stack of ITO:TiO_2 or ZnO:organic blend:Ag-electrode, show a significantly improved lifetime over the standard non-inverted geometry. One reason for this is the more stable metal top silver electrode. However, the exact role of the metal-oxide film and whether it contributes to degradation is not yet fully understood and needs to be investigated in more detail [312]. One major issue to ensure reasonable lifetime is the encapsulation of the solar cells. Only if oxygen and moisture can be kept outside a reasonable lifetime can be expected. This topic was briefly discussed previously in Section 5.1.1.5 concerning encapsulation layers.

The easiest way to measure the lifetime are outdoor tests. However, local weather conditions, the location, climate, and other factors will influence the lifetime of the cell. Also, such measurements are very time consuming or even unrealistic to perform due to desired lifetimes of 25 years or longer. Therefore, researchers try to find conditions for accelerated lifetime tests, where additional stress is induced to the samples and therefore the degradation mechanisms occur on faster time scales. Unfortunately it is very difficult to predict or extrapolate the real lifetime in outdoor conditions. The choice of aging parameters and test methods strongly influences the aging behavior. To choose the right aging parameters and test methods it is necessary to know the types of degradation, which are dominantly present in the device. In case of organic semiconductors, the degradation on molecular level should be known. It is then possible to select the laboratory aging conditions that allow to observe the same but accelerated degradation mechanisms as in outdoor tests. In general stress increase and time compression allow such aging tests. Stress increase can be induced by oxygen and moisture exposure, elevated temperatures, and increased light intensities. Often the combination of these parameters is the key issue. Time compression can be done for mechanical stress, such as high bending cycles of flexible cells. The selection of the right parameters depends strongly on the device type and its dominating degradation mechanisms. If devices are encapsulated the sensitivity to oxygen and moisture might be reduced drastically and other degradation mechanisms might become dominant. Looking at the main degradation types of donor and acceptor materials will help understand their intrinsic stability. In a donor–acceptor blend additional degradation mechanisms, such as reorganization at interfaces to the electrodes and phase separation in the bulk, will play a role. Overall, for each cell type the degradation mechanisms on microscopic scale should be understood. This will help understand property changes in function and appearance on a macroscopic level. Therefore, the right test protocol for aging needs to be designed for each specific system. To ensure comparability of devices also in terms of their lifetime, standardized test protocols are important. Reese et al. report about some testing procedures, which represent a consensus of the discussion and conclusions reached during the first 3 years of the international summit on OPV stability (ISOS) [313]. The procedures include directions for shelf life testing, outdoor testing, laboratory weathering testing, and thermal cycling testing, as well as guidelines for reporting data, which should help allow comparison between laboratories and make reported lifetimes more reliable.

6 Conclusion and Outlook

The development in the field of organic and hybrid solar cells has been very fast, and the understanding of the basic mechanisms has improved considerably over time. One example of the increased understanding is the influence of the morphology in bulk heterojunction solar cells. In the late 1990s the importance of the morphology had already been recognized. However, in the beginning, atomic force microscopy images had been one of the most important tools to look at morphology. This technique is still important, but it does not allow to look inside the structure. Therefore cross-section scanning electron microscopy images had been used, which became a standard tool. All these characterization tools have improved considerably over time, allowing much higher quality of the images. Nowadays many additional tools like wide-angle X-ray or advanced transmission electron microscopy have been employed to organic solar cells to investigate the structures. Only the combination of improved characterization tools, the chemistry to synthesize the desired molecules and polymers, and better understanding of the device physics have allowed the morphology to improve into a more systematic approach. The first pictures of the ideal donor–acceptor blend have been refined over time. Whereas in the beginning a morphology with phase separation of the exciton diffusion length and the direct percolation pathways to the electrodes had been seen as sufficient to create an ideal solar cell, we now know that a simple two-phase (donor–acceptor) image is too simplified to describe the structure accurately as also mixed phases in different aggregation states from amorphous to crystalline can exist, which influence the local energy landscape that is important for efficient charge separation and transport. Up to now there are many aspects uncovered, and the research in the field will ensure that our understanding and with it also the device performance will increase in future.

Additionally we have observed that with the sudden rise of perovskite solar cells, there are probably a number of "new" materials that have just not yet been discovered as solar cell materials. If we understand the physics of the perovskite solar cells in more detail, we will be able to define the most important properties that allow such efficient devices to be used in efficient photovoltaics, which again will allow for a systematic search for alternative materials. The current issues of toxicity and device stability of perovskite solar cells might be resolved with new materials. Several years ago probably nobody would have bet on a solution-processable semiconductor that allows to prepare 20% efficient solar cells. Now we have learned that it is possible and it remains exciting what other surprises in the field will excite the research community in future.

We hope that our textbook will draw the interest of the readers into this exciting field of research and encourage them to get deeper into this field by reading current research articles and will also motivate them to do research in the field of organic and hybrid solar cells. We are sure it is worth it.

Bibliography

[1] Core Writing Team. *Climate Change 2014: Synthesis Report: Contribution of Working Groups I, II and III to the Fifth Assessment Report of the Intergovernmental Panel on Climate Change.* IPCC, Geneva, Switzerland, 2014.

[2] William Grylls Adams and RE Day. The action of light on selenium. *Proceedings of the Royal Society of London*, 25(171–178):113–117, 1876.

[3] Daryl M Chapin, Calvin S Fuller, and Gerald L Pearson. A new silicon p-n junction photocell for converting solar radiation into electrical power. *Journal of Applied Physics*, 25:676–677, 1954.

[4] GA Chamberlain. Organic solar cells: A review. *Solar Cells*, 8(1):47–83, 1983.

[5] Hideki Shirakawa, Edwin J Louis, Alan G MacDiarmid, Chwan K Chiang, and Alan J Heeger. Synthesis of electrically conducting organic polymers: Halogen derivatives of polyacetylene, $(CH)_x$. *Journal of the Chemical Society, Chemical Communications*, (16):578–580, 1977.

[6] Ching W Tang. Two-layer organic photovoltaic cell. *Applied Physics Letters*, 48:183, 1986.

[7] Ching W Tang and Steven A VanSlyke. Organic electroluminescent diodes. *Applied Physics Letters*, 51(12):913–915, 1987.

[8] Jeremy H Burroughes, Donal DC Bradley, AR Brown, RN Marks, K Mackay, Richard H Friend, Paul L Burns, and Andrew B Holmes. Light-emitting-diodes based on conjugated polymers. *Nature*, 347(6293):539–541, 1990.

[9] Brian O'Regan and Michael Grätzel. A low-cost, high-efficiency solar cell based on dye-sensitized. *Nature*, 353:737–740, 1991.

[10] Shigenori Morita, Anvar A Zakhidov, and Katsumi Yoshino. Doping effect of buckminsterfullerene in conducting polymer: Change of absorption spectrum and quenching of luminescene. *Solid State Communications*, 82(4):249–252, 1992.

[11] Niyazi Serdar Sariciftci, L Smilowitz, Alan J Heeger, and Fred Wudl. Photoinduced electron transfer from a conducting polymer to buckminsterfullerene. *Science*, 258(1474), 1992.

[12] Niyazi Serdar Sariciftci, D Braun, C Zhang, VI Sradonov, Alan J Heeger, G Stucky, and Fred Wudl. Semiconducting polymer-buckminsterfullerene heterojunctions: Diodes, photodiodes, and photovoltaic cells. *Applied Physics Letters*, 62(6):585–587, 1993.

[13] Jonathan JM Halls, CA Walsh, Neil C Greenham, Elisabeth A Marseglia, Richard H Friend, SC Moratti, and Andrew B Holmes. Efficient photodiodes from interpenetrating polymer networks. *Nature*, 376(6540):498–500, 1995.

[14] Gang Yu, Jun Gao, Jan C Hummelen, Fred Wudl, and Alan J Heeger. Polymer photovoltaic cells: Enhanced efficiencies via a network of internal donor-acceptor heterojunctions. *Science-AAAS-Weekly Paper Edition*, 270(5243):1789–1790, 1995.

[15] Udo Bach, Donald Lupo, Pascal Comte, Jaques E Moser, F Weissörtel, J Salbeck, H Spreitzer, and Michael Grätzel. Solid-state dye-sensitized mesoporous TiO_2 solar cells with high photon-to-electron conversion efficiencies. *Nature*, 395(6702):583–585, 1998.

[16] Akihiro Kojima, Kenjiro Teshima, Yasuo Shirai, and Tsutomu Miyasaka. Organometal halide perovskites as visible-light sensitizers for photovoltaic cells. *Journal of the American Chemical Society*, 131(17):6050–6051, 2009.

[17] Michael M Lee, Joël Teuscher, Tsutomu Miyasaka, Takurou N Murakami, and Henry J Snaith. Efficient hybrid solar cells based on meso-superstructured organometal halide perovskites. *Science*, 338(6107):643–647, 2012.

[18] Hui-Seon Kim, Chang-Ryul Lee, Jeong-Hyeok Im, Ki-Beom Lee, Thomas Moehl, Arianna Marchioro, Soo-Jin Moon, Robin Humphry-Baker, Jun-Ho Yum, Jacques E Moser, Michael Grätzel, and Nam-Gyu Park. Lead iodide perovskite sensitized all-solid-state submicron thin film mesoscopic solar cell with efficiency exceeding 9%. *Scientific Reports*, 2, 2012.

[19] Siegfried Hunklinger. *Festkörperphysik*, volume 3. Oldenbourg Wissenschaftsverlag, 2011.
[20] Karlheinz Seeger. *Semiconductor Physics*, volume 9. Springer, 2004.
[21] Charles Kittel. *Introduction to Solid State Physics*, volume 8. John Wiley & Sons, 2004.
[22] Marius Grundmann. *The Physics of Semiconductors*, volume 2. Springer, 2010.
[23] Martin Pope and Charles E Swenberg. *Electronic Processes in Organic Crystals and Polymers*. Oxford, 2 edition, 1999.
[24] Wolfgang Brütting. *Physics of Organic Semiconductors*. John Wiley & Sons, 2006.
[25] Anna Köhler and Heinz Bässler. *Electronic Processes in Organic Semiconductors*. Wiley-VCH, 2015.
[26] Arthur George Milnes. *Heterojunctions and Metal Semiconductor Junctions*. Elsevier, 2012.
[27] Winfried Mönch. *Electronic Properties of Semiconductor Interfaces*, volume 1. Springer, 2004.
[28] Akiko Kobayashi, Otto F Sankey, Stephen M Volz, and John D Dow. Semiempirical tight-binding band structures of wurtzite semiconductors: AlN, CdS, CdSe, ZnS, and ZnO. *Physical Review. B, Condensed Matter*, 28(2):935–945, 1983.
[29] Peter Würfel. *Physics of Solar Cells*. Wiley-VCH, 2nd edition, 2009.
[30] Stuart Warren, Jonathan Clayden, and Nand Nick Greeves. *Organic Chemistry*. Oxford University Press, 2001.
[31] TW Graham Solomons and Craig B Fryhle. *Organic Chemistry*. John Wiley & Sons, Inc., USA, 2000.
[32] CK Chiang, CR Fincher, YW Park, Alan J Heeger, Hideki Shirakawa, EJ Louis, SC Gau, and Alan G MacDiarmid. Electrical conductivity in doped polyacetylene. *Physical Review Letters*, 39(17):1098–1101, 1977.
[33] WP Su, JR Schrieffer, and AJ Heeger. Solitons in polyacetylene. *Physical Review Letters*, 42(25):1698–1701, 1979.
[34] Roald Hoffmann. An extended hückel theory. I. hydrocarbons. *The Journal of Chemical Physics*, 39(6):1397–1412, 1963.
[35] Yuri Avlasevich, Chen Li, and Klaus Müllen. Synthesis and applications of core-enlarged perylene dyes. *Journal of Materials Chemistry*, 20(19):3814–3826, 2010.
[36] A Marletta, FEG Guimarães, and RM Faria. Line shape of emission spectra of the luminescent polymer poly (p-phenylene vinylene). *Brazilian Journal of Physics*, 32(2B):570–574, 2002.
[37] Markus Sauer, Johan Hofkens, and Jörg Enderlein. Basic principles of fluorescence spectroscopy. *Handbook of Fluorescence Spectroscopy and Imaging: From Single Molecules to Ensembles*, Wiley-VCH, Weinheim, 1–30, 2011.
[38] Martin Richter, Philipp Marquetand, Jesus Gonzalez-Vazquez, Ignacio Sola, and Leticia Gonzalez. Femtosecond intersystem crossing in the DNA nucleobase cytosine. *The Journal of Physical Chemistry Letters*, 3(21):3090–3095, 2012.
[39] Bruno Ehrler, Brian J Walker, Marcus L Böhm, Mark WB Wilson, Yana Vaynzof, Richard H Friend, and Neil C Greenham. In situ measurement of exciton energy in hybrid singlet-fission solar cells. *Nature Communications*, 3:1019, 2012.
[40] Daniel N Congreve, Jiye Lee, Nicholas J Thompson, Eric Hontz, Shane R Yost, Philip D Reusswig, Matthias E Bahlke, Sebastian Reineke, Troy Van Voorhis, and Marc A Baldo. External quantum efficiency above 100% in a singlet-exciton-fission–based organic photovoltaic cell. *Science*, 340(6130):334–337, 2013.
[41] Leon van Dommelen. *Quantum Mechanics for Engineers*. published online, 2013.
[42] Edward Condon. A theory of intensity distribution in band systems. *Physical Review*, 28:1182–1201, Dec 1926.
[43] J Franck and EG Dymond. Elementary processes of photochemical reactions. *Transactions of the Faraday Society*, 21:536–542, 1926.
[44] Edward U Condon. Nuclear motions associated with electron transitions in diatomic molecules. *Physical Review*, 32:858–872, Dec 1928.

[45] Fumitomo Hide, María A Díaz-García, Benjamin J Schwartz, and Alan J Heeger. New developments in the photonic applications of conjugated polymers. *Accounts of Chemical Research*, 30(10):430–436, 1997.

[46] Gregory D Scholes. Long-range resonance energy transfer in molecular systems. *Annual Review of Physical Chemistry*, 54(1):57–87, 2003.

[47] Alexander P Demchenko. *Introduction to Fluorescence Sensing*, volume 1. Springer, 2009.

[48] Paul R Selvin. The renaissance of fluorescence resonance energy transfer. *Nature Structural Biology*, 7(9): 730–734, 2000.

[49] David L Dexter. A theory of sensitized luminescence in solids. *The Journal of Chemical Physics*, 21(5):836–850, 2004.

[50] Paul E Shaw, Arvydas Ruseckas, and Ifor DW Samuel. Exciton diffusion measurements in poly (3-hexylthiophene). *Advanced Materials*, 20(18):3516–3520, 2008.

[51] H Najafov, B Lee, Q Zhou, LC Feldman, and V Podzorov. Observation of long-range exciton diffusion in highly ordered organic semiconductors. *Nature Materials*, 9(11):938–943, 2010.

[52] Heinz Bässler and Anna Köhler. Charge transport in organic semiconductors. In *Unimolecular and Supramolecular Electronics I*, 1–65. Springer, Berlin-Heidelberg, 2012.

[53] MB Prince. Drift mobilities in semiconductors. I. Germanium. *Physical Review*, 92(3):681, 1953.

[54] Valentin D Mihailetchi, HX Xie, Bert de Boer, L Jan Anton Koster, and Paul WM Blom. Charge transport and photocurrent generation in poly (3-hexylthiophene): Methanofullerene bulk-heterojunction solar cells. *Advanced Functional Materials*, 16(5):699–708, 2006.

[55] Serap Günes, Helmut Neugebauer, and Niyazi Serdar Sariciftci. Conjugated polymer-based organic solar cells. *Chemical Reviews*, 107(4):1324–1338, 2007.

[56] David C Coffey, Bryon W Larson, Alexander W Hains, James B Whitaker, Nikos Kopidakis, Olga V Boltalina, Steven H Strauss, and Garry Rumbles. An optimal driving force for converting excitons into free carriers in excitonic solar cells. *The Journal of Physical Chemistry C*, 116(16):8916–8923, 2012.

[57] Heinz Bässler. Charge transport in disordered organic photoconductors a monte carlo simulation study. *Physica Status Solidi (b)*, 175(1):15–56, 1993.

[58] Allen Miller and Elihu Abrahams. Impurity conduction at low concentrations. *Physical Review*, 120:745–755, Nov 1960.

[59] Carsten Deibel and Vladimir Dyakonov. Polymer – fullerene bulk heterojunction solar cells. *Reports on Progress in Physics*, 73(9):096401, 2010.

[60] Ronald Österbacka, Almantas Pivrikas, Gytis Juška, K Genevičius, K Arlauskas, and H Stubb. Mobility and density relaxation of photogenerated charge carriers in organic materials. *Current Applied Physics*, 4(5):534–538, 2004.

[61] Paul WM Blom, MJM de Jong, and MG van Munster. Electric-field and temperature dependence of the hole mobility in poly(p-phenylene vinylene). *Physical Review B*, 55:R656–R659, 1997.

[62] Damodar M Pai. Transient photoconductivity in poly(nvinylcarbazole). *The Journal of Chemical Physics*, 52(5):2285–2291, 1970.

[63] Ulf Larsen. New exact approach to models of diffusion in one-dimensional random systems and optimal network kinetics in spin glass models. *Physics Letters A*, 105(6):307–310, 1984.

[64] Jenny Nelson. *The Physics of Solar Cells*. Imperial College Press, 2007.

[65] J Tersoff. Theory of semiconductor heterojunctions: The role of quantum dipoles. *Physical Review B*, 30:4874–4877, Oct 1984.

[66] Arthur George Milnes and Donald Lee Feucht. *Heterojunctions and Metal-Semiconductor Junctions*. Academic Press New York and London, 1972.

[67] JX Tang, CS Lee, and ST Lee. Electronic structures of organic/organic heterojunctions: From vacuum level alignment to fermi level pinning. *Journal of Applied Physics*, 101(6):064504, 2007.

[68] IH Campbell, S Rubin, TA Zawodzinski, JD Kress, RL Martin, DL Smith, NN Barashkov, and JP Ferraris. Controlling schottky energy barriers in organic electronic devices using self-assembled monolayers. *Physical Review B*, 54:R14321–R14324, Nov 1996.
[69] Chiatzun Goh, Shawn R Scully, and Michael D McGehee. Effects of molecular interface modification in hybrid organic-inorganic photovoltaic cells. *Journal of Applied Physics*, 101(11):114503, 2007.
[70] Alexandre-Edmond Becquerel. Mémoire sur les effets électriques produits sous l'influence des rayons solaires. *Comptes Rendus*, 9(567):1839, 1839.
[71] Alexandre-Edmond Becquerel. Recherches sur les effets de la radiation chimique de la lumière solaire, au moyen des courants électriques. *CR Acad. Sci*, 9:145–149, 1839.
[72] solargis.info. http://solargis.info/doc/_pics/freemaps/1000px/ghi/solargis-solar-map-world-map-en.png, accessed 2015.
[73] NREL. http://rredc.nrel.gov/solar/spectra/am1.5/astmg173/astmg173.html, accessed 2016.
[74] William Shockley and Hans J Queisser. Detailed balance limit of efficiency of p-n junction solar cells. *Journal of Applied Physics*, 32(3):510–519, 1961.
[75] Alexis De Vos. Detailed balance limit of the efficiency of tandem solar cells. *Journal of Physics D: Applied Physics*, 13(5):839, 1980.
[76] Markus Hallermann, Stephan Haneder, and Enrico Da Como. Charge-transfer states in conjugated polymer/fullerene blends: Below-gap weakly bound excitons for polymer photovoltaics. *Applied Physics Letters*, 93:053307, 2008.
[77] Markus Hallermann, Ilka Kriegel, Enrico Da Como, Josef M Berger, Elizabeth Von Hauff, and Jochen Feldmann. Charge transfer excitons in polymer/fullerene blends: The role of morphology and polymer chain conformation. *Advanced Functional Materials*, 19(22):3662–3668, 2009.
[78] Lars Onsager. Initial recombination of ions. *Physical Review*, 54(8):554, 1938.
[79] M Tachiya. Breakdown of the onsager theory of geminate ion recombination. *The Journal of Chemical Physics*, 89(11):6929–6935, 1988.
[80] Kristofer Tvingstedt, Koen Vandewal, Abay Gadisa, Fengling Zhang, Jean Manca, and Olle Inganäs. Electroluminescence from charge transfer states in polymer solar cells. *Journal of the American Chemical Society*, 131(33):11819–11824, 2009. PMID: 19722595.
[81] L Jan Anton Koster, Sean E Shaheen, and Jan C Hummelen. Pathways to a new efficiency regime for organic solar cells. *Advanced Energy Materials*, 2(10):1246–1253, 2012.
[82] Jean-Luc Brédas, Joseph E Norton, Jérôme Cornil, and Veaceslav Coropceanu. Molecular understanding of organic solar cells: The challenges. *Accounts of Chemical Research*, 42(11):1691–1699, 2009.
[83] Eric Bittner, Vladimir Lankevich, Simon Gelinas, Akshay Rao, David S Ginger, and Richard Friend. How disorder controls the kinetics of triplet charge recombination in semiconducting organic polymer photovoltaics. *Physical Chemistry Chemical Physics*, 16:20321–20328, 2014.
[84] Akshay Rao, Philip CY Chow, Simon Gélinas, Cody W Schlenker, Chang-Zhi Li, Hin-Lap Yip, Alex K-Y Jen, David S Ginger, and Richard H Friend. The role of spin in the kinetic control of recombination in organic photovoltaics. *Nature*, 500(7463):435–439, 2013.
[85] Artem A Bakulin, Akshay Rao, Vlad G Pavelyev, Paul HM van Loosdrecht, Maxim S Pshenichnikov, Dorota Niedzialek, Jérôme Cornil, David Beljonne, and Richard H Friend. The role of driving energy and delocalized states for charge separation in organic semiconductors. *Science*, 335(6074):1340–1344, 2012.
[86] Giulia Grancini, Margherita Maiuri, Daniele Fazzi, Annamaria Petrozza, Hans-Joachim Egelhaaf, Daniele Brida, Giulio Cerullo, and Guglielmo Lanzani. Hot exciton dissociation in polymer solar cells. *Nature Materials*, 12(1):29–33, 2013.
[87] Koen Vandewal, Steve Albrecht, Eric T Hoke, Kenneth R Graham, Johannes Widmer, Jessica D Douglas, Marcel Schubert, William R Mateker, Jason T Bloking, George F Burkhard, et al. Efficient

charge generation by relaxed charge-transfer states at organic interfaces. *Nature Materials*, 13(1):63–68, 2014.

[88] Sarah Maria Falke, Carlo Andrea Rozzi, Daniele Brida, Margherita Maiuri, Michele Amato, Ephraim Sommer, Antonietta De Sio, Angel Rubio, Giulio Cerullo, Elisa Molinari, et al. Coherent ultrafast charge transfer in an organic photovoltaic blend. *Science*, 344(6187):1001–1005, 2014.

[89] David Kearns and Melvin Calvin. Photovoltaic Effect and Photoconductivity in Laminated Organic Systems. *The Journal of Chemical Physics*, 29(4): 950–951, 1958.

[90] Brian A Gregg. Excitonic solar cells. *The Journal of Physical Chemistry B*, 107(20):4688–4698, 2003.

[91] Khosrow Rahimi, Ioan Botiz, Natalie Stingelin, Navaphun Kayunkid, Michael Sommer, Felix Peter Vinzenz Koch, Ha Nguyen, Olivier Coulembier, Philippe Dubois, Martin Brinkmann, et al. Controllable processes for generating large single crystals of poly (3-hexylthiophene). *Angewandte Chemie International Edition*, 51(44):11131–11135, 2012.

[92] Adam J Moulé and Klaus Meerholz. Controlling morphology in polymer – fullerene mixtures. *Advanced Materials*, 20(2):240–245, 2008.

[93] Mukti Aryal, Krutarth Trivedi, and Wenchuang Hu. Nano-confinement induced chain alignment in ordered P3HT nanostructures defined by nanoimprint lithography. *ACS Nano*, 3(10):3085–3090, 2009.

[94] Kevin M Coakley, Bhavani S Srinivasan, Jonathan M Ziebarth, Chiatzun Goh, Yuxiang Liu, and Michael D McGehee. Enhanced hole mobility in regioregular polythiophene infiltrated in straight nanopores. *Advanced Functional Materials*, 15(12):1927–1932, 2005.

[95] Felix Peter Vinzenz Koch, Jonathan Rivnay, Sam Foster, Christian Müller, Jonathan M Downing, Ester Buchaca-Domingo, Paul Westacott, Liyang Yu, Mingjian Yuan, Mohammed Baklar, et al. The impact of molecular weight on microstructure and charge transport in semicrystalline polymer semiconductors–poly (3-hexylthiophene), a model study. *Progress in Polymer Science*, 38(12):1978–1989, 2013.

[96] Martin Brinkmann and Patrice Rannou. Effect of molecular weight on the structure and morphology of oriented thin films of regioregular poly (3-hexylthiophene) grown by directional epitaxial solidification. *Advanced Functional Materials*, 17(1):101–108, 2007.

[97] Edward JW Crossland, Kim Tremel, Florian Fischer, Khosrow Rahimi, Günter Reiter, Ullrich Steiner, and Sabine Ludwigs. Anisotropic charge transport in spherulitic poly (3-hexylthiophene) films. *Advanced Materials*, 24(6):839–844, 2012.

[98] Hajime Yamagata and Frank C Spano. Interplay between intrachain and interchain interactions in semiconducting polymer assemblies: The HJ-aggregate model. *The Journal of Chemical Physics*, 136(18):184901, 2012.

[99] Frank C Spano and Carlos Silva. H-and J-aggregate behavior in polymeric semiconductors. *Annual Review of Physical Chemistry*, 65:477–500, 2014.

[100] Fiona C Jamieson, Ester Buchaca Domingo, Thomas McCarthy-Ward, Martin Heeney, Natalie Stingelin, and James R Durrant. Fullerene crystallisation as a key driver of charge separation in polymer/fullerene bulk heterojunction solar cells. *Chemical Science*, 3(2):485–492, 2012.

[101] Eugene A Katz. Fullerene thin films as photovoltaic material. *Nanostructured Materials for Solar Energy Conversion*, 361–443, 2006.

[102] E Giudice, E Magnano, S Rusponi, C Boragno, and U Valbusa. Morphology of C60 thin films grown on Ag(001). *Surface Science*, 405(2):L561–L565, 1998.

[103] Peter Peumans, Soichi Uchida, and Stephen R Forrest. Efficient bulk heterojunction photovoltaic cells using small-molecular-weight organic thin films. *Nature*, 425(6954):158–162, 2003.

[104] Neil D Treat, Michael A Brady, Gordon Smith, Michael F Toney, Edward J Kramer, Craig J Hawker, and Michael L Chabinyc. Interdiffusion of PCBM and P3HT reveals miscibility in a photovoltaically active blend. *Advanced Energy Materials*, 1(1):82–89, 2011.

[105] Jun Zhao, Ann Swinnen, Guy Van Assche, Jean Manca, Dirk Vanderzande, and Bruno Van Mele. Phase diagram of P3HT/PCBM blends and its implication for the stability of morphology. *The Journal of Physical Chemistry B*, 113(6):1587–1591, 2009.

[106] Gordon J Hedley, Alexander J Ward, Alexander Alekseev, Calvyn T Howells, Emiliano R Martins, Luis A Serrano, Graeme Cooke, Arvydas Ruseckas, and Ifor DW Samuel. Determining the optimum morphology in high-performance polymer-fullerene organic photovoltaic cells. *Nature Communications*, 4:2867, 2013.

[107] Roman Gysel, Nichole Cates, Alex Mayer, Michael Toney, and Michael McGehee. Molecular self-ordering in organic solar cell materials. *Solar & Alternative Energy* SPIE, DOI: 10.1117/2.1200908.1769, 2009.

[108] Carmen Pérez León, Lothar Kador, Bin Peng, and Mukundan Thelakkat. Characterization of the adsorption of Ru-bpy dyes on mesoporous TiO_2 films with UV-Vis, Raman, and FTIR spectroscopies. *The Journal of Physical Chemistry B*, 110(17):8723–8730, 2006.

[109] Anders Hagfeldt, Gerrit Boschloo, Licheng Sun, Lars Kloo, and Henrik Pettersson. Dye-sensitized solar cells. *Chemical Reviews*, 110(11):6595–6663, 2010.

[110] Jonas Weickert and Lukas Schmidt-Mende. *Solid-State Dye-Sensitized Solar Cells*, chapter 15, pages 465–494. Wiley-VCH Verlag GmbH & Co. KGaA, 2014.

[111] Lukas Schmidt-Mende and Michael Gratzel. TiO_2 pore-filling and its effect on the efficiency of solid-state dye-sensitized solar cells. *Thin Solid Films*, 500(1-2):296–301, 2006.

[112] C Kyle Renshaw and Stephen R Forrest. Excited state and charge dynamics of hybrid organic/inorganic heterojunctions. I. Theory. *Physical Review B*, 90(4):045302, 2014.

[113] Stefan D Oosterhout, Martijn M Wienk, Svetlana S van Bavel, Ralf Thiedmann, L Jan Anton Koster, Jan Gilot, Joachim Loos, Volker Schmidt, and René AJ Janssen. The effect of three-dimensional morphology on the efficiency of hybrid polymer solar cells. *Nature Materials*, 8(10):818–824, 2009.

[114] Jonas Weickert and Lukas Schmidt-Mende. *Hybrid Solar Cells from Ordered Nanostructures*, chapter 14, pages 385–417. Wiley-VCH Verlag GmbH & Co. KGaA, 2013.

[115] Yana Vaynzof, Artem A Bakulin, Simon Gélinas, and Richard H Friend. Direct observation of photoinduced bound charge-pair states at an organic-inorganic semiconductor interface. *Physical Review Letters*, 108(24):246605, 2012.

[116] Dieter Weber. $CH_3NH_3PbX_3$, ein Pb(II)-system mit kubischer Perowskitstruktur/$CH_3NH_3PbX_3$, a Pb(II)-system with cubic perovskite structure. *Zeitschrift für Naturforschung B*, 33(12):1443–1445, 1978.

[117] Dieter Weber. $CH_3NH_3SnBrXI_{3-x}$ (x = 0-3), ein Sn(II)-system mit kubischer Perowskitstruktur/$CH_3NH_3SnBrXI_{3-x}$ (x = 0-3), a Sn(II)-system with cubic perovskite structure. *Zeitschrift für Naturforschung B*, 33(8):862–865, 1978.

[118] Cherie R Kagan, David B Mitzi, and Christos D Dimitrakopoulos. Organic-inorganic hybrid materials as semiconducting channels in thin-film field-effect transistors. *Science*, 286:945–947, 1999.

[119] Jeong-Hyeok Im, Chang-Ryul Lee, Jin-Wook Lee, Sang-Won Park, and Nam-Gyu Park. 6.5% efficient perovskite quantum-dot-sensitized solar cell. *Nanoscale*, 3(10):4088–4093, 2011.

[120] Jingbi You, Ziruo Hong, Yang Yang, Qi Chen, Min Cai, Tze-Bin Song, Chun-Chao Chen, Shirong Lu, Yongsheng Liu, Huanping Zhou, and Yang Yang. Low-temperature solution-processed perovskite solar cells with high efficiency and flexibility. *ACS Nano*, 8(2):1674–1680, 2014.

[121] In Chung, Jung-Hwan Song, Jino Im, John Androulakis, Christos D Malliakas, Hao Li, Arthur J Freeman, John T Kenney, and Mercouri G Kanatzidis. $CsSnI_3$: Semiconductor or metal? High electrical conductivity and strong near-infrared photoluminescence from a single material. High hole mobility and phase-transitions. *Journal of the American Chemical Society*, 134(20):8579–8587, 2012.

[122] Teruya Ishihara. Optical properties of PbI-based perovskite structures. *Journal of Luminescence*, 6061(0):269–274, 1994.
[123] Tom Baikie, Yanan Fang, Jeannette M Kadro, Martin Schreyer, Fengxia Wei, Subodh G Mhaisalkar, Michael Graetzel, and Tim J White. Synthesis and crystal chemistry of the hybrid perovskite (CH_3NH_3)PbI_3 for solid-state sensitised solar cell applications. *Journal of Materials Chemistry A*, 1(18):5628–5641, 2013.
[124] Constantinos C Stoumpos, Christos D Malliakas, and Mercouri G Kanatzidis. Semiconducting tin and lead iodide perovskites with organic cations: Phase transitions, high mobilities, and near-infrared photoluminescent properties. *Inorganic Chemistry*, 52(15):9019–9038, 2013.
[125] Jun Hong Noh, Sang Hyuk Im, Jin Hyuck Heo, Tarak N Mandal, and Sang Il Seok. Chemical management for colorful, efficient, and stable inorganic-organic hybrid nanostructured solar cells. *Nano Letters*, 13(4):1764–1769, 2013.
[126] RE Wasylishen, Osvald Knop, and JB Macdonald. Cation rotation in methylammonium lead halides. *Solid State Communications*, 56(7):581–582, 1985.
[127] A Poglitsch and Dieter Weber. Dynamic disorder in methylammoniumtrihalogenoplumbates (II) observed by millimeterwave spectroscopy. *The Journal of Chemical Physics*, 87(11):6373–6378, 1987.
[128] Wan-Jian Yin, Tingting Shi, and Yanfa Yan. Unique properties of halide perovskites as possible origins of the superior solar cell performance. *Advanced Materials*, 26(27):4653–4658, 2014.
[129] Stefaan De Wolf, Jakub Holovsky, Soo-Jin Moon, Philipp Löper, Bjoern Niesen, Martin Ledinsky, Franz-Josef Haug, Jun-Ho Yum, and Christophe Ballif. Organometallic halide perovskites: Sharp optical absorption edge and its relation to photovoltaic performance. *The Journal of Physical Chemistry Letters*, 5(6):1035–1039, 2014.
[130] Nam Joong Jeon, Jun Hong Noh, Woon Seok Yang, Young Chan Kim, Seungchan Ryu, Jangwon Seo, and Sang Il Seok. Compositional engineering of perovskite materials for high-performance solar cells. *Nature*, 517(7535):476–480, 2015.
[131] Woon Seok Yang, Jun Hong Noh, Nam Joong Jeon, Young Chan Kim, Seungchan Ryu, Jangwon Seo, and Sang Il Seok. High-performance photovoltaic perovskite layers fabricated through intramolecular exchange. *Science*, 348(6240):1234–1237, 2015.
[132] Qi Chen, Huanping Zhou, Ziruo Hong, Song Luo, Hsin-Sheng Duan, Hsin-Hua Wang, Yongsheng Liu, Gang Li, and Yang Yang. Planar heterojunction perovskite solar cells via vapor-assisted solution process. *Journal of the American Chemical Society*, 136(2):622–625, 2013.
[133] Mingzhen Liu, Michael B Johnston, and Henry J Snaith. Efficient planar heterojunction perovskite solar cells by vapour deposition. *Nature*, 501(7467):395–398, 2013.
[134] Giles E Eperon, Victor M Burlakov, Pablo Docampo, Alain Goriely, and Henry J Snaith. Morphological control for high performance, solution-processed planar heterojunction perovskite solar cells. *Advanced Functional Materials*, 24(1):151–157, 2014.
[135] Nam Joong Jeon, Jun Hong Noh, Young Chan Kim, Woon Seok Yang, Seungchan Ryu, and Sang Il Seok. Solvent engineering for high-performance inorganic-organic hybrid perovskite solar cells. *Nature Materials*, 13(9):897–903, 2014.
[136] Manda Xiao, Fuzhi Huang, Wenchao Huang, Yasmina Dkhissi, Ye Zhu, Joanne Etheridge, Angus Gray-Weale, Udo Bach, Yi-Bing Cheng, and Leone Spiccia. A fast deposition-crystallization procedure for highly efficient lead iodide perovskite thin-film solar cells. *Angewandte Chemie*, 126(37):10056–10061, 2014.
[137] Po-Wei Liang, Chien-Yi Liao, Chu-Chen Chueh, Fan Zuo, Spencer T Williams, Xu-Kai Xin, Jiangjen Lin, and Alex KY Jen. Additive enhanced crystallization of solution-processed perovskite for highly efficient planar-heterojunction solar cells. *Advanced Materials*, 26(22):3748–3754, 2014.
[138] Giles E Eperon, Severin N Habisreutinger, Tomas Leijtens, Bardo J Bruijnaers, Jacobus J van Franeker, Dane W deQuilettes, Sandeep Pathak, Rebecca J Sutton, Giulia Grancini, David S Ginger,

ReneAJ Janssen, Annamaria Petrozza, and Henry J Snaith. The importance of moisture in hybrid lead halide perovskite thin film fabrication. *ACS Nano*, 9(9), 9380–9393, 2015.

[139] Samuel D Stranks, Giles E Eperon, Giulia Grancini, Christopher Menelaou, Marcelo JP Alcocer, Tomas Leijtens, Laura M Herz, Annamaria Petrozza, and Henry J Snaith. Electron-hole diffusion lengths exceeding 1 micrometer in an organometal trihalide perovskite absorber. *Science*, 342(6156):341–344, 2013.

[140] Hui Yu, Feng Wang, Fangyan Xie, Wenwu Li, Jian Chen, and Ni Zhao. The role of chlorine in the formation process of $CH_3NH_3PbI_{3-x}Cl_x$ perovskite. *Advanced Functional Materials*, 24(45):7102–7108, 2014.

[141] Yaron Tidhar, Eran Edri, Haim Weissman, Dorin Zohar, Gary Hodes, David Cahen, Boris Rybtchinski, and Saar Kirmayer. Crystallization of methyl ammonium lead halide perovskites: Implications for photovoltaic applications. *Journal of the American Chemical Society*, 136(38):13249–13256, 2014.

[142] Silvia Colella, Edoardo Mosconi, Giovanna Pellegrino, Alessandra Alberti, Valentino LP Guerra, Sofia Masi, Andrea Listorti, Aurora Rizzo, Guglielmo Guido Condorelli, Filippo De Angelis, and Giuseppe Gigli. Elusive presence of chloride in mixed halide perovskite solar cells. *The Journal of Physical Chemistry Letters*, 5(20):3532–3538, 2014.

[143] David T Moore, Hiroaki Sai, Kwan W Tan, Detlef-M Smilgies, Wei Zhang, Henry J Snaith, Ulrich Wiesner, and Lara A Estroff. Crystallization kinetics of organicinorganic trihalide perovskites and the role of the lead anion in crystal growth. *Journal of the American Chemical Society*, 137(6):2350–2358, 2015.

[144] Wei Zhang, Michael Saliba, David T Moore, Sandeep K Pathak, Maximilian T Hörantner, Thomas Stergiopoulos, Samuel D Stranks, Giles E Eperon, Jack A Alexander-Webber, Antonio Abate, Aditya Sadhanala, Shuhua Yao, Yulin Chen, Richard H Friend, Lara A Estroff, Ulrich Wiesner, and Henry J Snaith. Ultrasmooth organic-inorganic perovskite thin-film formation and crystallization for efficient planar heterojunction solar cells. *Nature Communications*, 6, 6142, 2015.

[145] Julian Burschka, Norman Pellet, Soo-Jin Moon, Robin Humphry-Baker, Peng Gao, Mohammad K Nazeeruddin, and Michael Gratzel. Sequential deposition as a route to high-performance perovskite-sensitized solar cells. *Nature*, 499(7458):316–319, 2013.

[146] Yongzhen Wu, Ashraful Islam, Xudong Yang, Chuanjiang Qin, Jian Liu, Kun Zhang, Wenqin Peng, and Liyuan Han. Retarding the crystallization of PbI_2 for highly reproducible planar-structured perovskite solar cells via sequential deposition. *Energy & Environmental Science*, 7(9):2934–2938, 2014.

[147] Byeong Jo Kim, Dong Hoe Kim, Yoo-Yong Lee, Hee-Won Shin, Gill Sang Han, Jung Sug Hong, Khalid Mahmood, Tae Kyu Ahn, Young-Chang Joo, Kug Sun Hong, Nam-Gyu Park, Sangwook Lee, and Hyun Suk Jung. Highly efficient and bending durable perovskite solar cells: Toward a wearable power source. *Energy & Environmental Science*, 8(3):916–921, 2015.

[148] Dianyi Liu and Timothy L Kelly. Perovskite solar cells with a planar heterojunction structure prepared using room-temperature solution processing techniques. *Nature Photonics*, 8(2):133–138, 2014.

[149] Seong Sik Shin, Woon Seok Yang, Jun Hong Noh, Jae Ho Suk, Nam Joong Jeon, Jong Hoon Park, Ju Seong Kim, Won Mo Seong, and Sang Il Seok. High-performance flexible perovskite solar cells exploiting Zn_2SnO_4 prepared in solution below 100°C. *Nature Communications*, 6, 7410, 2015.

[150] Pablo Docampo, James M Ball, Mariam Darwich, Giles E Eperon, and Henry J Snaith. Efficient organometal trihalide perovskite planar-heterojunction solar cells on flexible polymer substrates. *Nature Communications*, 4, 2761, 2013.

[151] Henry J Snaith, Antonio Abate, James M Ball, Giles E Eperon, Tomas Leijtens, Nakita K Noel, Samuel D Stranks, Jacob Tse-Wei Wang, Konrad Wojciechowski, and Wei Zhang. Anomalous hysteresis in perovskite solar cells. *The Journal of Physical Chemistry Letters*, 5(9):1511–1515, 2014.

[152] Jeffrey A Christians, Joseph S Manser, and Prashant V Kamat. Best practices in perovskite solar cell efficiency measurements. Avoiding the error of making bad cells look good. *The Journal of Physical Chemistry Letters*, 6(5):852–857, 2015.

[153] Uwe Rau. Reciprocity relation between photovoltaic quantum efficiency and electroluminescent emission of solar cells. *Physical Review B*, 76(8):085303, 2007.

[154] Kristofer Tvingstedt, Olga Malinkiewicz, Andreas Baumann, Carsten Deibel, Henry J Snaith, Vladimir Dyakonov, and Henk J Bolink. Radiative efficiency of lead iodide based perovskite solar cells. *Scientific Reports*, 4:6071, 2014.

[155] Wolfgang Tress, Nevena Marinova, Olle Inganäs, Mohammad K Nazeeruddin, Shaik M Zakeeruddin, and Michael Graetzel. Predicting the open-circuit voltage of $CH_3NH_3PbI_3$ perovskite solar cells using electroluminescence and photovoltaic quantum efficiency spectra: The role of radiative and non-radiative recombination. *Advanced Energy Materials*, 5(3):1400812, 2015.

[156] Felix Deschler, Michael Price, Sandeep Pathak, Lina E Klintberg, David-Dominik Jarausch, Ruben Higler, Sven Hüttner, Tomas Leijtens, Samuel D Stranks, Henry J Snaith, Mete Atatüre, Richard T Phillips, and Richard H Friend. High photoluminescence efficiency and optically pumped lasing in solution-processed mixed halide perovskite semiconductors. *The Journal of Physical Chemistry Letters*, 5(8):1421–1426, 2014.

[157] Christian Wehrenfennig, Giles E Eperon, Michael B Johnston, Henry J Snaith, and Laura M Herz. High charge carrier mobilities and lifetimes in organolead trihalide perovskites. *Advanced Materials*, 26(10):1584–1589, 2014.

[158] Samuel D Stranks, Victor M Burlakov, Tomas Leijtens, James M Ball, Alain Goriely, and Henry J Snaith. Recombination kinetics in organic-inorganic perovskites: Excitons, free charge, and subgap states. *Physical Review Applied*, 2(3):034007, 2014.

[159] Christian Wehrenfennig, Mingzhen Liu, Henry J Snaith, Michael B Johnston, and Laura M Herz. Charge carrier recombination channels in the low-temperature phase of organic-inorganic lead halide perovskite thin films. *APL Materials*, 2(8):081513, 2014.

[160] M Hirasawa, T Ishihara, T Goto, K Uchida, and N Miura. Magnetoabsorption of the lowest exciton in perovskite-type compound $(CH_3NH_3)PbI_3$. *Physica B: Condensed Matter*, 201(0):427–430, 1994.

[161] Tom J Savenije, Carlito S Ponseca, Lucas Kunneman, Mohamed Abdellah, Kaibo Zheng, Yuxi Tian, Qiushi Zhu, Sophie E Canton, Ivan G Scheblykin, Tonu Pullerits, Arkady Yartsev, and Villy Sundström. Thermally activated exciton dissociation and recombination control the carrier dynamics in organometal halide perovskite. *The Journal of Physical Chemistry Letters*, 5(13):2189–2194, 2014.

[162] Valerio D'Innocenzo, Giulia Grancini, Marcelo JP Alcocer, Ajay Ram Srimath Kandada, Samuel D Stranks, Michael M Lee, Guglielmo Lanzani, Henry J Snaith, and Annamaria Petrozza. Excitons versus free charges in organo-lead tri-halide perovskites. *Nature Communications*, 5, 3586, 2014.

[163] Miao Hu, Cheng Bi, Yongbo Yuan, Zhengguo Xiao, Qingfeng Dong, Yuchuan Shao, and Jinsong Huang. Distinct exciton dissociation behavior of organolead trihalide perovskite and excitonic semiconductors studied in the same system. *Small*, 11(18):2164–2169, 2015.

[164] Emilio J Juarez-Perez, Rafael S Sanchez, Laura Badia, Germà Garcia-Belmonte, Yong Soo Kang, Ivan Mora-Sero, and Juan Bisquert. Photoinduced giant dielectric constant in lead halide perovskite solar cells. *The Journal of Physical Chemistry Letters*, 5(13):2390–2394, 2014.

[165] Qianqian Lin, Ardalan Armin, Ravi Chandra Raju Nagiri, Paul L Burn, and Paul Meredith. Electro-optics of perovskite solar cells. *Nature Photonics*, 9(2):106–112, 2015.

[166] Jin Hyuck Heo, Sang Hyuk Im, Jun Hong Noh, Tarak N Mandal, Choong-Sun Lim, Jeong Ah Chang, Yong Hui Lee, Hi-jung Kim, Arpita Sarkar, K NazeeruddinMd, Michael Gratzel, and Sang Il Seok. Efficient inorganic-organic hybrid heterojunction solar cells containing perovskite compound and polymeric hole conductors. *Nature Photonics*, 7(6):486–491, 2013.

[167] Roland Scheer and Hans-Werner Schock. *Chalcogenide Photovoltaics*. Wiley-VCH, 2011.

[168] Christopher Eames, Jarvist M Frost, Piers RF Barnes, Brian C O'Regan, Aron Walsh, and M Saiful Islam. Ionic transport in hybrid lead iodide perovskite solar cells. *Nature Communications*, 6, 2015.
[169] Stuart R Wenham and Martin A Green. Silicon solar cells. *Progress in Photovoltaics: Research and Applications*, 4(1):3–33, 1996.
[170] Simon Sze and Kwonk K Ng. *Physics of Seminconductor Devices*. John Wiley & Sons, Inc., USA, 2007.
[171] Guichuan Xing, Nripan Mathews, Shuangyong Sun, Swee Sien Lim, Yeng Ming Lam, Michael Grätzel, Subodh Mhaisalkar, and Tze Chien Sum. Long-range balanced electron- and hole-transport lengths in organic-inorganic $CH_3NH_3PbI_3$. *Science*, 342(6156):344–347, 2013.
[172] Vittoria Roiati, Edoardo Mosconi, Andrea Listorti, Silvia Colella, Giuseppe Gigli, and Filippo De Angelis. Stark effect in perovskite/TiO_2 solar cells: Evidence of local interfacial order. *Nano Letters*, 14(4):2168–2174, 2014.
[173] Jiangjian Shi, Xin Xu, Dongmei Li, and Qingbo Meng. Interfaces in perovskite solar cells. *Small* 11 (21), 2472–2486, 2015.
[174] Atsushi Wakamiya, Masaru Endo, Takahiro Sasamori, Norihiro Tokitoh, Yuhei Ogomi, Shuzi Hayase, and Yasujiro Murata. Reproducible fabrication of efficient perovskite-based solar cells: X-ray crystallographic studies on the formation of $CH_3NH_3PbI_3$ layers. *Chemistry Letters*, 43(5):711–713, 2014.
[175] Xu Dong, Hongwei Hu, Bencai Lin, Jianning Ding, and Ningyi Yuan. The effect of ald-zno layers on the formation of $CH_3NH_3PbI_3$ with different perovskite precursors and sintering temperatures. *Chemical Communications*, 50(92):14405–14408, 2014.
[176] Alberto Torres and Luis GC Rego. Surface effects and adsorption of methoxy anchors on hybrid lead iodide perovskites: Insights for spiro-MeOTAD attachment. *The Journal of Physical Chemistry C*, 118(46):26947–26954, 2014.
[177] Victor W Bergmann, Stefan AL Weber, F Javier Ramos, Mohammad Khaja Nazeeruddin, Michael Grätzel, Dan Li, Anna L Domanski, Ingo Lieberwirth, Shahzada Ahmad, and Rüdiger Berger. Realspace observation of unbalanced charge distribution inside a perovskite-sensitized solar cell. *Nature Communications*, 5, 2014.
[178] Antonio Guerrero, Emilio J Juarez-Perez, Juan Bisquert, Ivan Mora-Sero, and Germà Garcia-Belmonte. Electrical field profile and doping in planar lead halide perovskite solar cells. *Applied Physics Letters*, 105(13):133902, 2014.
[179] Eran Edri, Saar Kirmayer, Sabyasachi Mukhopadhyay, Konstantin Gartsman, Gary Hodes, and David Cahen. Elucidating the charge carrier separation and working mechanism of $CH_3NH_3PbI_{3-x}Cl_x$ perovskite solar cells. *Nature Communications*, 5, 3461, 2014.
[180] Zonglong Zhu, Jiani Ma, Zilong Wang, Cheng Mu, Zetan Fan, Lili Du, Yang Bai, Louzhen Fan, He Yan, David Lee Phillips, and Shihe Yang. Efficiency enhancement of perovskite solar cells through fast electron extraction: The role of graphene quantum dots. *Journal of the American Chemical Society*, 136(10):3760–3763, 2014.
[181] Agnese Abrusci, Samuel D Stranks, Pablo Docampo, Hin-Lap Yip, Alex KY Jen, and Henry J Snaith. High-performance perovskite-polymer hybrid solar cells via electronic coupling with fullerene monolayers. *Nano Letters*, 13(7):3124–3128, 2013.
[182] Konrad Wojciechowski, Samuel D Stranks, Antonio Abate, Golnaz Sadoughi, Aditya Sadhanala, Nikos Kopidakis, Garry Rumbles, Chang-Zhi Li, Richard H Friend, Alex KY Jen, and Henry J Snaith. Heterojunction modification for highly efficient organic-inorganic perovskite solar cells. *ACS Nano*, 8(12):12701–12709, 2014.
[183] Yuhei Ogomi, Atsushi Morita, Shota Tsukamoto, Takahiro Saitho, Qing Shen, Taro Toyoda, Kenji Yoshino, Shyam S Pandey, Tingli Ma, and Shuzi Hayase. All-solid perovskite solar cells

with HOCO-R-NH$_3^+$I$^-$ anchor-group inserted between porous titania and perovskite. *The Journal of Physical Chemistry C*, 118(30):16651–16659, 2014.

[184] Seigo Ito, Soichiro Tanaka, Kyohei Manabe, and Hitoshi Nishino. Effects of surface blocking layer of Sb$_2$S$_3$ on nanocrystalline TiO$_2$ for CH$_3$NH$_3$PbI$_3$ perovskite solar cells. *The Journal of Physical Chemistry C*, 118(30):16995–17000, 2014.

[185] Gill Sang Han, Hyun Suk Chung, Byeong Jo Kim, Dong Hoe Kim, Jin Wook Lee, Bhabani Sankar Swain, Khalid Mahmood, Jin Sun Yoo, Nam-Gyu Park, Jung Heon Lee, and Hyun Suk Jung. Retarding charge recombination in perovskite solar cells using ultrathin MgO-coated TiO$_2$ nanoparticulate films. *Journal of Materials Chemistry A*, 3(17):9160–9164, 2015.

[186] Haopeng Dong, Xudong Guo, Wenzhe Li, and Liduo Wang. Cesium carbonate as a surface modification material for organic-inorganic hybrid perovskite solar cells with enhanced performance. *RSC Advances*, 4(104):60131–60134, 2014.

[187] Yuhei Ogomi, Kenji Kukihara, Shen Qing, Taro Toyoda, Kenji Yoshino, Shyam Pandey, Hisayo Momose, and Shuzi Hayase. Control of charge dynamics through a charge-separation interface for all-solid perovskite-sensitized solar cells. *ChemPhysChem*, 15(6):1062–1069, 2014.

[188] Sandeep K Pathak, A Abate, P Ruckdeschel, B Roose, Karl C Gödel, Yana Vaynzof, Aditya Santhala, Shun-Ichiro Watanabe, Derek J Hollman, Nakita Noel, Alessandro Sepe, Ullrich Wiesner, Richard Friend, Henry J Snaith, and Ullrich Steiner. Performance and stability enhancement of dye-sensitized and perovskite solar cells by Al doping of TiO$_2$. *Advanced Functional Materials*, 24(38):6046–6055, 2014.

[189] Yaoming Xiao, Gaoyi Han, Yanping Li, Miaoyu Li, and Jihuai Wu. Electrospun lead-doped titanium dioxide nanofibers and the in situ preparation of perovskite-sensitized photoanodes for use in high performance perovskite solar cells. *Journal of Materials Chemistry A*, 2(40):16856–16862, 2014.

[190] Peng Qin, Anna L Domanski, Aravind Kumar Chandiran, Rudiger Berger, Hans-Jurgen Butt, M Ibrahim Dar, Thomas Moehl, Nicolas Tetreault, Peng Gao, Shahzada Ahmad, Mohammad K Nazeeruddin, and Michael Gratzel. Yttrium-substituted nanocrystalline TiO$_2$ photoanodes for perovskite based heterojunction solar cells. *Nanoscale*, 6(3):1508–1514, 2014.

[191] Juan Dong, Yanhong Zhao, Jiangjian Shi, Huiyun Wei, Junyan Xiao, Xin Xu, Jianheng Luo, Jing Xu, Dongmei Li, Yanhong Luo, and Qingbo Meng. Impressive enhancement in the cell performance of ZnO nanorod-based perovskite solar cells with Al-doped ZnO interfacial modification. *Chemical Communications*, 50(87):13381–13384, 2014.

[192] Yuchuan Shao, Zhengguo Xiao, Cheng Bi, Yongbo Yuan, and Jinsong Huang. Origin and elimination of photocurrent hysteresis by fullerene passivation in CH$_3$NH$_3$PbI$_3$ planar heterojunction solar cells. *Nature Communications*, 5, 5784, 2014.

[193] Kuo-Chin Wang, Jun-Yuan Jeng, Po-Shen Shen, Yu-Cheng Chang, Eric Wei-Guang Diau, Cheng-Hung Tsai, Tzu-Yang Chao, Hsu-Cheng Hsu, Pei-Ying Lin, Peter Chen, Tzung-Fang Guo, and Ten-Chin Wen. p-type mesoscopic nickel oxide/organometallic perovskite heterojunction solar cells. *Scientific Reports*, 4(4756):1–8, 2014.

[194] Antonio Abate, Michael Saliba, Derek J Hollman, Samuel D Stranks, Konrad Wojciechowski, Roberto Avolio, Giulia Grancini, Annamaria Petrozza, and Henry J Snaith. Supramolecular halogen bond passivation of organic – inorganic halide perovskite solar cells. *Nano Letters*, 14(6):3247–3254, 2014.

[195] Nakita K Noel, Antonio Abate, Samuel D Stranks, Elizabeth S Parrott, Victor M Burlakov, Alain Goriely, and Henry J Snaith. Enhanced photoluminescence and solar cell performance via lewis base passivation of organic-inorganic lead halide perovskites. *ACS Nano*, 8(10):9815–9821, 2014.

[196] T Swetha and Surya Prakash Singh. Perovskite solar cells based on small molecules hole transporting materials. *Journal of Materials Chemistry A*, 3, 18329–18344, 2015.

[197] Rebecka Schölin, Martin H Karlsson, Susanna K Eriksson, Hans Siegbahn, Erik MJ Johansson, and Håkan Rensmo. Energy level shifts in spiro-ometad molecular thin films when adding Li-TFSI. *The Journal of Physical Chemistry C*, 116(50):26300–26305, 2012.

[198] Qi Wang, Cheng Bi, and Jinsong Huang. Doped hole transport layer for efficiency enhancement in planar heterojunction organolead trihalide perovskite solar cells. *Nano Energy*, 15(0):275–280, 2015.

[199] Zhongwei Wu, Sai Bai, Jian Xiang, Zhongcheng Yuan, Yingguo Yang, Wei Cui, Xingyu Gao, Zhuang Liu, Yizheng Jin, and Baoquan Sun. Efficient planar heterojunction perovskite solar cells employing graphene oxide as hole conductor. *Nanoscale*, 6(18):10505–10510, 2014.

[200] Wenzhe Li, Haopeng Dong, Xudong Guo, Nan Li, Jiangwei Li, Guangda Niu, and Liduo Wang. Graphene oxide as dual functional interface modifier for improving wettability and retarding recombination in hybrid perovskite solar cells. *Journal of Materials Chemistry A*, 2(47):20105–20111, 2014.

[201] Long Hu, Jun Peng, Weiwei Wang, Zhe Xia, Jianyu Yuan, Jialing Lu, Xiaodong Huang, Wanli Ma, Huaibing Song, Wei Chen, Yi-Bing Cheng, and Jiang Tang. Sequential deposition of $CH_3NH_3PbI_3$ on planar NiO film for efficient planar perovskite solar cells. *ACS Photonics*, 1(7):547–553, 2014.

[202] Zonglong Zhu, Yang Bai, Teng Zhang, Zhike Liu, Xia Long, Zhanhua Wei, Zilong Wang, Lixia Zhang, Jiannong Wang, Feng Yan, and Shihe Yang. High-performance hole-extraction layer of sol-gel-processed NiO nanocrystals for inverted planar perovskite solar cells. *Angewandte Chemie International Edition*, 53(46):12571–12575, 2014.

[203] Zhiliang Ku, Yaoguang Rong, Mi Xu, Tongfa Liu, and Hongwei Han. Full printable processed mesoscopic $CH_3NH_3PbI_3$/TiO_2 heterojunction solar cells with carbon counter electrode. *Scientific Reports*, 3, 2013.

[204] Anyi Mei, Xiong Li, Linfeng Liu, Zhiliang Ku, Tongfa Liu, Yaoguang Rong, Mi Xu, Min Hu, Jiangzhao Chen, Ying Yang, Michael Grätzel, and Hongwei Han. A hole-conductor-free, fully printable mesoscopic perovskite solar cell with high stability. *Science*, 345(6194):295–298, 2014.

[205] Zhanhua Wei, Keyou Yan, Haining Chen, Ya Yi, Teng Zhang, Xia Long, Jinkai Li, Lixia Zhang, Jiannong Wang, and Shihe Yang. Cost-efficient clamping solar cells using candle soot for hole extraction from ambipolar perovskites. *Energy & Environmental Science*, 7(10):3326–3333, 2014.

[206] Yueyong Yang, Junyan Xiao, Huiyun Wei, Lifeng Zhu, Dongmei Li, Yanhong Luo, Huijue Wu, and Qingbo Meng. An all-carbon counter electrode for highly efficient hole-conductor-free organo-metal perovskite solar cells. *RSC Advances*, 4(95):52825–52830, 2014.

[207] Fuguo Zhang, Xichuan Yang, Haoxin Wang, Ming Cheng, Jianghua Zhao, and Licheng Sun. Structure engineering of hole-conductor free perovskite-based solar cells with low-temperature-processed commercial carbon paste as cathode. *ACS Applied Materials & Interfaces*, 6(18):16140–16146, 2014.

[208] Huawei Zhou, Yantao Shi, Qingshun Dong, Hong Zhang, Yujin Xing, Kai Wang, Yi Du, and Tingli Ma. Hole-conductor-free, metal-electrode-free TiO_2/$CH_3NH_3PbI_3$ heterojunction solar cells based on a low-temperature carbon electrode. *The Journal of Physical Chemistry Letters*, 5(18):3241–3246, 2014.

[209] Zhanhua Wei, Haining Chen, Keyou Yan, and Shihe Yang. Inkjet printing and instant chemical transformation of a $CH_3NH_3PbI_3$/nanocarbon electrode and interface for planar perovskite solar cells. *Angewandte Chemie International Edition*, 53(48):13239–13243, 2014.

[210] Zhen Li, Sneha A Kulkarni, Pablo P Boix, Enzheng Shi, Anyuan Cao, Kunwu Fu, Sudip K Batabyal, Jun Zhang, Qihua Xiong, Lydia Helena Wong, Nripan Mathews, and Subodh G Mhaisalkar. Laminated carbon nanotube networks for metal electrode-free efficient perovskite solar cells. *ACS Nano*, 8(7):6797–6804, 2014.

[211] Huiyun Wei, Jiangjian Shi, Xin Xu, Junyan Xiao, Jianheng Luo, Juan Dong, Songtao Lv, Lifeng Zhu, Huijue Wu, Dongmei Li, Yanhong Luo, Qingbo Meng, and Qiang Chen. Enhanced charge collection with ultrathin AlO_x electron blocking layer for hole-transporting material-free perovskite solar cell. *Physical Chemistry Chemical Physics*, 17(7):4937–4944, 2015.

[212] Jiang-Jian Shi, Wan Dong, Yu-Zhuan Xu, Chun-Hui Li, Song-Tao Lv, Li-Feng Zhu, Juan Dong, Yan-Hong Luo, Dong-Mei Li, Qing-Bo Meng, and Qiang Chen. Enhanced performance in perovskite organic lead iodide heterojunction solar cells with metal-insulator-semiconductor back contact. *Chinese Physics Letters*, 30(12):128402, 2013.

[213] Guangda Niu, Wenzhe Li, Fanqi Meng, Liduo Wang, Haopeng Dong, and Yong Qiu. Study on the stability of $CH_3NH_3PbI_3$ films and the effect of post-modification by aluminum oxide in all-solid-state hybrid solar cells. *Journal of Materials Chemistry A*, 2(3):705–710, 2014.

[214] Severin N Habisreutinger, Tomas Leijtens, Giles E Eperon, Samuel D Stranks, Robin J Nicholas, and Henry J Snaith. Carbon nanotube/polymer composites as a highly stable hole collection layer in perovskite solar cells. *Nano Letters*, 14(10):5561–5568, 2014.

[215] Jarvist M Frost, Keith T Butler, and Aron Walsh. Molecular ferroelectric contributions to anomalous hysteresis in hybrid perovskite solar cells. *APL Materials*, 2(8):081506, 2014.

[216] Yasemin Kutes, Linghan Ye, Yuanyuan Zhou, Shuping Pang, Bryan D Huey, and Nitin P Padture. Direct observation of ferroelectric domains in solution-processed $CH_3NH_3PbI_3$ perovskite thin films. *The Journal of Physical Chemistry Letters*, 5(19):3335–3339, 2014.

[217] Shi Liu, Fan Zheng, Nathan Z Koocher, Hiroyuki Takenaka, Fenggong Wang, and Andrew M Rappe. Ferroelectric domain wall induced band gap reduction and charge separation in organometal halide perovskites. *The Journal of Physical Chemistry Letters*, 6(4):693–699, 2015.

[218] SY Yang, J Seidel, SJ Byrnes, P Shafer, CH Yang, MD Rossell, P Yu, YH Chu, JF Scott, JW Ager, LW Martin, and R Ramesh. Above-bandgap voltages from ferroelectric photovoltaic devices. *Nature Nanotechnology*, 5(2):143–147, 2010.

[219] Huanping Zhou, Qi Chen, Gang Li, Song Luo, Tze-bing Song, Hsin-Sheng Duan, Ziruo Hong, Jingbi You, Yongsheng Liu, and Yang Yang. Interface engineering of highly efficient perovskite solar cells. *Science*, 345(6196):542–546, 2014.

[220] Wenzhe Li, Haopeng Dong, Liduo Wang, Nan Li, Xudong Guo, Jiangwei Li, and Yong Qiu. Montmorillonite as bifunctional buffer layer material for hybrid perovskite solar cells with protection from corrosion and retarding recombination. *Journal of Materials Chemistry A*, 2(33):13587–13592, 2014.

[221] Guangda Niu, Xudong Guo, and Liduo Wang. Review of recent progress in chemical stability of perovskite solar cells. *Journal of Materials Chemistry A*, 3(17):8970–8980, 2015.

[222] HY Fan. Infra-red absorption in semiconductors. *Reports on Progress in Physics*, 19(1):107, 1956.

[223] H Kim, CM Gilmore, A Pique, JS Horwitz, H Mattoussi, H Murata, ZH Kafafi, and DB Chrisey. Electrical, optical, and structural properties of indium–tin–oxide thin films for organic light-emitting devices. *Journal of Applied Physics*, 86(11):6451–6461, 1999.

[224] Leif AA Pettersson, Lucimara S Roman, and Olle Inganas. Modeling photocurrent action spectra of photovoltaic devices based on organic thin films. *Journal of Applied Physics*, 86(1):487–496, 1999.

[225] George F Burkhard, Eric T Hoke, and Michael D McGehee. Accounting for interference, scattering, and electrode absorption to make accurate internal quantum efficiency measurements in organic and other thin solar cells. *Advanced Materials*, 22(30):3293–3297, 2010.

[226] James M Ball, Michael M Lee, Andrew Hey, and Henry J Snaith. Low-temperature processed meso-superstructured to thin-film perovskite solar cells. *Energy & Environmental Science*, 6(6):1739–1743, 2013.

[227] William J Potscavage. Comment on high-efficiency panchromatic hybrid schottky solar cells. *Advanced Materials*, 25(35):4825–4825, 2013.

[228] Henry J Snaith. How should you measure your excitonic solar cells? *Energy & Environmental Science*, 5(4):6513–6520, 2012.

[229] Katsuhiko Takagi, Shinichi Magaino, Hidenori Saito, Tomoko Aoki, and Daisuke Aoki. Measurements and evaluation of dye-sensitized solar cell performance. *Journal of Photochemistry and Photobiology C: Photochemistry Reviews*, 14:1–12, 2013.

[230] Henry J Snaith, Robin Humphry-Baker, Peter Chen, Ilkay Cesar, Shaik M Zakeeruddin, and Michael Grätzel. Charge collection and pore filling in solid-state dye-sensitized solar cells. *Nanotechnology*, 19(42):424003, 2008.

[231] Eugen Zimmermann, Philipp Ehrenreich, Thomas Pfadler, James A Dorman, Jonas Weickert, and Lukas Schmidt-Mende. Erroneous efficiency reports harm organic solar cell research. *Nature Photonics*, 8(9):669–672, 2014.

[232] RA Street, M Schoendorf, A Roy, and JH Lee. Interface state recombination in organic solar cells. *Physical Review B*, 81(20):205307, 2010.

[233] Christoph J Brabec, Gerald Zerza, Giulio Cerullo, Sandro De Silvestri, Silvia Luzzati, Jan C Hummelen, and Serdar Sariciftci. Tracing photoinduced electron transfer process in conjugated polymer/fullerene bulk heterojunctions in real time. *Chemical Physics Letters*, 340(3):232–236, 2001.

[234] L Jan Anton Koster, VD Mihailetchi, H Xie, and Paul WM Blom. Origin of the light intensity dependence of the short-circuit current of polymer/fullerene solar cells. *Applied Physics Letters*, 87(20):203502, 2005.

[235] Paul Langevin. The recombination and mobilities of ions in gases. *Annales de Chimie et de Physique*, 28:433–530, 1903.

[236] L Jan Anton Koster, VD Mihailetchi, and Paul WM Blom. Ultimate efficiency of polymer/fullerene bulk heterojunction solar cells. *Applied Physics Letters*, 88(9):093511–093511, 2006.

[237] D Chirvase, Z Chiguvare, Martin Knipper, Jürgen Parisi, Vladimir Dyakonov, and Jan C Hummelen. Temperature dependent characteristics of poly (3 hexylthiophene)-fullerene based heterojunction organic solar cells. *Journal of Applied Physics*, 93(6):3376–3383, 2003.

[238] L Jan Anton Koster, ECP, Smits, VD Mihailetchi, and Paul WM Blom. Device model for the operation of polymer/fullerene bulk heterojunction solar cells. *Physical Review B*, 72(8):085205, 2005.

[239] William Shockley and WT Read Jr. Statistics of the recombinations of holes and electrons. *Physical Review*, 87(5):835, 1952.

[240] RN Hall. Recombination processes in semiconductors. *Proceedings of the IEE-Part B: Electronic and Communication Engineering*, 106(17S):923–931, 1959.

[241] M Magdalena Mandoc, FB Kooistra, Jan C Hummelen, Bert De Boer, and Paul WM Blom. Effect of traps on the performance of bulk heterojunction organic solar cells. *Applied Physics Letters*, 91(26):263505, 2007.

[242] M Magdalena Mandoc, Welmoed Veurman, L Jan Anton Koster, Bert de Boer, and Paul WM Blom. Origin of the reduced fill factor and photocurrent in MDMO-PPV:PCNEPV all-polymer solar cells. *Advanced Functional Materials*, 17(13):2167–2173, 2007.

[243] Pavel Schilinsky, Christoph Waldauf, and Christoph J Brabec. Recombination and loss analysis in polythiophene based bulk heterojunction photodetectors. *Applied Physics Letters*, 81(20):3885–3887, 2002.

[244] Ingo Riedel, Jürgen Parisi, Vladimir Dyakonov, Laurence Lutsen, Dirk Vanderzande, and Jan C Hummelen. Effect of temperature and illumination on the electrical characteristics of polymer – fullerene bulk-heterojunction solar cells. *Advanced Functional Materials*, 14(1):38–44, 2004.

[245] Sarah R Cowan, Anshuman Roy, and Alan J Heeger. Recombination in polymer-fullerene bulk heterojunction solar cells. *Physical Review B*, 82(24):245207, 2010.

[246] Brian C O'Regan, S Scully, AC Mayer, Emiliio Palomares, and James Durrant. The effect of Al_2O_3 barrier layers in TiO_2/dye/CuSCN photovoltaic cells explored by recombination and dos characterization using transient photovoltage measurements. *The Journal of Physical Chemistry B*, 109(10):4616–4623, 2005.

[247] Alison B Walker, Laurence M Peter, K Lobato, and PJ Cameron. Analysis of photovoltage decay transients in dye-sensitized solar cells. *The Journal of Physical Chemistry B*, 110(50):25504–25507, 2006.

[248] SC Jain and UC Ray. Photovoltage decay in p-n junction solar cells including the effects of recombinations in the emitter. *Journal of Applied Physics*, 54(4):2079–2085, 1983.

[249] MK Madan and VK Tewary. Measurements of the open-circuit photovoltage decay in a silicon solar cell. *Solar Cells*, 9(4):289–293, 1983.

[250] CB Shuttle, Brian O'Regan, AM Ballantyne, Jenny Nelson, Donal DC Bradley, John De Mello, and James R Durrant. Experimental determination of the rate law for charge carrier decay in a polythiophene: Fullerene solar cell. *Applied Physics Letters*, 92(9):3311, 2008.

[251] Gilles Dennler, Attila J Mozer, G Juška, Almantas Pivrikas, Ronald Österbacka, A Fuchsbauer, and Niyazi Serdar Sariciftci. Charge carrier mobility and lifetime versus composition of conjugated polymer/fullerene bulk-heterojunction solar cells. *Organic Electronics*, 7(4):229–234, 2006.

[252] Carsten Deibel, Andreas Baumann, and Vladimir Dyakonov. Polaron recombination in pristine and annealed bulk heterojunction solar cells. *Applied Physics Letters*, 93(16):163303, 2008.

[253] Attila J Mozer, Gilles Dennler, Niyazi Serdar Sariciftci, M Westerling, Almantas Pivrikas, Ronald Österbacka, and G Juška. Time-dependent mobility and recombination of the photoinduced charge carriers in conjugated polymer/fullerene bulk heterojunction solar cells. *Physical Review B*, 72(3):035217, 2005.

[254] Attila J Mozer, NS Sariciftci, Laurence Lutsen, Dirk Vanderzande, R Österbacka, M Westerling, and G Juška. Charge transport and recombination in bulk heterojunction solar cells studied by the photoinduced charge extraction in linearly increasing voltage technique. *Applied Physics Letters*, 86(11):112104, 2005.

[255] Juan Bisquert and F Fabregat-Santiago. Impedance spectroscopy: A general introduction and application to dye-sensitized solar cells. *Dye-Sensitized Solar Cells*, 457, 2010.

[256] Evgenij Barsoukov and J Ross Macdonald. *Impedance spectroscopy: Theory, experiment, and applications*. John Wiley & Sons, Inc., Hoboken, New Jersey. 2005.

[257] Francisco Fabregat-Santiago, Juan Bisquert, Germà Garcia-Belmonte, Gerrit Boschloo, and Anders Hagfeldt. Influence of electrolyte in transport and recombination in dye-sensitized solar cells studied by impedance spectroscopy. *Solar Energy Materials and Solar Cells*, 87(1):117–131, 2005.

[258] Francisco Fabregat-Santiago, Germà Garcia-Belmonte, Iván Mora-Seró, and Juan Bisquert. Characterization of nanostructured hybrid and organic solar cells by impedance spectroscopy. *Physical Chemistry Chemical Physics*, 13(20):9083–9118, 2011.

[259] Jing-Shun Huang, Tenghooi Goh, Xiaokai Li, Matthew Y Sfeir, Elizabeth A Bielinski, Stephanie Tomasulo, Minjoo L Lee, Nilay Hazari, and André D Taylor. Polymer bulk heterojunction solar cells employing forster resonance energy transfer. *Nature Photonics*, 7(6):479–485, 2013.

[260] Frederik C Krebs. Polymer solar cell modules prepared using roll-to-roll methods: Knife-over-edge coating, slot-die coating and screen printing. *Solar Energy Materials and Solar Cells*, 93(4):465–475, 2009.

[261] Roar Søndergaard, Markus Hösel, Dechan Angmo, Thue T Larsen-Olsen, and Frederik C Krebs. Roll-to-roll fabrication of polymer solar cells. *Materials Today*, 15(1):36–49, 2012.
[262] Frederik C Krebs, Nieves Espinosa, Markus Hösel, Roar R Søndergaard, and Mikkel Jørgensen. 25th anniversary article: Rise to power – OPV-based solar parks. *Advanced Materials*, 26(1):29–39, 2014.
[263] Azhar Fakharuddin, Rajan Jose, Thomas M Brown, Francisco Fabregat-Santiago, and Juan Bisquert. A perspective on the production of dye-sensitized solar modules. *Energy & Environmental Science*, 7(12):3952–3981, 2014.
[264] Frederik C Krebs (Editor). Stability and degradation of organic and polymer solar cells. *John Wiley & Sons, West Sussex, UK* 2012.
[265] Christoph Brabec, Ullrich Scherf, and Vladimir Dyakonov. *Organic photovoltaics: Materials, device physics, and manufacturing technologies*. 2nd edition, Wiley-VCH, Weinheim, 2014.
[266] Barry P Rand and Henning Richter (Editors). *Organic solar cells: Fundamentals, devices, and upscaling*. Pan Stanford Publishing, Singapore, 2014.
[267] David S Ginley, Hideo Hosono, David C Paine, et al. *Handbook of transparent conductors*. Springer, New York, 2010.
[268] John Scheirs and Jean-Luc Gardette. Photo-oxidation and photolysis of poly(ethylene naphthalate). *Polymer Degradation and Stability*, 56(3):339–350, 1997.
[269] Hans Zweifel, (Editor). *Plastic additives handbook*. Hanser Fachbuch, Munich, 2001.
[270] G Haacke. New figure of merit for transparent conductors. *Journal of Applied Physics*, 47(9):4086–4089, 1976.
[271] Cecilia Guillén and J Herrero. TCO/metal/TCO structures for energy and flexible electronics. *Thin Solid Films*, 520(1):1–17, 2011.
[272] Michael Layani, Alexander Kamyshny, and Shlomo Magdassi. Transparent conductors composed of nanomaterials. *Nanoscale*, 6(11):5581–5591, 2014.
[273] Robert M Pasquarelli, David S Ginley, and Ryan O'Hayre. Solution processing of transparent conductors: From flask to film. *Chemical Society Reviews*, 40(11):5406–5441, 2011.
[274] Eli Yablonovitch and George D Cody. Intensity enhancement in textured optical sheets for solar cells. *Electron Devices*, IEEE Transactions on, 29(2):300–305, 1982.
[275] Dechan Angmo and Frederik C Krebs. Flexible ITO-free polymer solar cells. *Journal of Applied Polymer Science*, 129(1):1–14, 2013.
[276] Ignasi Burgués-Ceballos, Florian Machui, Jie Min, Tayebeh Ameri, Monika M Voigt, Yuriy N Luponosov, Sergei A Ponomarenko, Paul D Lacharmoise, Mariano Campoy-Quiles, and Christoph J Brabec. Solubility based identification of green solvents for small molecule organic solar cells. *Advanced Functional Materials*, 24(10):1449–1457, 2014.
[277] Michel Schaer, Frank Nüesch, Detlef Berner, William Leo, and Libero Zuppiroli. Water vapor and oxygen degradation mechanisms in organic light emitting diodes. *Advanced Functional Materials*, 11(2):116–121, 2001.
[278] S Cros, R De Bettignies, S Berson, S Bailly, P Maisse, N Lemaitre, and S Guillerez. Definition of encapsulation barrier requirements: A method applied to organic solar cells. *Solar Energy Materials and Solar Cells*, 95:S65–S69, 2011.
[279] Jens A Hauch, Pavel Schilinsky, Stelios A Choulis, Sambatra Rajoelson, and Christoph J Brabec. The impact of water vapor transmission rate on the lifetime of flexible polymer solar cells. *Applied Physics Letters*, 93(10):103306, 2008.
[280] Sabine Amberg-Schwab, Ulrike Weber, Annette Burger, Somchith Nique, and Rainer Xalter. Development of passive and active barrier coatings on the basis of inorganic–organic polymers. *Monatshefte für Chemie/Chemical Monthly*, 137(5):657–666, 2006.

[281] Markus Hanika, H-C Langowski, Ulrich Moosheimer, and Wolfgang Peukert. Inorganic layers on polymeric films – influence of defects and morphology on barrier properties. *Chemical Engineering & Technology*, 26(5):605–614, 2003.

[282] Jakaria Ahmad, Kateryna Bazaka, Liam J Anderson, Ronald D White, and Mohan V Jacob. Materials and methods for encapsulation of OPV: A review. *Renewable and Sustainable Energy Reviews*, 27:104–117, 2013.

[283] Jau M Ding, Alejandro de la Fuente Vornbrock, Ching Ting, and Vivek Subramanian. Patternable polymer bulk heterojunction photovoltaic cells on plastic by rotogravure printing. *Solar Energy Materials and Solar Cells*, 93(4):459–464, 2009.

[284] P Kopola, T Aernouts, S Guillerez, H Jin, M Tuomikoski, A Maaninen, and J Hast. High efficient plastic solar cells fabricated with a high-throughput gravure printing method. *Solar Energy Materials and Solar Cells*, 94(10):1673–1680, 2010.

[285] Monika M Voigt, Roderick CI Mackenzie, Chin P Yau, Pedro Atienzar, Justin Dane, Panagiotis E Keivanidis, Donal DC Bradley, and Jenny Nelson. Gravure printing for three subsequent solar cell layers of inverted structures on flexible substrates. *Solar Energy Materials and Solar Cells*, 95(2):731–734, 2011.

[286] Arved Hübler, Bystrik Trnovec, Tino Zillger, Moazzam Ali, Nora Wetzold, Markus Mingebach, Alexander Wagenpfahl, Carsten Deibel, and Vladimir Dyakonov. Printed paper photovoltaic cells. *Advanced Energy Materials*, 1(6):1018–1022, 2011.

[287] Monika M Voigt, Roderick CI Mackenzie, Simon P King, Chin P Yau, Pedro Atienzar, Justin Dane, Panagiotis E Keivanidis, Ivan Zadrazil, Donal DC Bradley, and Jenny Nelson. Gravure printing inverted organic solar cells: The influence of ink properties on film quality and device performance. *Solar Energy Materials and Solar Cells*, 105:77–85, 2012.

[288] A Schneider, N Traut, and M Hamburger. Analysis and optimization of relevant parameters of blade coating and gravure printing processes for the fabrication of highly efficient organic solar cells. *Solar Energy Materials and Solar Cells*, 126:149–154, 2014.

[289] Junliang Yang, Doojin Vak, Noel Clark, Jegadesan Subbiah, Wallace WH Wong, David J Jones, Scott E Watkins, and Gerry Wilson. Organic photovoltaic modules fabricated by an industrial gravure printing proofer. *Solar Energy Materials and Solar Cells*, 109:47–55, 2013.

[290] Frederik C Krebs, Jan Fyenbo, and Mikkel Jørgensen. Product integration of compact roll-to-roll processed polymer solar cell modules: Methods and manufacture using flexographic printing, slot-die coating and rotary screen printing. *Journal of Materials Chemistry*, 20(41):8994–9001, 2010.

[291] D Deganello, JA Cherry, DT Gethin, and TC Claypole. Patterning of micro-scale conductive networks using reel-to-reel flexographic printing. *Thin Solid Films*, 518(21):6113–6116, 2010.

[292] Jong-Su Yu, Inyoung Kim, Jung-Su Kim, Jeongdai Jo, Thue T Larsen-Olsen, Roar R Søndergaard, Markus Hösel, Dechan Angmo, Mikkel Jørgensen, and Frederik C Krebs. Silver front electrode grids for ITO-free all printed polymer solar cells with embedded and raised topographies, prepared by thermal imprint, flexographic and inkjet roll-to-roll processes. *Nanoscale*, 4(19):6032–6040, 2012.

[293] Frederik C Krebs, Thomas Tromholt, and Mikkel Jørgensen. Upscaling of polymer solar cell fabrication using full roll-to-roll processing. *Nanoscale*, 2(6):873–886, 2010.

[294] Yulia Galagan, Jan-Eric JM Rubingh, Ronn Andriessen, Chia-Chen Fan, Paul WM Blom, Sjoerd C Veenstra, and Jan M Kroon. ITO-free flexible organic solar cells with printed current collecting grids. *Solar Energy Materials and Solar Cells*, 95(5):1339–1343, 2011.

[295] Matthieu Manceau, Dechan Angmo, Mikkel Jørgensen, and Frederik C Krebs. ITO-free flexible polymer solar cells: From small model devices to roll-to-roll processed large modules. *Organic Electronics*, 12(4):566–574, 2011.

[296] Bing Zhang, Heeyeop Chae, and Sung Min Cho. Screen-printed polymer: Fullerene bulk-heterojunction solar cells. *Japanese Journal of Applied Physics*, 48(2R):020208, 2009.

[297] Frederik C Krebs. Fabrication and processing of polymer solar cells: A review of printing and coating techniques. *Solar Energy Materials and Solar Cells*, 93(4):394–412, 2009.

[298] Andreas Kay and Michael Grätzel. Low cost photovoltaic modules based on dye sensitized nanocrystalline titanium dioxide and carbon powder. *Solar Energy Materials and Solar Cells*, 44(1):99–117, 1996.

[299] Tatsuo Toyoda, Toshiyuki Sano, Jyunji Nakajima, Syouichi Doi, Syungo Fukumoto, Atsushi Ito, Tomoyuki Tohyama, Motoharu Yoshida, Tetsuo Kanagawa, Tomoyoshi Motohiro, et al. Outdoor performance of large scale DSC modules. *Journal of Photochemistry and Photobiology A: Chemistry*, 164(1):203–207, 2004.

[300] R Sastrawan, J Beier, U Belledin, S Hemming, A Hinsch, R Kern, C Vetter, FM Petrat, A Prodi-Schwab, P Lechner, et al. A glass frit-sealed dye solar cell module with integrated series connections. *Solar Energy Materials and Solar Cells*, 90(11):1680–1691, 2006.

[301] Yongseok Jun, Jung-Ho Son, Dongwook Sohn, and Man Gu Kang. A module of a TiO_2 nanocrystalline dye-sensitized solar cell with effective dimensions. *Journal of Photochemistry and Photobiology A: Chemistry*, 200:314–317, 2008.

[302] Fabrizio Giordano, Andrea Guidobaldi, Eleonora Petrolati, Luigi Vesce, Riccardo Riccitelli, Andrea Reale, Thomas M Brown, and Aldo Di Carlo. Realization of high performance large area Z-series-interconnected opaque dye solar cell modules. *Progress in Photovoltaics: Research and Applications*, 21(8):1653–1658, 2013.

[303] Liyuan Han, Atsushi Fukui, Yasuo Chiba, Ashraful Islam, Ryoichi Komiya, Nobuhiro Fuke, Naoki Koide, Ryohsuke Yamanaka, and Masafumi Shimizu. Integrated dye-sensitized solar cell module with conversion efficiency of 8.2%. *Applied Physics Letters*, 94(1):013305, 2009.

[304] Fabrizio Giordano, Eleonora Petrolati, Thomas M Brown, Andrea Reale, and Aldo Di Carlo. Series-connection designs for dye solar cell modules. *Electron Devices, IEEE Transactions on*, 58(8):2759–2764, 2011.

[305] Lei Wang, Xiaoming Fang, and Zhengguo Zhang. Design methods for large scale dye-sensitized solar modules and the progress of stability research. *Renewable and Sustainable Energy Reviews*, 14(9):3178–3184, 2010.

[306] Songyuan Dai, Jian Weng, Yifeng Sui, Chengwu Shi, Yang Huang, Shuanhong Chen, Xu Pan, Xiaqin Fang, Linhua Hu, Fantai Kong, et al. Dye-sensitized solar cells, from cell to module. *Solar Energy Materials and Solar Cells*, 84(1):125–133, 2004.

[307] Kenichi Okada, Hiroshi Matsui, Takuya Kawashima, Tetsuya Ezure, and Nobuo Tanabe. 100 mm × 100 mm large-sized dye sensitized solar cells. *Journal of Photochemistry and Photobiology A: Chemistry*, 164(1):193–198, 2004.

[308] Tzu-Chien Wei, Jo-Lin Lan, Chi-Chao Wan, Wen-Chi Hsu, and Ya-Huei Chang. Fabrication of grid type dye sensitized solar modules with 7% conversion efficiency by utilizing commercially available materials. *Progress in Photovoltaics: Research and Applications*, 21(8):1625–1633, 2013.

[309] Jan M Kroon, NJ Bakker, HJP Smit, Paul Liska, KR Thampi, Peng Wang, Shaik M Zakeeruddin, Michael Grätzel, Andreas Hinsch, S Hore, et al. Nanocrystalline dye-sensitized solar cells having maximum performance. *Progress in Photovoltaics: Research and Applications*, 15(1):1–18, 2007.

[310] Songyuan Dai, Jian Weng, Yifeng Sui, Shuanghong Chen, Shangfeng Xiao, Yang Huang, Fantai Kong, Xu Pan, Linhua Hu, Changneng Zhang, et al. The design and outdoor application of dye-sensitized solar cells. *Inorganica Chimica Acta*, 361(3):786–791, 2008.

[311] Mikkel Jørgensen, Kion Norrman, Suren A Gevorgyan, Thomas Tromholt, Birgitta Andreasen, and Frederik C Krebs. Stability of polymer solar cells. *Advanced Materials*, 24(5):580–612, 2012.

[312] Ivan Litzov and Christoph J Brabec. Development of efficient and stable inverted bulk heterojunction (BHJ) solar cells using different metal oxide interfaces. *Materials*, 6(12):5796–5820, 2013.

[313] Matthew O Reese, Suren A Gevorgyan, Mikkel Jørgensen, Eva Bundgaard, Sarah R Kurtz, David S Ginley, Dana C Olson, Matthew T Lloyd, Pasquale Morvillo, Eugene A Katz, et al. Consensus stability testing protocols for organic photovoltaic materials and devices. *Solar Energy Materials and Solar Cells*, 95(5):1253–1267, 2011.

Index

β-carotene 38
π-π stacking 138
π-electrons 34

Absorbance 104, 194
Absorber nanoparticles 163
Absorption 47, 56, 104, 194
Absorption spectroscopy 193
− Haze measurement 211
− Total absorption 208
− Transfer matrix simulations 210
Active area 212
− Current density 214
− Masking 214
AM 1.5G 97
Anderson's rule 89
Annealing 145
Auger recombination 222

Bässler model 75
Ball-and-stick model 30
Band bending 82
Band structure 6, 10
− Conduction band 7
− Valence band 7
− Bandgap 7
Bandgap 7, 17
− Direct bandgap 17
− Indirect bandgap 17
Base 115
$BaSO_4$ 208
Beer–Lambert's law 194, 199
Bias-dependent charge extraction 128
Bilayer OPV 130
Bimolecular recombination 223
Black body radiation 99
Bloch wave 14
Born-Oppenheimer approximation 49
Brillouin zone 12
Built-in voltage 83
Bulk heterojunction 132
− Annealing 145
− Demixing 144
− External quantum efficiency 135
− Glass transition 146
− Internal quantum efficiency 134

− Metal oxide–polymer 159
− Phase separation 134, 144
− Polymer–fullerene mixing 135

C_{60}
− Crystallization 141
Carbon atom
− Carbon orbitals 26
− Hybridization 26
− single/double/triple bond 28
Carboxylic acid 152
CdSe 164
CdTe 164
Characteristics of solar cells 106
Charge diffusion 79
Charge extraction probability 225
Charge generation
− Charge separated state 119
− Charge transfer state 118
− Excess energy 128
− Exciton splitting 117
− Organic solar cell 117
Charge hopping 67
Charge separated state 119
Charge separation probability 223
Charge transfer exciton 62
Charge transfer state 118
− CTS_0 124
− CTS_n 124
− Electroluminescence 122
− Energetics 124
− Excess energy 128
− Hot charge transfer state 125
− Pump-push spectroscopy 125
Charge transport 67
− Bässler model 75
− Charge transport Hamiltonian 68
− Continuity equation 78
− Diffusion 78
− Drift 78
− Electrical mobility equation 79
− Gaussian disorder model 74
− Macroscopic parameters 78
− Miller-Abrahams hopping 72
Charge transport Hamiltonian 68
Chemical structure 30

– Mesomerism 33
– Polymer 32
Complex refractive index 198
Condon approximation 55
Conduction band 7
Conjugation 34
Continuity equation 78
Crystal 11
Crystal structure
– Bloch wave 14
– Brillouin zone 12
– Lattice vector 11
$CuInS_2$ 164
Current density 214
Current density–voltage measurement 179, 212
Current through p–n junction 85

Davydov splitting 60
Density of states 19
Depletion approximation 86
Depletion region 82, 84
Detailed balance limit 101
Dexter energy transfer 64
Diffusion coefficient 79
Dipole operator 54
Doping 22
– n-type doping 23
– Organic semiconductor 79
– p-type doping 25
Double bond 28
Drift-diffusion equation 78
Dye bath 152
Dye-sensitized solar cell 151
– Assembly 151
– Dye bath 152
– Dye-sensitized solar cells 5
– Liquid electrolyte 151, 153
– Nanoparticle film 151
– Quantum dot sensitization 153
– Redox couple 155
– Solid state DSC 156
– Time scales 154

Efficiency 112
Electrical mobility equation 79
Electroluminescence 122, 197
Electron 8
Electron current 78

Electron density 21
Electron–hole pair 118
Electrostatic potential 85
Ellipsometry 198
Emission 47, 56
Emission spectroscopy 195
– Electroluminescence 197
– Photoluminescence 195
Emitter 115
Energy band 6, 10
Energy transfer 62
EQE 104
EQE measurement 218
– Internal quantum efficiency 218
– White light-bias 220
Exciton 60, 117
– Binding energy 119
– Charge transfer exciton 62
– Dexter energy transfer 64
– Diffusion equation 65
– Energy transfer 62
– Exciton diffusion length 66, 131
– Exciton diffusion 60, 65
– Exciton splitting 117
– Frenkel exciton 62
– FRET 62
– Wannier exciton 62
Exciton diffusion 65
Extended Hückel theory 44
External quantum efficiency 104, 135, 218
Extinction coefficient 198
Extremely thin absorber 163–164

Förster resonance energy transfer (FRET) 62
– Förster radius 63
– Overlap integral 63
Fermi energy 20
Fermi level 20
– Temperature dependence 20
Fermi–Dirac statistics 20
FF 112
Fill factor 112
Flat heterojunction OPV 130
Four-point probe 204
Franck–Condon factor 55
Franck–Condon principle 52
– Absorption 56
– Condon approximation 55
– Emission 56

– Franck–Condon factor 55
Frenkel exciton 62
Fresnel equations 199
Fullerene
– Crystallization 141

Gaussian disorder model 72
Geminate recombination 124, 222
Generalized Shockley equation 111
Generalized Shockley model 106, 108, 111
Generation rate 222
Global irradiance 98

H-/J-aggregate 60
H-aggregate 60
Hückel theory 44
Haacke's figure of merit 206
Haze measurement 211
Heterojunction 89, 116
– Solar cell 116
HJ-aggregate model 140
Hole 8
Hole current 78
Hole density 21
Hole transport material 158
HOMO 39
Hybrid heterojunction 95
Hybrid junction solar cell 150
Hybrid solar cell 150, 158
– Charge separation 161
– Dye-sensitization 160
– Extremely thin absorber 164
– Inorganic absorber 163
– Metal oxide–polymer 159
– Pump-push spectroscopy 163
– ZnO-P3HT 160
Hybridization 26
Hysteresis 217

Ideality factor 88, 108
Inorganic absorber hybrid solar cell 163
Inorganic heterojunction 89, 116
– Anderson's rule 90
– Solar cell 116
Inorganic junction solar cell 113
Inorganic–organic PV 158
Insulator 8
Integrating sphere 208
Internal quantum efficiency 104, 134, 218

Intersystem crossing 48
Intrinsic charge density 24, 223
Inverted Marcus region 72
Iodine electrolyte 155
IPCE 104
IQE 104

J-aggregate 60
J–V measurement 179, 212
– Hysteresis 217
– Light intensity dependence 221
– Masking 214
– Sweep time 217
J_{SC} 112
– Light intensity dependence 225
Jablonski diagram 47
Junction 81
– Anderson's rule 90
– Inorganic heterojunction 89
– Junction types 89
– Organic–inorganic junction 95
– Organic–organic junction 93
– p–n homojunction 81
– Schottky Junction 92

Kasha's rule 48

Langevin recombination 223
Lattice vector 11
Lewis structure 30
Light intensity–dependent J–V measurement 221
– Charge extraction probability 225
– J_{SC} 225
– V_{OC} 222
Light polarization 199
Linear combination of atomic orbitals 43
Liquid electrolyte 153
LUMO 39

Marcus theory 71
– Inverted Marcus region 72
– Reorganization energy 72
MEH-PPV 128
Mesomerism 33
Mesoporous TiO_2 151
Metal 8
Metal complex dye 36
Metal oxide nanoparticles 159

Metal oxide–dye–polymer PV 160
Metal oxide–polymer PV 158–159
Miller-Abrahams hopping 72
Mixed films 143
Molecular orbital 37
– Extended Hückel theory 44
– HOMO 39
– Linear combination of atomic orbitals 43
– LUMO 39
– Molecular orbital theories 42
– Optical transition 45
– Particle in a box approximation 37
– Perimeter free electron orbital theory 41
Molecular orbital Hückel theory 44
Molecular orbital theory 42
Monomolecular recombination 224

n-type semiconductor 22–23
Nanoporous alumina 139
Non-geminate recombination 124

Onsager–Braun model 120
– Dissociation rate 122
– Onsager radius 121
Open-circuit voltage 112
Optical cavity 209
Optical density 104
Optical transition
– 0-0/0-1/... transition 55
– Aggregates 58
– Born–Oppenheimer approximation 49
– Franck–Condon factor 55
– Franck–Condon principle 52
– Physical dimer 59
– Singlet/Triplet 48
– Transition dipole moment 54–55
– Transition probability 55
Optical transitions 45
– Jablonski diagram 47
– Kasha's rule 48
OPV 117
– Bulk heterojunction 132
– Donor–acceptor heterojunction 130
– External quantum efficiency 135
– Flat heterojunction 130
– Internal quantum efficiency 134
– Mixed films 143
Organic heterojunction 93

Organic semiconductor 26, 34
– Absorption 47
– Charge hopping 67
– Charge transport 67
– Conjugation 34
– Doping 79
– Electronic states 34
– Emission 47
– Molecular orbital 37
– Optical transitions 45
– Organic heterojunction 93
– Polyacetylene 35
– Polythiophene 35
Organic solar cell 117
– Charge generation 117
– Charge transfer state 118
– Exciton splitting 117
– Polymer-fullerene 120
Organic–Inorganic Heterojunction 95

p–i–n structure 117
p–n junction 81, 114
– Band bending 82, 115
– Built-in voltage 83
– Charge generation 115
– Current 85
– Depletion approximation 86
– Depletion region 82–84
– Electric field 84
– Electron and hole current 86
– Electrostatic potential 85
– p–n junction solar cell 114
– Poisson equation 84
– Quasi-Fermi level 116
– Shockley equation 88
– Space charge region 82
p–n junction solar cell 114
p-type semiconductor 25
P3HT
– Crystallization 136, 139
– HJ-aggregate model 140
– Regioregularity 137
– Spherulites 140
P_{max} 108
Parasitic resistances 108, 112
Particle in a box approximation 37
PBDTTPD 128
PbS 164

PCBM 123
PCE 112
PEDOT:PSS 81
– HOMO/LUMO levels 44
Perimeter free electron orbital theory 41
Perovskite solar cell 165
Phase separation 145
Phase velocity 199
Photocurrent 89, 102
Photoluminescence spectroscopy 195
– Emission measurements 197
– Excitation measurements 197
Physical dimer 59
Physical oligomer 59
Planck's law 99
Point lattice 11
Poisson equation 84
Polarization 199
Polaron 8
Polaron pair 119
Polyacetylene 35
Polymer
– Structural Formula 32
Polymer crystallization 136
– Extended chains 139
– Spherulites 140
Polymer-fullerene 120
– Structural Formula 32
Polymer–fullerene co-crystal 149
Polymer–fullerene mixing 135
Polythiophene 35
Power conversion efficiency 99, 112
Power law 225
PPV 123
Predicted J_{SC} 219
Pump-push spectroscopy 125
– Differential photocurrent 127

Quantum efficiency 134
Quasi-Fermi level 116

R_{sh} 112
R_s 112
Radiative recombination 87
Recombination
– Geminate recombination 124
– Non-geminate recombination 134
Recombination mechanisms 222
– Langevin recombination 223
Recombination rate 224

Redox couple 155
Refractive index 198
Regioregularity 137
Resistivity 203
Reverse saturation current 88
Rotating analyzer 201
Rotational state 53
Rubrene 132
Rylene dye 46

Sb_2 165
Schottky Junction 92
Self-assembled monolayer 152
Semiconductor 8
– Band structure 10
– Charge density 26
– Density of states 19
– Direct semiconductor 17
– Doping 22
– Fermi–Dirac statistics 20
– Indirect semiconductor 17
– Intrinsic charge density 24
– Intrinsic Semiconductor 22
– n-type semiconductor 23
– p-type semiconductor 25
Semiconductor Junctions 184
Series resistance 108, 112
Shadow mask 212, 214
Sheet resistance 202
Shockley equation 88
Shockley model 106
Shockley–Queisser limit 101
Shockley-Read-Hall recombination
    87, 224
Short circuit current 112
Shunt resistance 108, 112
Single bond 28
Singlet 48
Solar cell
– Characteristics 106
– Charge generation 115
– Efficiency 112
– Equivalent circuit 106
– External quantum efficiency 104
– Generalized Shockley model 111
– Heterojunction solar cell 116
– Inorganic junction 113
– Internal quantum efficiency 104

- Open-circuit voltage 115
- Optical density 104
- Organic solar cell 117
- Parasitic resistances 108, 112
- Photocurrent 102
- p–i–n structure 117
- p–n junction solar cell 114
- Power conversion efficiency 165
- Quasi-Fermi level 116
- Shockley–Queisser limit 101
- Short circuit current 112
- Spectral response 102
Solar cell geometry 213
Solar energy conversion 97
Solar simulator 212
Solar spectrum 97
Solid state DSC 156
sp/sp²/sp³-hybridization 26
Space charge region 82
Spectral response 102
Sperulites 140
Spiro-OMeTAD 157
Standing electromagnetic wave 209
Structural formula 30

TCO 202
- Haacke's figure of merit 206
- Sheet resistance 202
- Transmission 204
TiO_2 151
- Charge transport 154
- Nanoparticle film 151
Total absorption 208
Transfer matrix simulations 210
Transition dipole moment 54
Transmission 194, 204
Transmission mode 193
Transparent conducting oxide 202
Triple bond 28
Triplet 48

UV-vis spectroscopy 193

V_{OC} 112
- Light intensity dependence 222
Valence band 7
Vibrational state 53
Vibronic energy levels 53

Wannier exciton 62

ZnO 159
ZnO–P3HT solar cell 160